5/86

W9-BLF-838

BRANCHES WITHOUT ROOTS

BRANCHES
W I T H O U T
ROOTS

*Genesis of the Black Working
Class in the American
South, 1862–1882*

Gerald David Jaynes

New York   Oxford
OXFORD UNIVERSITY PRESS
1986

Oxford University Press

Oxford   New York   Toronto
Delhi   Bombay   Calcutta   Madras   Karachi
Singapore   Hong Kong   Tokyo
Nairobi   Dar es Salaam   Cape Town
Melbourne   Auckland

and associated companies in
Beirut   Berlin   Ibadan   Nicosia

Published by Oxford University Press, Inc.,
200 Madison Avenue, New York, New York 10016

Library of Congress Cataloging in Publication Data
Jaynes, Gerald David.
Branches without roots,
Includes index.
1. Afro-Americans—Employment—Southern States—History—19th century.
2. Share-cropping—Southern States—History—19th century.
I. Title.
HD8081.A65.J39   1985   331.6'3'96073075   85-5116
ISBN 0-19-503619-0

Printing (last digit): 9 8 7 6 5 4 3 2 1
Printed in the United States of America

FOR MY WIFE

Ann Shepherd Jaynes

Preface

This book owes its being to Yale's Afro-American Studies program's request that I participate in the team teaching of an introduction to the social sciences for Afro-American Studies. In grappling with the question of what an economist could teach a group of first and second year students, none of whom could be presumed to have had any economics, I decided that since the course was subtitled "Patterns of Social Change" it would be appropriate for me to lecture on economic history, developing my lectures on three periods of significant social transition for the Afro-American community: emancipation and the great northern migrations of the two world war periods. With alacrity I began to read the literature on these subjects produced by historians and economic historians. From the beginning I felt that something very essential was lacking in the accounts of the economic transformation of southern agriculture during Reconstruction. The more I read and prepared lectures, the more convinced I became that a study of the incipient development of the free Black worker was greatly needed.

Initially this book was to be a short monograph presenting an economic analysis of the origins of sharecropping and family tenancy in southern agriculture in the aftermath of American slavery. That intention succumbed as I became aware that a full understanding of the problem encompassed issues that go far beyond the boundaries usually self-imposed by economic historians and encroached upon areas best covered by social historians. Not being blessed by those trepidations which the ancients inform us abound among angels I decided to try my hand at a fuller study.

This work attempts to do a number of tasks. And I am painfully aware from the friendly criticism of some economists whom I dearly respect that a few of these things obstruct a simple reading of "the" main economic argument. Be that as it is, I hope readers will find the textual structure interesting enough to put in the extra work required to synthesize the main lines of argument.

A major desire was to attempt to create a work in the tradition of Afro-American scholarship paying strict attention to shades of meaning and perspective in the narrative structure. For a period whose historical interpretation is still as controversial as Reconstruction this tradition seems very appropriate. The historian who searches blindly for absolute truths seems to me to be destined to play a role something like the straight man to Socrates in a Platonic Dialogue. Yet if we believe that historical knowledge is "true belief," then controversial history must be explored from many perspectives as if one were viewing the spectrum of a magnificent kaleidoscope sensing the futility of deciding which of the many shades were the central color while still recognizing an ordered structure.

In the chapters which follow, I have striven to provide a spectrum of perspectives on the agrarian transformation of the American South. To draw an analogy from jazz—these chapters are best understood as linked but distinct improvisations upon a central theme with planters, freedpeople, and others each contributing distinct but compatible solos.

My debts to others who made this project possible are enormous. John Blassingame and William Parker unknowingly, at least at first, provided an ignorant economist with postgraduate courses in history and economic history during seemingly impromptu sessions in department corridors and coffee houses. Both read various stages of the manuscript, listened endlessly to my arguments and offered valuable criticism and encouragement; Parker always insisting upon high standards of economic analysis and Blassingame never refraining from informing me at crucial spots of the text that economic logic was fine in and of itself but that "real" historians would remain unconvinced without more hard evidence. Henry Louis Gates read and red-inked some of the more "literary" chapters and he and Anthony Appiah joined John Blassingame in discussing my work in our near daily meetings over coffee. Chapters Four and Five, which in ways perhaps evident only to myself are central to the entire book, were conceived and outlined one rainy afternoon in the office of Charles T. Davis, whose untimely death prevented his seeing the completion of a project he greatly encouraged. Stanley L. Engerman read the entire manuscript and dashed off an incredibly detailed and cogent criticism of thirteen single spaced pages with unbelievably unselfish lightning speed that would draw the envy of a NASA computer. Of many others who offered advice or criticism, David Levine of the University of Denver, George Grantham of McGill University, and Gavin Wright of Stanford made criticisms that caused me to do some rewriting. Julie Saville, a graduate student in the Yale History Department, gave me invaluable aid in constructing a bibliography of historical sources.

During the research and writing I was fortunate to receive a junior faculty research fellowship from Yale for the academic year 1979–80 and a fellowship from the Rockefeller Foundation for 1982–83. The Afro-American Studies Program at Yale provided me with student research assistants to do legwork, computer work, and aid in the codification of early Reconstruction labor contracts. At various stages, I received assistance from Richard Ligon, Edmondo Corpus, and Lynn Carter. I am especially indebted to Eric Arnesen, Lisa George, and Kimla Wilkins, who worked with me for long stretches of time and gave service beyond the call of pay. Two former economic graduate students Marcellus Andrews now of the University of Denver and Frederick McKinney of Brandeis University lent me their expertise at the Yale Computer Center. Finally, all of the typing and various other duties were performed superbly by Lorraine O'Donnell and Norma Larson.

My introduction to the topics slavery and emancipation occurred prior to the beginning of my formal education. Frequently on Saturday or Sunday mornings my parents would take my sister and myself to Pontiac, Illinois, to visit my father's uncle and aunt [brother and sister]. I still recall, vividly, the photograph, sitting conspicuously upon the piano in the parlor, of the old blackskinned man with a violin in his hands. This photograph always held a strange fascination over me and I cannot remember just when I finally asked who he was and what were the words inscribed upon the photograph. I was told that the picture was of my great grandfather and subsequently I have read hundreds of times the inscription which read:

William Alonzo Janes, born a
slave, Paris, Tennessee, 1850.

This book responds to some of my early inquisitions about old William.

Gerald David Jaynes

New Haven, Connecticut
September 1984

Contents

Part I
The Management
of Colored Races

Us colored folks is branches without roots and that makes things come round in queer ways.

Zora Neale Hurston,
Their Eyes Were Watching God

1

The question is not whether you can obtain free labour at all, but whether you can obtain it so as to keep up the same state of society as exists at present.

Earl of Stanley,
Colonial Secretary of Great Britain

At no time during the existence of the race, have the freedmen as a mass, been called upon to exercise care, economy and forecast. It cannot be expected they should, without cultivation, suddenly become possessed of these qualities.

General Davis Tillson

The Moral Economy of Slavery

Negro slavery's most valuable benefit was seldom more apparent than on the eve of an emancipation. In December 1832 Viscount Howick, lately under secretary for the colonies and destined to become the 3rd Earl Grey, drafted a memorandum on the future of the British economy: "the great problem to be solved in drawing up any plan for the emancipation of the slaves in our colonies is to devise some mode of inducing them when relieved from the fear of the driver and his whip, to undergo the regular and continuous labour which is indispensable in carrying on the production of sugar."[1] The British government's solution to this problem was the labor apprenticeship system contained in the emancipation plan proposed by Colonial Secretary Stanley to the House of Commons. This system, the third part of a five-point plan, declared "that all persons now slaves be entitled to be registered as apprenticed labourers, and to acquire thereby all rights and privileges of freedom; subject to the restriction of laboring" for their former owners three-quarters of each working day for a period of twelve years.

---

[1] Eric Williams, *From Columbus to Castro: The History of the Caribbean, 1492–1969* (London, 1970), p. 328. Howick had resigned from office in protest of the government's plans for solving this problem.

On the floor of the House of Commons, Viscount Howick opposed this plan on the grounds that it was "neither more nor less than a continuance of the whole principle of slavery." But Howick's earlier memorandum fore-shadowed his subsequent admission that true freedom meant "the cultivation of sugar" must end. "The negroes would find so little difficulty in procuring land in order to maintain themselves with more comfort than they have been used to," he warned, that they "would generally refuse to work for wages." This showed that he not only shared Stanley's fears of economic disaster but also desired to devise some means of avoiding the disaster.[2]

Howick's own emancipation plan took sustenance from the knowledge that after freedom the condition of the British West India colonies would not be unlike that experienced by colonizers in other areas of the world. In Asia and many of the Pacific Islands, in contrast to the predominantly hunter-gatherer cultures of the Americas, Europeans found the indigenous populations settled upon the land as experienced agricultural cultivators. In such colonies the major problem was not to supplement a scarce laboring population but to induce the available peoples to abandon their traditional way of life. An emphasis on the growing of subsistence food crops would have to be replaced by the production of commercial staples for the world market and the profit of the colonizers. In the Americas, the colonial system's desire for large quantities of commercial agricultural products led to the introduction of the African slave and the development of the slave plantation, but lifelong chattel slavery was only one solution to the problem of harnessing productive powers for the development of commercial agriculture.

An instructive example is the island of Java. Beginning in the seventeenth century the Dutch East India Company, through the auspices of the home country, took advantage of the existing feudalistic institutions on the island by contracting with Indonesian chiefs. Under this contract the chiefs agreed to force the peasant population to deliver a portion of their crops as a tax tribute. In the early nineteenth century the British took control of the island. Their administrator, Thomas Stamford Raffles, attempted to destroy the feudal institutions by giving the individual peasant the autonomy to freely dispose of his own produce subject to the payment of a rent in the form of a share of the rice harvest. This policy of allowing a degree of freedom to the peasantry failed to provide what the colonizers considered adequate quantities of commercial crops.[3] In 1830, fourteen years after the British restored control of Java to the Dutch, the new governor-general, Johannes Van den Bosch, decided that free trade and voluntary labor would continue to produce insufficient commercial crops. Van den Bosch introduced the "culture system." This policy made commercial agriculture compulsory by mandating that peasant landholders

[2] *Hansard's Parliamentary Debates*, vol. 17 (London, 1833), pp. 1230, 1234, 1254.
[3] W. F. Wertheim, *Indonesian Society in Transition* (The Hague, 1956), p. 55.

cultivate designated commercial crops on one-fifth of their land and deliver them to the government.[4]

Viscount Howick's emancipation plan was constructed to produce the same effect as Van den Bosch's culture system. Howick was ingenious enough to maintain the appearance of voluntary trade by applying compulsion not directly but through the market system. After emancipation, island legislatures would be encouraged to pass laws promoting the labor of freed Africans by taxing all subsistence crops and imported foodstuffs and prohibiting vagrancy and, above all, illicit squatting upon waste lands.[5] Under these policies, the free market choices of the black laboring population would inevitably lead to plantation wage labor to escape the alternatives of starvation or imprisonment.

In fairness to Viscount Howick and the British Parliament, the debates over the emancipation issue also show evidence of a concern with the educational development and welfare of the emancipated blacks. Emancipation, for the British and later for Americans, raised a complex of political and moral issues that had probably never before been confronted as sincerely by a group of emancipators. In the United States, where the slave population was contained in a region that was not a colony of the central government but would ultimately be represented in that government on an equal footing, these political and moral problems were far more important. Nonetheless, these issues in the United States and in Great Britain were secondary to immediate concerns with the state of the economy. We shall see that irrespective of one's philosophical orientation, any attempt to understand American Reconstruction must face this central issue: after the abolition of slavery, how was an industrial order dedicated to commercial production of agricultural products for a market system to be maintained? The essential problem about which this issue revolved was defined by a member of Britain's Parliament as "the management of colored races."[6]

Thaddeus Stevens, in February 1850, during the debates over the extension of slavery into United States territories, affirmed the basic principles which over a decade later would underpin radical reconstruction proposals:

> That republic must be feeble, both in peace and war, that has not an intelligent and industrious yeomanry, equally removed from luxury and from poverty. The middling classes who own the soil, and work it with their own hands, are the main support of every free government.[7]

[4] Ibid., pp. 61, 243.

[5] Lowell Joseph Ragatz, *A Guide for the Study of British Caribbean History, 1763–1834* (Washington, D.C., 1932).

[6] Sir George Campbell, *Black and White: The Outcome of a Visit to the United States* (New York, 1879, reprinted N.Y., 1969), "The Management of Colored Races," pp. 111–25. Campbell espouses what for 1879 was a very liberal view, which makes his essay all the more interesting.

[7] *Congressional Globe*, 30th Cong., 2nd sess., Appendix, p. 103. These three principles formed a package. It is not difficult to find Republicans who would disavow one or two if the third could not be satisfied. Indeed, we shall see below that the absence of education among the black masses was a central obstacle to radical Reconstruction plans.

We find implicit in this statement three defining characteristics of nineteenth-century radical Republicanism: universal education, universal male suffrage, and universal small proprietorship. While all Republicans shared one or more of these beliefs, inclusion in that select club known as radicals should be restricted to those who espoused all three. To the radicals these three principles were the foundation upon which an ideal or good society, in the Platonic sense of these terms, must be built. For the radical mind the proper course for reconstruction was not difficult to envision, because the antebellum South, an important constituent of the United States, failed to satisfy each of these fundamental criteria.

"How," asked Stevens fifteen years later, "can republican institutions, free schools, free churches, free social intercourse, exist in a mingled community of nabobs and serfs; of the owners of twenty thousand acre manors with lordly palaces and the occupants of narrow huts inhabited by 'low white trash' "? Stevens was not the kind of man who could wait for others to suggest the proper answer. He supplied his own: " 'If the South is ever to be made a safe republic let her lands be cultivated by the toil of the owners or the free labor of intelligent citizens.' "[8] These views rested squarely in the center of the Republican Party's ideology. Eric Foner's deservedly acclaimed study of the mid-nineteenth-century Republican Party effectively argues that a unifying foundation for the party was that southern society and its people were presumed industrially and socially backward precisely because the institution of slavery was inherently inferior to free labor as a system of social organization.[9] For this reason, in 1865 most Republicans could agree with Stevens that "the whole fabric of southern society must be changed" by breaking up and relaying "the foundations of their institutions."[10]

The crucial issue was how such a reordering was to be achieved. For Stevens, the quickest and surest way to install republicanism was to drive the southern "nobility into exile" by confiscating their estates and creating an independent black yeomanry in one stroke. On this point Stevens parted company with the heart of his party. The explanation for Stevens's failure to persuade his party, and therefore the Congress, to deliver land to the freedmen is much more complicated than the traditional emphasis upon the Republican reverence for the sanctity of private propterty implies. The Republican belief in the inferiority of the southern slave economy and the generally shared assessment that the economy should be reformed was a commitment to the need for extensive agrarian reform. The debate over the proper course of reconstruction was therefore intricately connected with all

[8] Quoted in W. E. B. Du Bois, *Black Reconstruction* (New York, 1935), p. 197.

[9] Eric Foner, *Free Soil, Free Labor, Free Men* (London, 1970), especially chapters 1, 2, and 9. *See also* W. R. Brock, *An American Crisis* (London, 1963), and David Montgomery, *Beyond Equality* (New York, 1967). The role of Republicans in the larger context of the free labor ideology of nineteenth-century civilization is yet to be adequately addressed. However, for clues see David Brion Davis, *The Problem of Slavery in the Age of Revolution, 1770–1823* (Ithaca, 1975).

[10] Du Bois, p. 197.

of the issues which must be faced in any political confrontation over an attempt to institute an extensive reform of agrarian institutions that includes land redistribution.

The view of Reconstruction as a problem of agrarian reform isolates three basic issues which commanded the attention of all protagonists in the debates: what economic structure for southern agriculture would be most efficient for spurring economic growth and American economic development, what political structure was most conducive to the strength of the Republic, and, lastly, what role should and could the emancipated slaves assume in providing for the success of the first two factors? Opposing sides split over classical lines that as of today remain unresolved issues for the bulk of the developing and underdeveloped nations of the modern world. What is the optimal size producing unit for agricultural enterprise, and who should control the allocation decisions of these units? It is clear that for radicals, and perhaps many moderate Republicans, the answer was small- to medium-sized farms operated by independent self-cultivators. But even among Republicans this position was not unanimously held.

The radicals' attack upon the land-labor monopoly power of southern slaveholders could just as well have been directed toward large manufacturing and commercial interests in general. Within the logic of the radical philosophy, these interests posed just as much of a threat to the development of democratic independence and the egalitarianism of equal opportunity (as opposed to equal results). Thus radicals such as Henry Wilson of Massachusetts warned the country of the danger from "the power of wealth, individual and associated, concentrated and diffused." Large manufacturers, corporations, and especially publicly subsidized railroads, characterized by Elihu Washburn of Illinois as "that bloated and pampered and merciless monopoly," had no place in a democratic society. George Julian of Indiana, a relentless crusader against monopoly power, favored land redistribution to the Negroes and small white farmers in order to destroy the land-labor monopoly of the slaveholders and to prevent northern capitalists from creating a new land monopoly whose control over southern labor might become "more galling than slavery itself."[11]

This distaste for concentrated economic power had its roots not only in the warnings of classical writers like Adam Smith, whose dislike for monopoly power and joint-stock companies is well known, but also in an empirical condition of the nineteenth century. To the radicals, what made the United States prosperous was not merely the drive and energy of its inhabitants but the fact that, unlike Europe, its land and institutions were not controlled or influenced by the presence of a hereditary class of nobles.[12] Egalitarian opportunity and upward mobility required a large class of small-proprietors. Every instance of monopoly control decreased the

---

[11] Brock, p. 224; Kenneth M. Stampp, *The Era of Reconstruction* (New York, 1965), p. 128; Foner, pp. 21–23.

[12] E. L. Godkin, "The Labor Crisis," *North American Review*, vol. 216 (1867). Godkin was editor of the *Nation*.

available opportunities for independent ownership. Here the radicals were in many ways just as anachronistic as the slaveholders. Their conception of liberty was, as David Montgomery reminds us, associated "with the ownership of productive property; its opposite—lack of property—was thus a form of slavery."[13] The analogy went further. If economic growth and widespread prosperity were more enhanced in a free than a slave economy, the reasons seemed clear. The free laborer, striving to emancipate himself into the employing class, had the incentive to work hard, to save, and by accumulating, to spread the wealth of society horizontally in the process. The mere hireling, if he had little hope of being anything else, would perform little better than a slave, and consequently industrial development would suffer.

The radical philosophy epitomized both a desire for political equality and a program for optimal economic development. It was inconsistent with the needs of big business and the opposing philosophy of those who believed that the proper course of economic development required large enterprises operated by talented entrepreneurs. Such men, utilizing the capitalist mode of production, were to bring about capital accumulation on a scale so vast that prosperity, through the vertical process of trickle-down, would ultimately be universal. As long as those of influence, such as the *New York Times* and E. L. Godkin of the *Nation*, insisted upon terming "manufacturing capitalists" the "masters" of "white laborers who are to be slaves," they posed an ideological threat to an industrial structure which was becoming increasingly the province of those who in fact worked for wages. The radical ideology represented an obstacle not only to the capitalists and believers in capitalistic growth like William H. Seward, but also to the actual course of economic growth and the conversion to an economy of wage laborers. But in 1865, the radicals' idea of freedom and independence had not yet succumbed to the social philosophy of factory production which would give to the wage laborer a respectability beyond mere lip service.[14]

In the fall of 1865 the Boston Board of Trade, under the influence of a speech by merchant Edward S. Tobey, resolved that the federal government secure financial assistance to the southern states for the purpose of stimulating commercial enterprise. Tobey considered the problem of finding "the best mode in which industry in the southern states can be most effectually secured and rapidly promoted" one "of immediate importance." It was "perhaps paramount at the present moment to any other having reference to the condition and future welfare of [the] country." For Tobey, whose speech was phrased in terms of "laboring classes,", "employers," and "capitalists," the optimal position for freed laborers was as wage laborers upon either private or government-controlled plantations.[15]

---

[13] Montgomery, p. 30.

[14] Montgomery, p. 26; Godkin, loc. cit.

[15] Edward S. Tobey, *The Industry of the South: Its Immediate Organization Indispensable to the Financial Security of the Country* (Boston, Nov. 27, 1865), p. 5.

The logic of Tobey's argument was a flawless exercise in the political economy of the Atlantic commercial system and the mechanics of an international monetary system based upon the gold standard. The burden of a wartime economy had effectively forced the United States Treasury off the gold standard for internal commerce. A large federal debt had been incurred to finance the war and this had been accommodated with a large issuance of paper currency. The postwar northern economy was beset with price inflation, and many business interests were clamoring for a contraction of the quantity of paper currency. They argued that a decreased money supply and restoration of a gold-based hard money policy was the best procedure for combating inflation and stabilizing the economy. The opponents of this hard money policy feared that a contraction of the money supply would lead to high interest rates, falling prices, unemployment, and bankruptcy for small merchants and farmers.

Edward Tobey was a member of a select group of financial experts who saw in the cotton industry a painless solution to the growing controversy over these issues. Tobey, and others, reasoned that since cotton was the number one export commodity of the United States, its supply essentially controlled the flow of gold to and from the states. "Cotton *is* gold anywhere in Europe," argued Thomas W. Conway; "by its extensive production our currency will be more rapidly reduced to a gold basis than in any other manner." Mr. Tobey elaborated upon the point. The United States' imports of European goods, and thus the "balance of trade, must be paid in specie [gold], unless paid in cotton, which is to Europe the same as specie; and if the additional amount of specie could be retained at home it would tend greatly to strengthen the paper currency of the government, and probably avoid altogether the much-dreaded financial crisis."[16] The Tobey argument was simple. Not only would large cotton exports allow gold to be retained at home, but a surplus crop would induce a flow of gold to the United States. The resulting increase in the domestic supply of money would allow a contraction of the paper currency sans the debilitating effect of a depression, buoy the economy, and allow payment of the national debt.

The Tobey argument was impeccable, but it did not prove that a large cotton crop required plantation production units operated with hired Negro labor. Thaddeus Stevens, reading the mood of the country, proposed a

---

[16] T. W. Conway, "Communication to the Chamber of Commerce of New York," *Eighth Annual Report of the Chamber of Commerce of New York State* (New York 1866), Dec. 14, 1865, p. 64. Colonel Conway had been General Nathaniel P. Banks's number one aide in the Bureau of Free Labor, Department of the Gulf, during the war. At the time of this communication he was preparing to depart for Europe with the intention of inducing English capitalists to invest in southern cotton lands "for the purpose of benefitting the people of the South, both black and white." Tobey, p. 15. The argument is stated in full by Alanson Penfield in a "Communication" to the New York Chamber of Commerce, (*see above*), pp. 60–61. Penfield was at the time employed by the U.S. Treasury Department. For the ins and outs of the paper money controversy during Reconstruction, see W. R. Brock, chap. 6, and also Robert P. Sharkey, *Money, Class and Party: An Economic Study of Civil War and Reconstruction* (Baltimore, 1959).

confiscation plan that was calculated to deliver all that the Tobeyites promised. Stevens proposed to confiscate the land of the 70,000 largest landowners in the South. Forty million acres were to be given to adult freedmen, each to receive a forty-acre tract. The remaining estimated 354 million acres were to be divided into farms and sold in pieces to the highest bidder. He calculated "an average price of ten dollars an acre, making a total of three billion five hundred forty million in six per cent bonds; the income of which should go towards the payment of pensions to deserving veterans and the widows and orphans of soldiers and sailors who had been killed in the War. Two hundred million dollars should be appropriated to reimburse loyal men in both North and South whose property had been destroyed or damaged during the War. With the remaining three billion, forty million dollars, he would pay the national debt." The plan was calculated to create a class of independent farmers whose free and self-interested labour would produce an enormous crop of cotton and democratize the South. In answer to those who worried about the sanctity of private property, Stevens reasoned "that since all this property which has to be confiscated was owned by 70,000 persons, the vast majority of the people in the South would not be affected by this policy. These 70,000 were the arch-traitors; and since they had caused an unjust war, they should be made to suffer the consequences." [17]

The Stevens argument was ingenious. It had been constructed to meet both the requirements of the Tobeyites and the reservations of those closest to his own position. In 1862, during early discussions of confiscation, the inviolability of private property had been a major obstacle. Senator Charles Sumner spoke for many when he confessed that he looked "with more hope and confidence to liberation than to confiscation. To give freedom is nobler than to take property." He added the caveat that "there is in confiscation, unless when directed against the criminal authors of the rebellion, a harshness inconsistent with that mercy which it is always a sacred duty to cultivate."

By 1865 the practical reality that political reform could not be achieved without agrarian reform which encompassed land redistribution had been impressed upon the minds of the best thinkers among the abolitionists. John Stuart Mill wrote to his American friends that if slavery was not to be resurrected by "an aristocracy of ex-slaveholders" regaining control "of the state legislatures," it would be "absolutely necessary to break altogether the power of the slaveholding caste." Mill outlined the strategy which radicals would soon carry into Congress. The slaveholders "and their dependents must be effectually outnumbered at the polling places; [this] can only be effected by the concession of full equality of political rights to negroes and by a large immigration of settlers from the North; both of them being made independent by the ownership of land."

Two months earlier Sumner had written the Englishman John Bright stressing the point that confiscation would only affect the southern aris-

[17] Du Bois, p. 198.

tocracy, who had forfeited their rights by violating the Declaration of Idependence:

> Can emancipation be carried out without using the lands of the slave-masters? We must see that the freedmen are established on the soil, and that they may become proprietors. From the beginning I have regarded confiscation only as ancillary to emancipation. The great plantations, which have been so many nurseries of the rebellion, must be broken up, and the freedmen must have the pieces.[18]

Why did not more northerners come round to this position? Confiscation of the property of the leaders of an unsuccessful insurrection was hardly a serious threat to the preservation of private ownership of property. Indeed, such confiscation was deeply rooted in Anglican tradition. That point was implicitly appealed to by both Stevens and Sumner, whose own strong beliefs in private property rights required justification of any violation of those rights.

The Tobey argument carried the debate because of an inescapable dilemma which the Radicals had created for themselves. Across the Atlantic, the London-published *Economist* entered the contest with a new perspective. European commercial concerns were just as anxious as those in the United States about the forthcoming supply of cotton, for identical reasons. European holders of United States public securities and "European capitalists" counseled the United States to "return as soon as practicable to specie payments" through the mechanism of large cotton exports. This would "reduce a portion of the volume of the precious metals which flows to Egypt and Asia in payment for cotton, and engulfed in that maelstrom, never returns to the commerce of American and European nations." It would also ensure prompt payment to the buyers of U.S. securities, and prevent a continuation of the recession caused by the reduced supply of cotton to Europe.

Support for this position was sought by appealing to the cotton textile industry, which had undergone a harsh recession during the cotton famine period of the American Civil War. "We have stated repeatedly, what indeed is obvious enough, that until industry and production in the southern states of the union shall be once more *settled*, there can be no *certainty* nor much comfort for either spinners or importers." For the *Economist*, there was only one method by which these disasters to the Atlantic economy could be avoided: "*How* the negroes will be 'regimented,' as Mr. Carlyle calls it, into industrial gangs or squadrons again, does not seem clear. That in some way or other they will be so 'regimented'—and either induced to work, or 'persuaded' to work, we entertain little doubt."[19]

---

[18] Edward L. Pierce, *Memoirs and Letters of Charles Sumner* (Boston, 1893), vol. 4, p. 76. *National Anti-Slavery Standard* vol. 26, no. 6 (June 17, 1865); Mill's letter was dated May 13. Pierce, p. 229.

[19] "Communication of Alanson Penfield" (see n. 16 above), p. 60. *Economist* (London), Nov. 17, 1865, pp. 1396, 1397.

The meaning of the reference to Thomas Carlyle is unmistakable. After emancipation in the British West Indies the fears of the government and Lord Howick had been realized. Vast numbers of blacks had deserted the sugar plantations to take up subsistence farming on waste lands. The consequent fall in the level of sugar production and rise in wages required to induce labor onto the plantations seemed to threaten the ruin of the planters and commercial interests. In 1853 Carlyle published his infamous "Occasional Discourse on the Nigger Question" in which he argued "that no Black man who will not work according to what ability the gods have given him for working, has the smallest right to eat pumpkin, or to any fraction of land that will grow pumpkin, however plentiful such land may be; but has an indisputable and perpetual *right* to be compelled, by the real proprietors of said land, to do competent work for his living." Basing his argument on the inherent inferiority of blacks to whites, Carlyle maintained that the natural and best condition of affairs would have all blacks in the British empire bound to white masters under lifetime contracts. This solution was not only good for the blacks, he argued, it was necessary for sound commercial development and the growth of civilization. The islands were good for growing many agricultural products other than food for home consumption, and to Carlyle the crucial question was "the right of chief management in cultivating those West-India lands." The answer was not difficult to ascertain: "if Quashee will not honestly aid in bringing-out those sugars, cinnamons and nobler products of the West-Indian Islands, for the benefit of all mankind, then I say neither will the powers permit Quashee to continue growing pumpkins there for his own lazy benefit." The state wanted "sugar from these islands" and, by one way or another, "means to have it."[20]

Mr. Carlyle's recourse to racism was an excess which only those with large personal interests at stake need indulge. For others, less personally involved or more politic, an argument based upon class would just as easily serve the essential purpose. The rallying cry was taken up on both sides of the Atlantic. How could the radicals, the very same abolitionists who had recently decried the debilitating effects of slavery upon the independence and very humanity of the blacks, advocate the ruin of the Atlantic economy by turning over to these debased people the finest cotton lands in the world? Hard-working, commercially oriented free landholders might well produce more cotton than slaves, but what would these recently made freedmen do?

Mr. Tobey's speech had alluded to reports on conditions in the South, and it is likely his information came from Bostonians who had been work-

[20] Thomas Carlyle, "Occasional Discourse on the Nigger Question" (London, 1853), in Eugene R. August, ed., *Carlyle: The Nigger Question, Mill: The Negro Question* (New York, 1971), pp. 9, 27, 29, 31. The article was originally published as "Occasional Discourse on the Negro Question" in *Fraser's Magazine*, vol. 40 (Dec. 1849), pp. 670–79. The 1853 expanded version was published in response to critics, chiefly J. S. Mill. Abolitionists and other free labor proponents blamed the condition of the post-emancipation West Indies upon the poor labor-management techniques of planters and the unjust apprentice system of the British Parliament. See Sir Morton Peto in the same issue of the *Economist*, pp. 1398–99.

ing with the freedpeople in the South Carolina and Georgia Sea Islands for a number of years. The young Boston attorney Edward Pierce, super-intendent of these government controlled lands since 1862 and a true friend of the blacks, reported what he had observed with all candor: "I have seen these people where they are said to be lowest, and sad indeed are some features of their lot; yet with all earnestness and confidence I enter my pro-test against the wicked satire of Carlyle." In the same article, he reported that "they were beginning to plant corn in their patches, but were dis-inclined to plant cotton, regarding it as a badge of servitude." In his 1862 report to Secretary of the Treasury Salmon P. Chase, Pierce reported that "without the system here put in operation [plantations with centrally super-vised wage labor] the mass of the laborers, if left to themselves and properly protected from depradations and demoralization by white men, would have raised on their negro patches corn and potatoes sufficient for their food, though without the incentives and moral inspirations thereby applied, they would have raised no cotton, and had no exportable crop."

When Edward Philbrick, a Boston engineer and undoubtedly the ablest manager and businessman in Pierce's group, testified before Congress in 1865 he was more explicit. The typical attitude of the Sea Island freedpeople was expressed by the following declaration: "Cotton is no good for nigger. Corn good for nigger; ground nuts good for nigger; cotton good for massa; if massa want cotton he may make it himself, cotton do nigger no good; cotton make nigger perish."[21]

In a letter to a friend discussing possible ways of providing land to the freedmen, Philbrick confessed that it made his "orderly bones ache" to think that future generations of black landholders would subdivide the land among their inheritors. This would duplicate "the minute subdivision of lands among the French peasantry," where "a man has not land enough to live on or work economically, and hence a vast amount of time and energy is wasted in France for lack of organization." Philbrick was a member of a modern school of economic development ideas, but he was not without company. William B. Dana, editor of the *Commercial and Financial Chronicle*, disseminating the interests of New York merchants, was in philosophical agreement with President Johnson's reconstruction plans but questioned the decision to deny amnesty to secessionists who possessed more than $20,000 in property. This policy would disenfranchise practically every real "capitalist in the South," he warned, and thereby jeopardize the flow of investment to the region. Earlier in the year the *New York Times* had made a similar point, arguing that the blacks' "industry will merely seek the most immediate gratification as its reward." "White ingenuity and enterprise ought to direct black labor," the *Times* advised; more to the

21 A. W. Stevens, ed., *Addresses and Papers of E. L. Pierce* (Boston, 1896), p. 123, from "The Freedmen at Port Royal," *Atlantic Monthly*, Sept. 1863, p. 93. "Second Report of Edward L. Pierce to Salmon P. Chase," June 2, 1862, in Frank Moore, ed., *The Rebellion Record* (New York, 1864), pp. 317–18. Edward Philbrick, "Testimony, Special Report No. 3," *House Executive Document No. 34, 39th Cong., 1st sess., 1865–66.,* p. 18.

point, cotton production required "the white brain employing the black labor." [22]

Corroboration of these points came in from all over the South. General Nathaniel P. Banks, defending his administration of the Free Labor Bureau in wartime Louisiana against charges of injustice and harsh treatment of Negroes, argued that the closely supervised contract system he authorized was a necessary condition to maintain order in occupied territory during wartime. Affirming his committment to the proposition that land should be put into the hands of the laborer, Banks advised the nation not to proceed too quickly: "It has required the legislation of four centuries to give white labor the freedom it enjoys in contracting for employment." Edward Pierce was in basic agreement: "As it is said to take three generations to subdue a freeman completely to a slave, so it may not be possible in a single generation to restore the pristine manhood." [23]

By early 1866 the general public was receiving information concerning the prospects that freedmen farmers would provide the needed export crop. Correspondents of northern newspapers were anxious to learn if the Negroes would produce an export crop as large as that of 1860. One newsman, generally impressed with the progress and prospects of freedmen who had acquired land, expressed the ultimate reservation: "To make their titles [to the land] perpetual is to give over to uncertain cultivation, by a race supposed to exist only for the convenience of another, the most valuable cotton lands;" and this was "a conclusion deemed inadmissable and monstrous, especially by the rebel owners of the lands." [24]

If the nature of Negro slavery's most valuable benefit was ever in doubt, that doubt was dispelled during the debates attendant upon the making of federal commercial and political policy for the reconstructed Union. The attempt to make land reform a governmental function was one early aspect of these debates. Argument on this issue would continue, but the basic requirements of a commercially oriented society were an insurmountable obstacle to such a revolutionary basis for agrarian reform. Discussion of the failure of land distribution to the blacks is a crucial step in the examination

[22] Elizabeth Ware Pearson, *Letters from Port Royal* (Boston, 1906), p. 274. *Commercial and Financial Chronicle*, July 1, 1865, Aug. 26, 1865. *New York Times*, Feb. 26, 1865, p. 4.

[23] Nathaniel P. Banks, *Emancipated Labor in Louisiana* (Boston, Oct. 30, 1864), p. 16. Stevens, p. 128. Other officers in charge of freedmen's affairs throughout the South were of like mind. See General Davis Tillson, National Archives, Record Group 105, Freedmen's Bureau, film M798, roll 32, Georgia, p. 15; General Hurlbutt, ibid., film M752, roll 15, Louisiana, p. 9. Laurence N. Powell, *New Masters* (New Haven, 1980), p. 2. In 1832 British Colonial Secretary Stanley had defended his proposed apprenticeship system for the emancipated slaves upon the same grounds: "The slaves should not be made free by one hasty step—that the shackles should not be burst at once—that they should not be flung forth suddenly from slavery to freedom, for which they may be unfit." (*Hansard's Parliamentary Debates*, vol. 17, p. 1222). These positions should not be hastily interpreted as hypocritical rationalizations of unstated ulterior motives.

[24] J. T. Trowbridge, *The South: A Tour of Its Battlefields and Ruined Cities* (Hartford, 1866), p. 536. Banks, p. 24. *New York Tribune*, Jan. 30, 1865. *Boston Daily Advertiser*, June 15, 1865. Whitelaw Reid, *After the War* (New York, 1866), pp. 115, 287.

of the postwar development of the southern economy and the evolution of the free black worker. Such examination reveals that the federal government was willing to push agrarian reform only up to a point, and that the issues that defined this boundary were a national propensity to accept the thesis of black inferiority and the extreme fiscal conservatism of a ruling political party facing a large public debt. Within the context of these key issues, the course of the policies adopted seems almost inevitable. We shall see that these two considerations virtually guaranteed that the development of the southern economy and the black labor force would be largely left to the fluctuations of the market. The slave population was not to enter freedom as a yeomanry, nor, ironically, would their former masters receive the financial aid necessary to provide for a rebuilding of the economy upon firm capitalistic grounds.

Limited federal aid to the postwar south further damaged its ruined financial system and in a direct way led to the much maligned sharecropping system of agriculture. The roots of black and white southern poverty lie in this failure to construct a "Marshall Plan" for the reorganization of the cotton economy on the basis of free labor.

In the fight over the questions, of confiscation and the economic emancipation of the Negro, the weapons utilized were a complex mixture of racial and economic ideologies and the class-based commercial politics that underlie all attempts at agrarian reform. In the final analysis, economic independence for black freedmen was seen as a poor gamble whose stakes encompassed the entirety of West Atlantic civilization. Like the French philosophe Abbé Raynal, most men in power understood "that without this labor these lands" must "remain uncultivated." Few of them could contemplate so incredible a sacrifice as did Raynal nearly one hundred years earlier: "Let them lie fallow, if it means that to make these lands productive man must be reduced to brutishness."[25] The moral economy of managing colored races had not changed that significantly.

[25] David Brion Davis, *The Problem of Slavery in Western Culture* (Ithaca, 1966), p. 16.

2

Let the Bureau set apart one or more plantations as may be required in each state or county, to be called Industrial colonies. . . . Let small but comfortable tenements, workshops, schoolhouses, and chapels, be built on them. . . . If they refuse to labor, the Scripture rule may be applied, 'If any will not work neither should he eat' . . . for society has long since claimed and exercised the right to protect the industrious portion of the community from the burden of supporting the idle and the vicious and it is the part of true humanity and practical justice.

Edward S. Tobey

The Benthamite Economy

Leading radicals were most conspicuous in their attempts to grapple with the dilemma enveloping their reconstruction policy. Charles Sumner, in advocating the creation of a federal bureau of freedmen in his famous "bridge from slavery to freedom" speech, had predicated the argument upon the awesome and debilitating effects of slavery upon the blacks. Sumner spent much time, even at this relatively early stage of the debate (1864), arguing that freed Negroes, under proper conditions, would respond to market incentives. And the proper conditions included external guidance and protection: "The curse of slavery is still upon them. Somebody must take them by the hand,—not to support them, but simply to help them obtain the work which will support them."[1]

The creation of the Bureau of Refugees, Freedmen, and Abandoned Lands on March 3, 1865, was for many people signal evidence of fatal flaws in radical reconstruction plans. Ironically, those plans might never have survived had it not been for developments in the South.

In the fall of 1865, after Lincoln's assassination, reconstituted southern states under Andrew Johnson's conciliatory reconstruction policy had legislated the infamous "black codes." These statutes, built around

---

[1] Edward Pierce, ed., *The Works of Charles Sumner*, vol. 3 (Boston, 1873), p. 479.

stringent vagrancy and apprenticeship laws which would immob'lize the emancipated laborer, were not merely the South's attempt to reestablish the old order but also its solution to what Viscount Howick characterized as the key problem of emancipation. This is an interpretation that should not be overlooked, but historians who have correctly pointed out the similarities between these codes and contemporary northern vagrancy laws have missed an essential point. The South was, as W. E. B. Du Bois put it, "looking backward." Not only were basic civil rights for Negroes unrecognized, but the southern states seemed to be saying that for blacks Reconstruction would begin and end with the codes.

Such proscription of an entire race was incompatible with the Republican ideal of a free labor society. During the presidential campaign of 1860, Abraham Lincoln had given an unequivocal statement of the minimal conditions which the Republican Party stood upon: "I want every man to have his chance—and I believe a black man is entitled to it—in which he can better his condition—when he may look forward and hope to be a hired laborer this year and the next, work for himself afterward, and finally to hire men to work for him. That is the true system." [2] Under the "true system" all men had the opportunity to prove themselves—to upgrade their condition by individual effort. A well-enforced vagrancy code, applied indiscriminately against all men, would not affect those who willingly worked for a living but was supposed to protect society from would-be drones. In this sense the Republicans could, with no thought of hypocrisy, attack the "black codes," which, as Henry Wilson of Massachusetts proclaimed, made "the colored people of South Carolina serfs, a degraded class, the slaves of society." Senator Lyman Trumbull of Illinois asked, What would the proclamation of emancipation mean? "It is idle to say that a man is free who cannot go and come at pleasure, who cannot buy and sell, who cannot enforce his rights. I give notice that, if no one else does, I shall introduce a bill and urge its passage through Congress that will secure to those men every one of these rights: they would not be freemen without them." [3]

In the parliamentary debates over West Indian emancipation Viscount Howick had attacked the government's apprenticeship program on the basis of this same free labor ideology. The proposed plan "did not introduce the principle of competition" and therefore could "not in reality change the nature of the system," for "competition was the essence of free labor," and "there is no intermediate state between slavery and freedom." Despite its similarities to the "black codes" that the southern states would enact years later, the government plan which Howick excoriated was put forth with a different objective in mind.

British emancipators confronted the same dilemma that Americans would face. If emancipation was to enable its architects to "get rid of every trace and vestige of the existing system" and allow them to "re-organize society upon better and sounder principles," much depended upon the resulting

[2] Du Bois, p. 17. See also Foner, pp. 11–39.
[3] *Congressional Globe*, 39th Cong., 1st sess., pt. 1, pp. 39, 43.

behaviour of the freedpeople. Free labor advocates like the Englishman Howick and the American Lyman Trumbull, who argued that freedom must be immediate and absolute, had much more faith in the market system as a school for freed laborers than did most men.

Secretary Stanley had proposed what he believed to be "a safer and a middle course which [would] give to the slave all the essentials of freedom," while it left them "still subject to such regulations as [would] operate as an incentive to the acquisition of industrious habits."[4] The transitory state offered by Stanley drew upon the social philosophy of a group of deceased Anglo-American philosophers and would shortly thereafter serve as a blueprint for American reconstruction plans. The safer and middle course promised that "stipendiary Magistrates" would "be appointed by the Crown." These magistrates, "unconnected with the colonies [and] having no local or personal prejudices," would be "paid by this country, for the purpose of doing justice between the negro and the planter, of watching over the negro in his state of new-born freedom, and of guiding and assisting his inexperience in the contracts into which he may enter with his employer." The Crown was "bound to see that they are fitted for the enjoyment" of freedom by establishing "a religious and moral system of education." This would "imbue them with feelings calculated to qualify them for the adequate discharge of their duties." Howick, who at this time (1833) proposed a compromise which differed from Stanley's plan only to the extent that freed laborers would be paid, was extremely skeptical of the feasibility of the experiment.[5]

In 1865, the turmoil caused by the political truculence of the readmitted states who enacted "black codes" served to strengthen the tenuous bonds of agreement which existed between radicals and some northern business interests. Existing conditions between labor and capital within the South and northern attitudes toward the South were not conducive to the quick resumption of staple production desired by businessmen. Edward Silas

---

[4] *Hansard's Parliamentary Debates*, vol. 17, pp. 1248 (Howick), 1222 (Stanley).

[5] Ibid., p. 1228. Stanley's plan required that freedpeople labor for three-fourths of their time without wages; he argued that the forgone wages would help pay for their emancipation and thereby give them incentives to work for their own freedom. Howick countered that their freedom was a natural right and if they must be under guardianship they should at least receive the full value of their labor (ibid., pp. 1235–36). The Howick-Trumbull position was endorsed by the leading abolitionists in their respective legislatures, but for different reasons. For a good statement of the various motivations of British abolitionists and emancipators that holds up as a fair characterization of Americans too, consult the speech of Mr. Buckingham, an abolitionist who stood on the far left of this political issue (ibid., vol. 18, pp. 227–30). A new study of the tensions between these various groups is sorely needed. With respect to British and United States emancipation, historians from the right and left have downplayed the noneconomic arguments of emancipators and focused upon the arguments of centrists, who usually were most interested in economics, although not solely for reasons of self-interest. For example, Eric Williams, in *Capitalism and Slavery* (Chapel Hill, 1944), argues that economic interest, as opposed to humanitarianism, led to British emancipation. There were, of course, elements of both, but a more general interpretation of economic interest than Williams's has more validity than some recent revisionist critiques of Williams imply.

Tobey, referring to the Freedmen's Bureau bill of March 1865 in his speech before the Boston Board of Trade, sounded the keynote which would ultimately lead to the compromise that could be accepted by moderates and radicals. "Happily for the country, Congress at its last session created an agency and a power which in the hands of the President may do much towards accomplishing these most important objects," he said. The bureau, "with enlarged powers which may be granted early in the next Congress," could be so organized as to "secure to the laborer the rewards of industry, and to the national wealth the results of that industry."[6] The primary requirement need to seal the compromise was a precise delineation of the bureau's enlarged powers and its organization.

Thaddeus Stevens, abandoning all concern for the pitfalls inherent in the radical dilemma, moved to seize the initiative in the wake of these developments:

> We have turned, or are about to turn, loose four million slaves without a hut to shelter them or a cent in their pockets. The infernal laws of slavery have prevented them from acquiring an education, understanding the commonest laws of contract, or of managing the ordinary business of life. This Congress is bound to provide for them until they can take care of themselves. If we do not furnish them with homesteads, and hedge them around with protective laws; if we leave them to the legislation of their late masters, we had better have left them in bondage.

Stevens's insistence that slavery had not prepared the blacks for freedom was for most people a weak antecedent for a conclusion which included provision of homesteads. The core of any compromise was contained in Stevens's last warning. Moderate Lyman Trumbull represented the centrist mood of Congress when he said, "We may, if deemed advisable, continue the Freedmen's Bureau, clothe it with additional powers, and if necessary, back it up with a military force."[7] A point conspicuously absent in the proposals of centrists was the provision of homesteads.

In taking up the challenge of civil rights for Negroes, Charles Sumner had done his homework. The Proclamation of Emancipation pledged the republic "to maintain the emancipated slave in his freedom." He argued that the sanctity of such a pledge could hardly "be entrusted to the old slavemasters" but "must be performed by the National Government." This was "according to reason" and "to the examples of history."

> In the British West Indies we find this teaching. Three of England's greatest orators and statesmen; Burke, Canning, and Brougham, at successive periods, united in declaring, from the experience in the British West Indies, that

---

6 Tobey, pp. 8, 9.

7 *Congressional Globe*, 39th Cong., 1st sess., pt. 1, p. 74, Dec. 18, 1865; ibid., p. 43. Sir Robert Peel, who thought that immediate "abolition of the present system under the notion that wages would induce labor would be a most dangerous experiment" but considered Secretary Stanley's aprenticeship plan "equally hazardous," serves as a good barometer of the centrist position in the British debates (*Hansard's Parliamentary Debates*, vol. 19, p. 1064).

whatever the slavemasters undertook to do for their slaves was always "arrant trifling," and that, whatever might be its plausible form, it always wanted the executive principle.[8]

A weakness in this position, from the radical perspective, was that the executive principle could be the basis for a wide variety of policies towards the freed slaves. The conservative Edmund Burke had combined this principle with a "Negro Code" for the British West Indies. Burke's emancipation plan served as Secretary Stanley's blueprint for a policy far removed from a desire to create homesteads for emancipated slaves.

If a freedmen's bureau was to be the primary vehicle of reconstruction, its organization and purpose was fundamental in determining the direction and extent of reconstruction. Two contemporaries, New England's Edward L. Pierce and the Americanized Scotsman Robert Dale Owen, played leading roles in determining the lines of debate which developed around the organization and function of the Bureau of Refugees, Freedmen, and Abandoned Lands. However, it was the intellectual legacy of the English philosopher Jeremy Bentham that defined those lines.

The entire issue could be summed up in one question, which Owen considered to be "one of the gravest social problems ever presented to a government." Edward Pierce had asked it, and it was on the lips of every informed member of Euro-American civilization. Could the habits of generations of a race of men and women, even a race of slaves, be so modified, so altered as to permit "the possibility of lifting it to civilization"? Was it really possible, as the Frenchman DeGasparin claimed, "to transform slaves into peasants, to show by facts that free negroes are not monsters; to preserve the cultivation of cotton."?[9]

It is not surprising that men and women of this period, when confronted with a social question of this nature, were affected by the ideas of one of the most influential social philosophers in Western thought. But in this particular case the intellectual presence of Bentham is so evident and so powerful as to be almost personal. Jeremy Bentham had died just thirty years earlier, and he had devoted a major portion of the last years of his life to attempting to put into practice a theoretical schema created precisely for the kind of problem the Americans faced at the dawn of Reconstruction. The problem Bentham had originally sought to solve was to design an optimal prison system which, among other things, would lead to complete reformation of the characters of the inmates. The reformer began his treatise by listing the advantages of this method: "Morals reformed—health preserved—industry invigorated—instruction diffused—public burthens lightened—economy seated, as it were, upon a rock—the gordian knot

[8] Quoted in Du Bois, p. 272.

[9] Robert Dale Owen, "Report of Freedmen's Inquiry Commission," *Senate Executive Doc. No. 53*, 38th Cong., 1st sess., p. 13. Stevens, pp. 102, 130. Sir Robert Peel had made the point earlier: the objective of emancipation "was to produce an industrious class of cultivators, willing to labor and to reap the profits of their industry" (*Hansard's Parliamentary Debates*, vol. 18, p. 354.

of the poor-laws not cut, but untied—all by a simple idea in architecture!"[10]

Bentham adroitly named this paradigm of Skinnerian architecture "the panopticon," for it was a circular or cylindrical structure so constructed as to allow a centrally located inspector or "inspective force" to observe the actions of all subjects in a single glance, "without any change of situation." This was the essential point of the plan. Without it, the reformers could have no assurance that their teachings and instructions were being followed and put to good effect. Bentham was quite convinced that behaviour modification on so profound a scale required absolute control:

> If it were possible to find a method of becoming master of everything which might happen to a certain number of men, to dispose of everything around them so as to produce on them the desired impression, to make certain of their actions, of their connections, and of all the circumstances of their lives, so that nothing could escape, nor could oppose the desired effect, it cannot be doubted that a method of this kind would be a very powerful and a very useful instrument which governments might apply to various objects of the utmost importance.[11]

Objects considered to be of sufficient importance included houses of correction, workhouses, madhouses, schools, manufactories, and the "*training* [of] *the rising race* in the path of *education.*"

In addition to the "inspection principle," Bentham recommended free enterprise, and cited three rules as imperative for successful management: the rules of lenity, severity, and economy. First, the condition of subjects should not lead to bodily suffering or endanger health. But second, they should not enjoy more than "the poorest class of subjects in a state of innocence and liberty." Third, "no public expense" should be incurred "or profit or saving rejected." To guarantee observation of these rules, Bentham argued, the best organizational structure for the system was "contract-management." Panopticons should be leased to profit-seeking private individuals, not administrated under "trust-management" by a board of public officials.[12]

The debates over the Freedmen's Bureau centered on the issues implied by Bentham's three rules and on the choice between contract and trust management. The first plans to gain wide public notice were put forth by Pierce and Owen. Pierce's plan "to reorganize the laborers, prepare them to become sober and self-supporting citizens, and secure the successful culture of a cotton-crop, now so necessary to be contributed to the markets of the

---

[10] John Bowring, ed., *Works of Jeremy Bentham* (Edinburgh, 1843), vol. 4, "Panopticon; or, The Inspection House," p. 39.

[11] Ibid., p. 44. See also Elie Halevy, *The Growth of Philosophical Radicalism* (New York, 1965), p. 85.

[12] Bowring, pp. 40 (my italics), 122–23. Interestingly, the architectural idea of the panopticon was invented by Bentham's brother to assist in employing "Mujiks, or peasantry of Russia." Bentham conceived the plan of adapting and improving the scheme while visiting his brother in Crichoff, Russia (Bowring, vol. 10, p. 50).

world" reads like a catechism of Benthamite social engineering. The principle of economy recommended that the organization "should not be a burden on the treasury." The recommendation for lenity demanded that they who were "as readily spoiled as children should not be treated with weak and injurious indulgence" but be placed under conditions which would put them upon a material equality with the family of "a white farm laborer in [the] vicinity." As for physical severity, Bentham had counseled against physical whipping and advocated solitary confinement and "a denial of food" as more humane and effective punishments for disobedience and refusal to work. Pierce advocated "the milder and more effective punishments of deprivation of privileges, isolation from family and society, the workhouse or even the prison." [13]

Both plans were consistent in minute detail with the organizational and internal management specified by Bentham, with one important exception: Edward Pierce favored "trust-management" over Bentham's absolute insistence upon the desirability of "contract management." Pierce's primary argument was that contract leasing of "the plantations and the people-using them" would lead to an influx of "doubtful men, offering the highest price" out of the desire "to obtain a large immediate revenue—perhaps to make a fortune in a year or two." He considered this system "beset with many of the worst vices of the slave system, with one advantage in favor of the latter, that it is to the interest of the planter to look to permanent results." He thought "the better course could be to appoint superintendents for each large plantation," clothing them with an "adequate power to enforce a paternal discipline" upon the laborers.[14]

Robert Dale Owen favored the transference of plantations to loyal and respectable owners or lessees of plantations as quickly as possible. His basic motivation was that Bentham's principle of lenity was most likely to be put into efficient operation under contract management with free labor. The freedpeople should be made to rely upon themselves as soon as possible, he warned, for otherwise "the risk is serious that under the guise of guardianship, slavery, in a modified form, may be practically restored." Like Viscount Howick thirty years earlier, Owen believed that emancipation "should be unconditional and absolute." Freedmen, "like whites," were "to be self-supporting," and all this required was that they be given "a fair chance." [15]

This position had other uses that gained rapid support in the North. Men like Owen believed that the freedmen should ultimately become owners of southern land, but the requirements of northern industry for southern crops argued against an indiscriminate distribution of land to untested laborers. Some in the North argued that the free labor market would serve as a filter. Industrious and frugal freedmen would soon acquire land and become a commercial benefit to the country. Those who could not survive the test

[13] Moore, pp. 310, 311. Owen, pp. 15, 20. Bowring, vol. 4, p. 164. Moore, p. 311.

[14] Moore, pp. 310–11.

[15] Owen, pp. 14, 109, 110.

should not be given the opportunity to waste valuable property before the fact of their incompetence had been demonstrated. Edward Philbrick was a staunch supporter of this position. He saw it as "the wise order of Providence as a means of discipline" from which the "negroes should not be excepted" until they proved they had "aquired the necessary qualifications to be benefited" from such acquisition. The *New York Times* maintained that if "northern capital should flow into those rich cotton lands" and "make them bloom with wealth, intelligence and civilization," certain "individuals of [the Negroe race] rising to an equality with the superior race," after being constantly elevated for "centuries," would join "a great prosperous, increasingly intelligent peasantry, in the southern country."[16]

This argument, attractive as it was to most northerners, placed a poor second to another, which destroyed the quest for government-managed panopticons as thoroughly as it had killed the movement for land redistribution. Even Edward Pierce, when arguing for trust-management of government plantations, had admitted that leasing to private contractors "might yield to the treasury a larger immediate revenue," even if "it would be sure to spoil the country and its people in the end." After this admission, there was little of import left to be said. Northern commercial interests and small would-be northern capitalists looking to lease plantations combined with southern planters to form an unlikely political coalition. The *New York Times* came out in favor of Andrew Johnson's policy of restoring lands to owners because it would pay the national debt faster than the radical plan could.

Bentham's principles of lenity and economy, bedrocks of the science of moral economy, triumphed for those who argued for contract leasing. Hadn't the father of social engineering counseled that "economy, I have said, should be the leading object; and it is principally because the contract plan is the most favorable to economy, that it is so superior to every other plan"? The Bureau of Refugees, Freedmen, and Abandoned Lands was organized to watch over the interests of the country and the emancipated slaves. The blacks, in the fairest filter the world had yet devised, would, in the words of Henry Ward Beecher, have to "root hog or die." Bentham as J. H. Clapham has suggested in another context, would most likely "have criticized each new thing" in the final plan, but overall Jeremy would have been pleased.[17]

[16] Pearson, p. 276. *New York Times*, Feb. 26, 1865. *New York Tribune*, Jan. 30, 1865. Before Parliament Colonial Secretary Stanley had supported the apprenticeship plan on the basis of a similar argument: *Hansard's Parliamentary Debates*, vol. 18, p. 140. But see also Pierce in Moore, p. 312. Edward Tobey's plan, which was a mixture of contract and trust management, advocated the filtering argument.

[17] Moore, p. 311. *New York Times*, Sept. 6, 1866. Bowring, vol. 4, p. 128. J. H. Clapham, *An Economic History of Modern Britain*, 3 vols., vol. 2 (Cambridge, 1932), p. 387.

3

Resolved, that in view of the fact that efforts by private enterprise to draw capital and intelligent labor to the Southern States, must in their results be remote and contingent, and however ultimately useful they must be, cannot be relied upon materially to increase the agricultural products of next year, or even for several years to come, it is vitally important to the interests of the nation that the Government should, within the sphere of its legitimate powers, aid forthwith in sustaining and organizing such portion of the laboring classes now in the Southern States, as cannot be reached by private capital, not only on the ground of a considerate humanity, but also to save their labor to themselves and to the country.

Boston Board of Trade

Preconditions of Failure

At the inception of the postbellum period of American history, the United States stood poised to embark upon a period of economic growth which would enable it to emerge a half century later as the premier industrial nation in the world. Optimistic Americans were confident about that future and the prospect that the South would not only aid the industrial expansion but also share in its benefits. The staunchest advocates of the free labor ideology were certain that the Civil War had been fought for the express purpose of fulfilling this destiny. As early as 1845, William H. Seward, political leader of the movement for business expansion and industrial growth, argued that the abolition of slavery would determine "whether impartial public councils shall leave the free and vigorous North and West to work out the welfare of the country, and drag the reluctant South up to participate in the same glorious destinies."[1] Events would prove otherwise.

The data in table 3.1, demonstrating regional disparities in per capita income and its growth during the nineteenth century, have been collected and ably analyzed by Richard A. Easterlin and expounded on in an admirable essay by Stanley L. Engerman.[2] Some controversy exists concerning the interpretation of these data. There is, however, a consensus that the significant collapse in the relative performance of the southern economy

[1] Foner, p. 51.
[2] Richard A. Easterlin, "Regional Income Trends, 1840–1950," in *American Economic History*, ed. Seymour E. Harris (New York, 1961). Stanley L. Engerman, "Some Economic

24

Table 3.1 Personal Income Per Capita in the South as a Percentage of United States Average, 1850–1950

	1840	1860	1880	1900	1920	1930	1940	1950
United States	100	100	100	100	100	100	100	100
South	*76*	*72*	*51*	*51*	*62*	*55*	*65*	*73*
South Atlantic	70	65	45	45	59	56	69	74
East South Central	73	68	51	49	52	48	55	62
West South Central	144	115	60	61	72	61	70	80

SOURCE: Richard Easterlin, "Regional Income Trends, 1840–1950," in *American Economic History*, ed. Seymour E. Harris (New York, 1961).

NOTE: South Atlantic, all dates: Delaware, Maryland, and the District of Columbia excluded. West South Central, 1840 and 1860: Oklahoma and Texas excluded. 1880: Oklahoma excluded. For 1920–50 the personal-income figure used in computing per capita income was an average over the period of a business cycle: 1920, average of 1919–21; 1930, average of 1927–32; 1940, average of 1937–44; 1950, average of 1948–53.

immediately after the war and its failure to regain its postwar position for ninety years is highly suggestive of the extraordinary costs of the Civil War to the South. To explain what went wrong is a primary task of any economic study of the post–Civil War southern economy. Students of United States history have for some time recognized that the problem involves complicated issues concerning the course of economic development and growth. Adopting this line of analysis, and without committing ourselves to his concept of "take-off," we may, for descriptive purposes, consider W. W. Rostow's fairly exhaustive and representative taxonomy of the internal factors which influence economic growth.

Professor Rostow has written that "the level of output of an economy is a function of the size and quality of the working force and the size and quality of the capital stock." The rate of growth of output can then be seen to be a complex function of these two factors. Rostow argues that these two fundamental variables are determined by a list of "major subvariables" encompassing social and technical factors:

Related to Changes in the Size and Productivity of the Work Force
1. The birth rate.
2. The death rate.
3. The role of women and children in the working force.
4. The degree of effort put forward by the working force.

Related to Changes in the Size and Productivity of the Capital Stock
5. The yield from additions to the capital stock (including natural resources, fundamental and applied science, organizational techniques, etc.).

Factors in Southern Backwardness in the Nineteenth Century," in *Essays in Regional Economics,* ed. John F. Kain and John R. Meyer (Cambridge, Mass., 1970).

6. The (prior) volume of resources devoted to the pursuit of fundamental science.
7. The (prior) volume of resources devoted to the pursuit of applied science.
8. The proportion of the flow (and pool) of potential innovations accepted.
9. The volume of resources allocated to current investment.
10. The appropriateness of the desired level of consumption in relation to 5, above.[3]

At this level of generality there should be little disagreement. The primary sources of disagreement among students of development are the social and economic contexts within which these factors can or should be determined and their relative importance. If we add to the list such external factors as the growth of the terms of trade in product and financial markets, our list will encompass the explanatory variables of every major study of the postbellum southern economy from the racial theories of early twentieth-century historians (items 3 and 4) to the class-based Marxist critiques of more recent years (items 5, 8, and 9).[4]

My choice of Professor Rostow's delineation of the factors which determine economic growth is motivated by the fact that Rostow explicitly recognized that an understanding of the major factors contributing to economic growth should also enable an understanding of the absence of significant growth. Assuming that this proposition has some validity, it should be clear that any practical use of the idea must be approached within the context of a comparative analysis. One could hardly expect to learn why significant growth did not occur in a socioeconomic system without comparing it to some alternative where substantial growth did take place. This method of analysis is common among students of development economics, and I shall be so conventional as to focus most of my comparative analysis upon the classical case of the British economy in the eighteenth and nineteenth centuries. A major difference from most studies will be a primary focus upon the role of industrial organization as the key element in southern economic performance.

The early postbellum southern economy was being transmuted from a network of centralized plantation factories—operated with slave gang labor—to a network of decentralized *domestic* production units. In the eighteenth and nineteenth centuries, many British industries underwent a mirrorlike transmutation from a system of decentralized domestic production units to a network of centralized factories operated with free labor. The domestic or "putting out" system which dominated British textiles prior to the emergence of the textile factory centered production in the homes of

[3] W. W. Rostow, *The Process of Economic Growth* (New York, 1952), pp. 56–57.
[4] The two most complete analyses that stress racial factors are Matthew B. Hammond, *The Cotton Industry* (New York, 1897, reprint 1966), and Robert P. Brooks, *The Agrarian Revolution in Georgia, 1865–1912* (Madison, Wisc., 1914). Two recent Marxist studies are Jay R. Mandle, *The Roots of Black Poverty* (Durham, N.C., 1978), and Jonathan M. Wiener, *Social Origins of the New South: Alabama, 1860–1885* (Baton Rouge, 1978).

individual weavers. These weavers, with their families and sometimes additional hired help, produced cloth for either a local market or a merchant, who often supplied them with the essential materials needed for production.

Because there were regional and even local differences care must be taken when attempting a detailed description of the organizational structure of the domestic system. The primary differentiations are concerned with the types of direct producers and the relationships between them and other agents involved in the process of production and marketing. In general, we may single out five types of producers distinct from (a) the master weaver who employed workmen and their families:[5]

(b) The customer-weaver producing in an independent status, sometimes selling directly to a market and sometimes filling orders for merchant-factors;
(c) The weaver self-employed and working at home on a *piece* rate for a choice of masters;
(d) The journeyman weaver working at home for a master;
(e) The journeyman weaver working in the shop of his master full time;
(f) The part-time weaver, often a farmer.

The exact relationships between these producers and their masters and merchant-factors might easily involve a mixture of these categories in many individual cases. However, for theoretical clarity we shall keep them in their purest form. The independence of those in category (b) could exist in gradations, from those producing only for the open market in complete independence to those selling only to merchant-factors. We will insist that producers in this category own their fixed capital (looms and so on), although they may be in debt to a merchant-factor or a master. The last three categories will invariably not own their fixed capital. This capital, as well as such working capital as raw materials (yarn, for example) and consumption subsistence, was often supplied at a price or rental fee by a master or merchant-factor. In many cases the cottages occupied by these three types were leased to them by the employer as partial payment for their services.

William J. Ashley shows that these same distinctions also existed in the industrial history of other countries. The German word for the last three categories is *Verlagsystem*, Ashley explains: "*Verlager* is a term still in common use for a merchant who gives out work to be done in the employee's own workshop or workroom: the English term which most nearly covers the same meaning is *factor*." The primary functions of the merchant-factor

---

[5] The following draws upon William James Ashley, *The Economic Organization of England* (London, 1914); T. S. Ashton, *An Economic History of England: The Eighteenth Century* (London, 1955); Paul Mantoux, *The Industrial Revolution in the Eighteenth Century* (London, 1928, reprint New York, 1961); E. P. Thompson, *The Making of the English Working Class* (New York, 1963); Abbot Payson Usher, *An Introduction to the Industrial History of England* (Boston, 1920; and J. H. Clapham, *Economic History of Modern Britain*, 3 vols. (Cambridge, 1926–38).

were to perform the role of a middleman between producers and buyers and to act as a lender of credit. The last function often placed him into the role of the first category, (a), the master who employed those in categories (c) through (f). A schematic representation of these categories and their relationships is provided in figure 3.1. The arrows represent directions of product flows. The broken arrows represent secondary flow paths connecting the same agents as primary paths but used less often. The height of the various categories is meant to capture any possible relationships of dependence, such as borrower-lender and authoritative control.

Figure 3.2 is a schematic representation of the organizational structure of the postbellum agricultural economy during the period 1880–1940. The producer relationships may be described as follows:

(a') The planter who employed those in categories c' through f';
(b') The small independent farmer owning his land and other fixed capital and producing for the market or selling to a merchant-factor;
(c') The sharecropper living in a house and working on land, both provided by a planter, for a share of the crops;
(d') The renter self-employed and working at home on leased land;
(e') The full-time wage hand working on the land of and under the direct supervision of a planter;
(f') The part-time wage hand or day worker who often leaves nonagricultural employment during peak agricultural seasons.

The obvious similarity between these two economic structures suggests what modern mathematicians define as two systems or structures with an *isomorphic* relationship. Each category of the domestic system can be associated with a unique category in the agricultural system. The two

Figure 3.1   Market-Buyers and Sellers

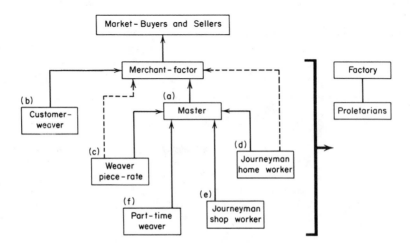

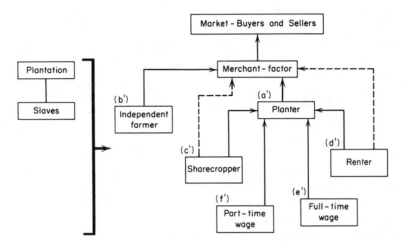

Figure 3.2   Market-Buyers and Sellers

categories have the same hierarchical position, and agents in the two categories perform the same function. For example, the customer-weaver, category (b) in the domestic system, has an equivalent hierarchical relationship of independence or dependence with (a), (c), (d), (e), and (f) as the small independent farmer (b') has with (a'), (c'), (d'), (e') and (f'). Furthermore, the customer-weaver and the small independent farmer in their respective systems perform identical functions; that is, both produce for the market with their own hands and enjoy the full fruit of their labor. Without belaboring the point, the same can be said for each of the "categorical pairs" in the two systems. This isomorphism is highly suggestive of further similarities in the socioeconomic systems of the two structures. We shall discover that such similarities exist and use that fact to gain a deeper understanding of the postbellum southern economy.

One significant difference between figures 3.1 and 3.2 is that in 3.1 the domestic system is evolving into the free labor factory system, while in 3.2 the slave labor gang system is evolving into the domestic tenancy system. Was the movement from gang factories to domestic tenancy an anachronism? The isomorphic relationship suggests that the answers to a variety of questions that may be posed to the two economies must stand or fall together. The explication of the forces which effected this transmutation of the organization of southern agricultural production units is the main technical task of this book. Completion of this task involves dealing with a large set of social and economic questions which cannot be discussed in isolation.

The most obvious empirical difference between the British and southern United States economies during the periods under consideration is the fact that in 1865 not only were the South's wealth and capital stocks not entering a period of rapid growth, as had these two stocks in Britain during the nineteenth century, but the South's wealth and capital had actually suffered a calamitous decline.[6] The primary burden of this study is to demonstrate that the comparative inversion in industrial organization just discussed is largely explained by the inversion in the direction of change in wealth and capital stocks at the two respective time periods. A much less comprehensive variant of this explanation, intended only to explain the long period required for the region to regain its relative income position (table 3.1), has maintained that the once-and-for-all wartime devastation of southern physical capital was responsible for the South's prolonged failure to regain the position of 1860.

In their recent book, Roger Ransom and Richard Sutch offer a fresh modification of an older argument. They claim that the decrease in labor supplied by the free black population was the chief reason for the South's reduced relative income. They present strong evidence that the magnitude of the loss of nonslave wealth has been greatly exaggerated, and argue effectively that the quick recovery of other post-war economies suggests that the impact of the Civil War is insufficient to explain the problem.[7] Below it will be shown that both Ransom and Sutch's argument and the one they critique overlook crucial differences in the early postbellum southern economy and the economies of post–World War II Europe, to which Ransom and Sutch make comparisons.

In the destitute conditions of the southern economy in 1865, laborers and landowners had to join in an effort to resume production and economic life. For the landowner, this meant that he had to gain command over the use of labor services and working capital such as seed, tools, work stock, and subsistence consumption for his own family as well as his work force.[8] The underlying destitution was the major exogenous condition which affected the historical outcome. For the landlord endowed with greatly reduced resources, the two requirements of physical capital and labor services produced an abnormally high demand for credit. In what position was the planter to secure such credit? When the anticipated profit of an investment venture is uncertain and there is a chance that the borrower's future receipts will not cover the amount of the loan, the lender must concern himself with the borrower's underlying asset or wealth position. With the *uncompensated emancipation* of slaves came two important consequences:

---

[6] Ralph Andreano, ed., *The Economic Impact of the American Civil War* (Cambridge, Mass., 1967). James L. Sellers, "The Economic Incidence of the Civil War in the South," *Mississippi Valley Historical Review*, Sept. 1927.

[7] Roger L. Ransom and Richard Sutch, *One Kind of Freedom: The Economic Consequences of Emancipation* (Cambridge, 1977). Brooks and Hammond stress the reduction in black labor. Ransom and Sutch, however, do not build their model on the thesis of race inferiority, as do Brooks and Hammond.

[8] My use of the term *labor service* rather than *labor* is intentional. The ideology which the planter class, and most other southern whites, brought into the free labor regime maintained that the employer must have control of the black laborer per se in order to control his services. This has important implications both for the present chapter and later, when we seek to understand what some writers have termed "postbellum planter paternalism."

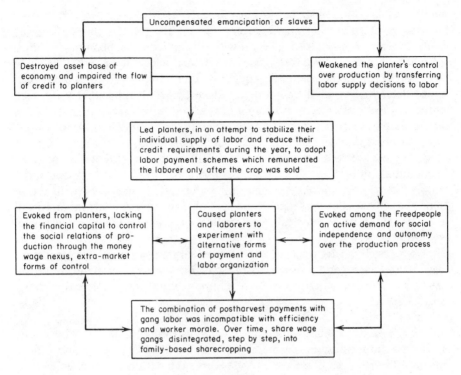

Figure 3.3

(1) The transfer of labor supply decisions to the laborer weakened the landlord's control of the production process *and* increased the real cost of
labor services;

(2) The wealth collateral of planters decreased by the value of their slaves.
This destroyed the asset base of the antebellum credit system.

Both of these lowered the credit-worthiness of planters. The first reduced
the expected profit possibilities at each set of prices on goods other than
labor services. This increased the possibility that returns may fall below the
value of current debt, and the borrower's asset position, which played the
role of insurance for the lender, became all the more important in case of a
loan default. The uncompensated loss of slave wealth, coupled with the
depreciated value of other property, all but eliminated this insurance collateral. Given these conditions, even without the related collapse of
southern banking, the postbellum credit system could not equal the antebellum system.

My chronicle of the evolution of family based sharecropping begins with
observations 1 and 2 and traces them along the interconnected paths illustrated in figure 3.3. Each entry in the chart is a simplified summary of complex phenomena. All of these issues are treated at some length, and from a
variety of perspectives, in different sections of the book. An oversimplified
exegesis might stress the chain of events flowing down the center of the

chart. A reader interested in an account that strays least from a "pure" economic interpretation might, after finishing the present chapter, simply read Chapters 9 through 12. There it will be found that the planter's recourse to what I term "postharvest payment schemes" led, in the wake of three successive crop failures and consequent planter bankruptcies and labor payment defalcations, to a demand for more security and less risk by both employer and employee. For reasons that are entirely historical, and involve activities of the Bureau of Refugees, Freedmen, and Abandoned Lands, both parties opted for share payments in lieu of money wages. Crucial to an understanding of the development of family sharecropping is the fact that this mutual adoption of share-wage *payments* occurred in the context of the gang system of labor. As an economic institution, gang labor with group share payments is an inefficient and unstable organization of labor for all but the most paternalistically controlled or collectively oriented social groups.

With a logic all its own, the centralized organization of the plantation with share-wage gang labor inexorably disintegrated over time under the pressure of its incompatibility with economically efficient incentive structures.

This explanation, while a central theme of this book and logically consistent within itself, begs some important questions by making a number of implicit assumptions. Foremost of these is the actual behavior and motivation of the freedpeople. Even if one accepts the notion that collective payment work organizations is incompatible with social harmony and economic efficiency among an independent and individualistically oriented population, these characteristics must still be shown to be true of the freedpeople. With respect to independence, a large literature, from Stanley Elkins's dependent Sambo to Eugene Genovese's statement of reciprocity and paternalism between master and slave, suggests that the postemancipation behavior of ex-slaves would be easily compatible with institutions that could not survive the presence of socially independent people.[9] After all, what long-term institution was based more upon a collective payment work system than plantation slavery?

Why was the dominant planter class unable to maintain more continuity with the social relations of production inherited from slavery? This question is examined in Chapters 4, 5, and especially 6 and 7, and in Chapters 11 and 14. As the left side of figure 3.3 indicates, an important determinant of this failure can be traced to the planters' loss of wealth and economic power. In chapters 6 and 7 examination of the changing social relations between planter and field laborer leads to a discussion of the connection between free markets and social deference.

Much of the American slaves' experience and possibly their African heritage might lead to a strong collective world view regarding economic as

[9] Stanley Elkins, *Slavery: A Problem in American Institutional and Intellectual Life* (Chicago, 1959). Eugene Genovese, *Roll Jordan Roll: The World The Slaves Made* (New York, 1974).

well as other social phenomena. The absence of this world view in certain forms and its occurrence in others must be examined. In the chapters just listed these topics are discussed. In Chapters 11 and, particularly, 14 an attempt is made to elucidate the freedpeoples' own independent effort to revolutionize the agrarian structure of southern society through collective action. Why this effort failed and, indeed, why so little is still known about it are questions closely related to the book's other themes.

Chapters 4 and 5 are a continuation of 1 and 2. In these next two chapters I examine the implementation of the Benthamite approach to creating a black yeomanry. The failure of contract leasing and the interface between northern lessees, southern planters, Freedmen's Bureau labor policies, and the black laborer are discussed within the framework of the interpretation of the central problem of Reconstruction developed in the first two chapters. Chapter 12 after examination of these issues, returns to an assessment of the questions raised in the present chapter.

II

Roger Ransom and Richard Sutch have argued that the financial losses to slaveholders brought about by uncompensated emancipation "cannot be generalized and used as evidence of the destructive impact of the Civil War" on the productiveness of the southern economy. The basis of their argument is that

> the outlawing of slavery did not destroy the "capital" embodied in the black population. The apparent disappearance of nearly one-half of the southern capital stock represented, not a loss to the South, but a transfer of the owner-ship of "capitalized labor" from the slaveholders to the ex-slaves themselves.[10]

This argument is logically fallacious.[11] It is true that emancipation represented a transfer of the ownership of labor services from master to freedperson: the slaveowner's loss of the capitalized value of those services was canceled out by the freedperson's gain. Since ex-master and freedperson were members of the same society, the net loss to that society incurred by the transfer of labor services would equal zero.

To the ex-slaveholder the loss was not just of labor services, however, but of the laborer as property. The slave, being a capital asset, could be sold on

---

[10] Ransom and Sutch, p. 52. Charles Nordhoff made the same argument in 1875: "The emancipation of the slaves destroyed at a blow for the slaveowners, the greater part of the accumulated capital of these states. The labor is still there . . . but in the redistribution of this wealth the former wealthy class is reduced to moderate means. It is by no means a public calamity" (*The Cotton States in the Spring and Summer of 1875* [New York, 1875], p. 24). Interestingly, the same error was made by J. S. Mill; see his *Principles of Political Economy* (London, 1862), p. 8.

[11] The non sequitur is of a classic type studied by philosophers of language. If the argument is to be valid the phrases "capital stock" and "capitalized labor" must have the same semantical meaning. They do not, however, since the first refers to slaves as physical objects or property and the second refers not to that physical object, but to one specific and intangible aspect of the object.

the market for human capital or offered as insurance collateral for a loan. For Ransom and Sutch's argument above to be valid, it would be necessary that the master's loss of slave capital be no loss to the economic system as a whole. The freedperson must be able to sell complete control of her or himself, for any fixed length of time, as a negotiable asset on a market at the *same* price terms as the former master could have obtained. But free laborers cannot sell or, what is essentially the same, offer their bodies as collateral upon the same terms as those on which a master can negotiate a slave. The primary reason for this is explained by the fact that there exist obvious incentives for free laborers who have sold themselves into bondage to renege upon their contracts at some future date. To prevent this the legal system must sanction the sale of persons as property and enforce, through the courts, contracts of this nature. Thus ancient Roman law mandated that "any *liber homo* . . . who knowingly allows himself to be sold as a slave, in order to share the price, is enslaved, or as it is more usually expressed, is forbidden *proclamore in libertatem*, i.e., to claim his liberty."[12]

Such a sanction is anathema in a nonslave society. Indeed, by definition, any society that did legally sanction such contracts would be a slave society. With uncompensated emancipation, the property and collateral aspect of the black population became a dead-weight loss to the economy. This was a point well understood by planters or creditors who inherited antebellum debts from loans which had been secured by slave collateral.

The argument that emancipation had no social costs has become so entrenched that it will be wise to dwell upon it a moment longer. Some have argued that since the ownership of slaves enhanced borrowing power solely because of the slaves' productive labor, the fact that the labor was still in the South left overall credit-worthiness unaltered. This argument, simply a rephrasing of Ransom and Sutch's, falls victim to a fallacy of composition, holding that what is true for a whole must necessarily be true for each of the constituent parts of that whole. Thus while it is true that the emancipated labor was available to employers as a whole, no single employer had control over the future disposition of any labor services, and therefore no single employer could hypothecate a loan by promising someone else's labor as surety in the event of crop failure, as could have been done if the laborer

---

[12] W. W. Buckland, *The Roman Law of Slavery* (Cambridge, 1908), pp. 427–28. In economics and financial literature, the phenomenon of the erstwhile bondswoman who reneges on her contract is called "the moral hazard." This generally can be defined as the existence of incentives for one party of a contractual agreement to alter future behaviour to the detriment of another party who lacks important information required to prove or anticipate *intent*. A simple example is the financial incentive to commit arson after purchasing a fire insurance policy. Under Roman law, the putative slave could still claim liberty if it could not be shown that he had actually shared in the price of his sale. See Buckland, p. 428, and R. H. Barrow, *Slavery in the Roman Empire* (London, 1928), pp. 11–12. In recent years, formal recognition of the importance of moral hazard in affecting resource allocation and market arrangements has dramatically changed economists' thinking. See Kenneth Arrow and Frank H. Hahn, *General Competitive Analysis* (San Francisco, 1971), p. 126, and their citations on p. 128.

were owned. At least one contemporary understood this problem and its relationship to the credit system:

> The change that has passed over the slave states has in the meantime dislocated the conditions of credit. When the planter was an owner of slaves, and had along with that ownership and *fund of property*, now swept away, an *unlimited control over the labor* . . . he enjoyed great credit in the rivertowns and seaports. That is now gone.[13]

According to Ransom and Sutch's own estimate, slave capital represented $1.6 billion of the assets in the five major cotton-producing states alone. Their estimate makes this 45.8 percent of the total wealth of these states.[14] The uncompensated loss of such a significant fraction of the region's real wealth dealt the financial system a severe blow. We shall see that this loss of social wealth created serious inefficiencies in southern credit markets which permeated the entire economic system. These inefficiencies became crucial determinants of the legal status of laborers in an economic organization whose basic structure was to come into fundamental contradiction with the democratic ideals of a free labor society.

One fortunate circumstance for cotton production was that the high cotton prices caused by curtailed production during the war initially attracted capital from the North. But the extreme credit requirements dictated by a heavily indebted and war-torn economy were not met because of prevailing high interest rates and northern doubts about the social stability of the southern economy. Much of the immediate capital flowing from the North seems to have settled prewar debts and provided working capital for current operations. This left few funds for the heavy fixed capital investments required to revive the plantation economy.[15]

Planters large and small desperately searched for means of securing credit and seemed willing to meet extraordinary terms. A group of planters organized the American Cotton Planters' Association in New York City in September 1865 for the purpose of borrowing $60 million from northern and European financiers. There is no evidence that this attempt to raise such an enormous sum of centrally controlled capital was successful, but the terms upon which members volunteered to pay creditors are indicative of the financial difficulties planters were experiencing.

The association members resolved to pay creditors one-quarter of their net cotton revenues in addition to interest of 10 percent per annum. According to the very detailed calculations of the association's treasurer, Robert V.

---

[13] Robert Somers, *The Southern States since the War* (New York, 1871), p. 241 (my emphasis). For a discussion of the fallacy of composition in an economic context see Paul A. Samuelson, *Economics*, 8th ed. (New York, 1970), pp. 11–12. A very general discussion is contained in George Edward Moore, *Principia Ethica* (Cambridge, 1908), pp. 27–36, especially the discussion of the "organic whole," pp. 30–34.

[14] Ransom and Sutch, pp. 52–53.

[15] Harold D. Woodman, *King Cotton and His Retainers* (Lexington, Ky., 1968), p. 250. Woodman's study of the ante- and postbellum southern financial system is an invaluable aid to research on the southern economy.

Richardson, a $1 million loan was expected to bring the lender a gross return of $4,283,302 at the end of five years. The estimated payment at the end of the first year was to earn the lender a net return of 33 percent on his entire loan, with all payments following being pure profit, the principal having been returned the first year. In addition, borrowers pledged to secure the loans with mortgages on their plantations, a lien on the growing crop, and a promise to ship all cotton to the factors designated by lenders. These terms were to be restricted to "the pick and choice of both planters and plantations to select the best lands, and the most industrious, skillful, successful, and reliable planters."[16]

If the "pick and choice" of planters were willing to meet terms as adverse as these, smaller planters were indeed in dire need. Journalist Whitelaw Reid gave an eyewitness account of the fervent search for credit among planters in the cotton belt stretching from Memphis to New Orleans:

> Scores of planters were already announcing their anxiety to borrow money on almost any conceivable terms, to carry on operations for the next year. . . . Small planters from the interior of Mississippi, proposed to a heavy capitalist, in considerable numbers, to borrow severally ten to fifteen thousand dollars, to mortgage their plantations as security for the loan and give the consignment of one-half the crop as interest for the year's use of the money.

Few students of the era have doubted that credit was exceedingly expensive, but expensive credit need not cause structural rearrangements in business operations. The only certain consequence of high interest rates would be a curtailment of borrowing and planting operations accompanied by an increase in the volume of complaints about the exorbitant cost of credit. The severe difficulties confronting planters were exacerbated by the fact that credit was not only expensive but, for many, unavailable in desired amounts even if a prospective borrower were willing to pay the prices being asked. J. D. B. DeBow, the most informed contemporary authority on southern industry, assessing the situation in February 1866, wrote that "the planters find it impossible to produce advances to work the estates. Capital is too cautious to seek such adventures. The capital and labor which was to have come from the North or Europe have not yet appeared."[17]

In the antebellum period, the cotton factors were the middlemen who held the southern credit system together. Factorage houses, often situated in such centres of southern trade as New Orleans, Mobile, and Charleston, would obtain financial credit from northern and European financial firms and southern banks. With the obtained line of credit, they would extend

[16] Mary Wilkin, "Some Papers of the American Cotton Planters' Association, 1865–1866," *Tennessee Historical Quarterly*, vol. 7 (Dec. 1948), vol. 8 (March 1949). R. V. Richardson was also vice-president of the United States Cotton Company, and according to Mary Wilkin had acquired 100,000 acres of Arkansas land prior to the Civil War. The document from which these calculations come is a good example of the business ability and profit-estimating techniques of the educated large planter (see ibid., pp. 344–47).

[17] Whitelaw Reid, *After the War*, pp. 414–15. J. D. B. DeBow, Editorial, *DeBow's Review*, Feb. 1866, p. 332.

loans to planters desirous of obtaining the working capital needed to produce a crop during the year. Often the planter receiving credit pledged the
harvest of cotton to the factor extending finance. The factor would then sell
the crop. In the early postwar period, many of these factors found
themselves unable to provide the services required by planters. Some factors
had been bankrupted by the war. The northern blockage of southern shipping and confiscation of cotton stocks, combined with the failure of
planters to repay loans secured by slaves, broke some factorage houses and
weakened the base of others' credit lines from the North and from Europe.

The factorage firm of Linton and Doughty in Augusta illustrated the
problems to which DeBow had referred: "An experience of 15 years with a
large list of cotton planters is sufficient to enable us to know very well the
standing, means, and integrity of each one of them, and we do not hesitate
to say there are many among them that we would not require the 'scratch of
the pen' from, if we had the ready means at hand to supply them what is
necessary to make the crops."[18] Uncertainty concerning future events and
the state of the economy frightened those with capital away from offering
credit. The fear of bankruptcy was not only in the air but visible. For many
merchants no terms of credit, irrespective of how high, were acceptable
given the risks involved. The owner of a large furniture store in Mobile
refused to reopen his business:

> Everybody, he said, wanted to buy and nobody had any money. When they
> began to sell their lands, or . . . cotton and get money, he was ready to resume
> business; but till then it would ruin him to have his store open. If he refused
> credit, he would make all his old customers enemies; if he gave credit, he
> would soon be bankrupt.[19]

The southern economy was trapped in a perfect conundrum: no money or
cotton meant little or no credit, which meant little money or cotton. The
credit trap could be broken only if the South could raise a large and profitable crop.

The gods did not smile upon southern producers during 1866–68.
Disastrous crop shortages served only to make conditions worse. The credit
situation became so bleak that southerners who had been preaching the
blessings of laissez-faire and states' rights for decades were forced to pocket
their convictions and ask the federal government for financial assistance.[20]

---

18 Theodore C. Peters, *A Report upon Conditions of the South, with Regard to Its Needs
for a Cotton Crop and Its Financial Wants in Connection Therewith as Well as the Safety of
Temporary Loans* (Baltimore, 1867), pp. 16–17. Also see Woodman, chaps. 21 and 22.

19 Reid, pp. 206–7. Merchants who were open for business were forced to change their
methods of doing business. A group of wholesalers in Charleston published a joint notice to
the public that their antebellum methods of extending credit would have to be altered due to
the "cash and short-term" credit procedures adopted in the North (*Charleston Daily Courier*,
March 28, 1866).

20 In February 1868 the Georgia Constitutional Convention asked the federal government to
make a $30 million loan "to the impoverished planters of the South" (*House Miscellaneous
Documents No. 65*, 40th Cong., 2nd sess.).

All this suggests that one crucial key to understanding the postbellum southern economy lies in a detailed analysis of the structure of its financial system. The credit systems in the antebellum and post–Civil War periods are exhibited in figures 3.4 and 3.5.[21] Arrows indicate sources and directions of credit flows to and within the southern agricultural economy. Dashed arrows again indicate secondary, less frequently used credit flows, and solid arrows indicate primary, more frequently used flows. What strikes the eye is the structural similarity of the two systems. A closer inspection of the figures reveals that after the Civil War there is a new class of *borrower-lenders*, farm tenants; the relative importance of factors and merchants has changed; and the flow of credit from merchants has not only broadened but become a primary source for planters.

We shall find that the forces which altered the credit system cannot be separated from the broader questions concerned with the evolution of the tenancy system. The explanation requires a departure from current discussions of the emergence of the post–Civil War agricultural credit and tenancy systems. The most common procedure is to offer an explanation of how sharecropping tenancy developed, independent of any discussion of the credit system, and then to explain the credit system as a consequence of the tenancy system. A logical implication of this line of analysis leads to Robert

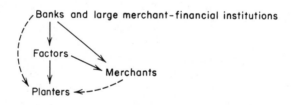

Figure 3.4   Antebellum financial structure

Figure 3.5   Postbellum financial structure

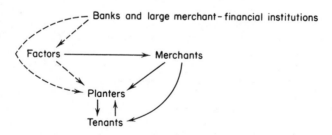

[21] Figures 3.4 and 3.5 draw heavily upon Harold D. Woodman's *King Cotton and His Retainers*, and Alfred Holt Stone, "The Cotton Factorage System of the Southern States," *American Historical Review*, vol. 20 (April 1915).

Higgs's conclusion that "the crop-lien system was merely incidental."[22] The analysis and evidence presented in this book will show that the causal forces in the development of these two institutions were preeminently the collapse of the antebellum credit base and the consequent *simultaneous* development of the new credit and tenure formations.[23]

The adoption of a federal reconstruction policy that did not demolish the plantation system by confiscation and redistribution of land insured that an immediate agrarian revolution, converting the commercially oriented plantation economy into a small freeholder or peasant subsistence economy, would not occur. But the announced intention of enforcing a democratic and relatively decentralized labor policy, along with the refusal to deliver federal aid to a region struggling with a crippled credit market, made it unclear whether the plantation economy, spared from legislative annihilation from without, would be secure from internal disintegration.

In April 1866 Charles G. Baylor put forth a dire prediction for the future of the southern economy:

> Time will show that the cultivation of cotton at the south, upon the principle of vast estates and large gangs of hired laborers, will fail. It will bring poverty to the country, disease and extermination to the African race, and increase the distress of the white people themselves—it will fail.

Baylor's prediction was based upon one crucial assumption. It was an assumption shared with the radical Republicans. The radicals' almost religious belief that the abolition of slavery would result in the breakup of the plantation system and the development of an independent black and white yeomanry was based upon their presumption that the planter class, dispossessed of their human capital, would be unable to obtain the financial credit necessary to maintain a large centralized organization of production.

The analysis underlying the radical presumption was ably summarized by Baylor, a former slave owner, who had served as commercial commissioner for the state of Georgia:

> The cultivation of cotton at the south before the emancipation of the slaves was capitalized. A large amount of capital was committed irretrievably to the growth of cotton and nothing else . . . so that if the planter made a crop at a price so low that it would not pay, his factor, by the aid of banks, would carry him through. . . . that condition of things gave a steadily increasing production of cotton, and it gave that stability and security to the manufacturing interests which justified the investment of capital in the improvement of machinery, and the enlargement of its operations. Now that capital basis at the south is, of course, destroyed through the abolition of slavery. The basis represented by the value of the negroes held as chattel property devoted to this interest has gone, and for the future the lands themselves will have to become

---

22 Referred to here are Hammond, Brooks, and Ransom and Sutch. Robert Higgs, *Competition and Coercion* (Cambridge, 1977), quotation p. 59.

23 A general equilibrium analysis of these developments is crucial to a proper understanding.

the basis upon which money can be loaned in safety in the event of the reduction of the price of cotton to a point below that which would justify its production. The consequence will be, that the attempt to carry on cotton cultivation in the south by vast estates, worked by large gangs of laborers, will be
liable to all the fluctuations and shiftings of the market; and if the price of
middling cotton at New York, Boston, and Liverpool, at the beginning of the
season, is such as to discourage the planter and to induce him to believe that he
cannot afford to pay the established rate of wages to grow it, the consequence
will be that the production will thereby be diminished, or be discontinued
altogether.

This analysis explains the opposition of some free-laborites to the granting
of government loans to southern planters. Edward Atkinson, a fervent
believer in the ideology of free labor, looked upon the attempt "to obtain a
loan for the south," in 1868, "as the last struggle of the plantation system,
which in [his] judgement was doomed by the war."[24]

Charles Baylor's analysis demonstrated a sophisticated understanding of
the financial basis of a slave economy. Its only flaw was its underestimation
of the ability of a wounded market system to sustain its existence indefinitely
while functioning very inefficiently.

With planters no longer able to obtain a flow of credit upon the security
of slave capital, new methods had to be adopted. An immediate awareness
of the required changes is signified by the rapid response of state legislatures
to the concern of citizens like South Carolinean Henry Ravenel that "unless
some means are devised" to enable planters to receive "an advance of funds
or rations, hypothecated on the coming crop,—or the lands are sold to
capitalists who can furnish this advance, they must remain idle." By 1870
every southern state had legalized the contractual right of a lender to attach
a lien upon the crops of agricultural borrowers.[25]

In the attempt to reorganize the antebellum factorage system the legal
formalization of the crop lien was merely the most obvious of the changes
which occurred. Its necessity was indicative of other changes, all of which
were based upon the reduced wealth of the planters. Prior to the war,
planters were able to obtain credit from factors who based the extent of
loans upon easily obtainable information concerning the value of the borrowers' land and slaves. In the prosperous antebellum period land and
slaves were assets which could easily be liquidated to pay off debts that
could not be covered by the value of the year's crops. In the postbellum
period all slave capital had been liquidated at a zero price, and land values
were enormously depressed. But more importantly, with respect to the col-

---

[24] *Report of the Joint Committee on Reconstruction*, 39th Cong., 1st sess., "Testimony of
Charles G. Baylor," pp. 176–77, Georgia. Charles Baylor served as the secretary of the
American Cotton Planters' Association. Atkinson quoted in George R. Woolfolk, *The Cotton
Regency* (New York, 1958), p. 80. Atkinson's views on slavery, cotton, and the world economy
can be found in his early essay, *Cheap Cotton by Free Labor* (Boston, 1861).

[25] Arney Robinson Childs, ed., *Private Journal of Henry William Ravenel* (Columbia,
S. C., 1947), p. 256.

lateral value of land, the large number of landowners attempting to sell land in an effort to raise capital put the real estate market in a state of continual short-run excess supply. More to the point was that although a piece of land could always be sold in time, the expected waiting period became so long that its value to a foreclosing lender was greatly reduced.

Lenders require liquid assets, those which can be sold quickly, as collateral. In the postbellum South the most liquid assets were storable staples like cotton and tobacco.[26] Lenders who once had been willing to grant credit to landholding slave owners on the basis of a gentleman's word could no longer afford to be so amenable. When cotton became the only liquid collateral, lenders found it imperative that they know not only how much land a borrower had but how much of it was in cotton. Creditors also needed to know the condition of the crops at various stages throughout the season. Since labor was mobile the creditor had to know if the planter was able to retain enough hands throughout the season to satisfactorily harvest the entire crop.

Planters were well aware of these problems. "I've got a good plantation, and not a cent of money. The war has stripped me," complained a cotton planter. "The man who loans me the money ought to come right down on the plantation with me or send down an agent. Then he can keep accounts of everything, and see that he is not cheated; he can keep accounts with the hands, and look after things generally."

Even the large planters who formed the American Cotton Planters' Association saw the necessity of accepting such arrangements. In their attempt to establish a large line of credit for members, the board of directors informed prospective lenders that the association would appoint agents to "examine the plantations" and "verify their values and productiveness." These two precautions were neither unusual nor unexpected. However, commercial negotiations with prospective creditors may have played some part in awakening planters of the association to the *new* informational concerns of lenders. The nature of these concerns is outlined in the following letter, dated November 8, 1865, from the association's treasurer to General O. O. Howard.

Sir:
    I have been appointed Treasurer of the "Executive Committee of the Cotton Planters' Association." And as such, it was made my duty, with the Secretary— G. G. Baylor to negotiate, in America and Europe, a loan of money to Cotton Planters, secured by Bonds and Mortgages created upon their plantations. We initiated negotiations, through responsible sources, with capitalists in France. Our proposition has been favourably received and is still under consideration; but a reference has been made to us, by them, for additional information upon the reorganization of free colored labor in the southern states. Your Bureau, having the whole subject of the Freedmen under its care, we desire to propose

---

[26] Somers, *Cotton States since the War*, pp. 21–23. J. C. Hemphill, "Agriculture in South Carolina and Georgia," *Department of Agriculture Special Report No. 47*, 1882, p. 39.

the following points; and do solicit, your earliest possible answer, as fully, as you may be able to give it.

1. Please to state the powers, and duties of your Bureau, upon the subject of Freedmen!

2. Please to state your general system of management of Freedmen! Are they required to seek employment; and are they generally employed, and at labor, throughout the southern states.

3. Does your Bureau encourage them to seek employment under fair contracts? And, do they do so? And do you protect them in their just rights, but require them, at the same time, to fulfill their contracts?

4. Please to state if it is your opinion that the Freedmen will make efficient and reliable laborers in the cultivation of cotton, in the southern states? Please to add any general information within your possession, showing the efficiency, and reliability of Freedmen as agricultural laborers, in the cotton producing states.

You will confer a great favour upon our committee and association, by forwarding your answer to me at 179 Broadway, New York City, as soon as practicable.

I have the honor to be your obedient servant.

R. V. Richardson, Treasurer

Irrespective of their motivations, the cotton planters sought to introduce new practices to supply the desired information. Further *"duties of agents"* were to "supervise the operations of planters, to see that money borrowed is faithfully used in the production of cotton; that the crops are properly planted; cultivated, gathered, prepared, & sent to the Factors appointed by the Capitalists of the Association and to make regular reports on these points."[27] The procedures were identical to the activities which would soon be undertaken by local merchants.

Harold Woodman has noted that technological improvements in transportation and communications systems in the interiors of many southern states had marked the incipient stages of the demise of the factorage system before the War. A large planter explained how these changes would affect small planters and cotton commerce:

This zone has a well planned, and complete system of railroads connecting every part of it with . . . the seaboard. There is now being introduced into these states a new, improved cotton press, very cheap, and of great power. Every planter will have one. . . . thus, the bales will be made on the plantations for shipment to Paris, or St. Petersburg.

These improvements made it *possible* for northern and European financial houses and wholesalers to trade directly with interior planters, bypassing the costs imposed by middlemen factors.[28]

[27] John R. Dennett, *The South as It Is* (1866, reprint New York, 1965), p. 102. Richardson letter in *BRFAL*, film, M752, roll 23, p. 632. Wilkin, "Some Papers of the American Cotton Planters' Association," p. 58.

[28] Woodman, *King Cotton*, pp. 269–76. Wilkin, p. 51. Hammond, *Cotton Industry*, pp. 144–45. On the spread of centralized cotton gins to service the small farmer see Edward King, *The Great South* (Hartford, 1875), p. 11.

The freeing of the slaves, giving them the right to make their own purchasing decisions, made direct trade *profitable*. This change had been anticipated, and the expectation of stronger local markets in the southern economy explains the large influx of northern merchants which preceded the tenancy system. Wendell Phillips, of Massachusetts, argued that his land confiscation plan to deliver both capital and acres to freedmen would solve the "financial condition" of the nation, because "5 million blacks [who] never bought $2.00 apiece before, under [his] plan" would "buy $100 million [in] manufactured goods." A merchant who continued his antebellum business in Alabama verified the predicted effect of unleashing millions of working consumers upon a local market system: Negroes "give an impetus to trade that we never before had. I had sold Jack Peters' negroes more goods this year and last year than I ever sold Peters, and he owned four hundred and fifty negroes. So you see the free negro system is working well with us."

The stimulus of private enterprise also affected local economies' capacity to produce less salubrious goods. In 1865 a Freedmen's Bureau agent in Darlington, South Carolina, reported, "Already numerous distillerys have been established throughout the District; through whose authority I do not know,—the owners of which doubtless anticipate the opportunity of purchasing corn of the negro, at a very low figure."[29] The fact that the alteration in the mode of production from slave to free labor created an enlarged demand for merchandise is often overlooked by those arguing that share tenancy created a need for the local merchant.

Better interior transportation systems made it possible for the local merchant to usurp the factor's business, and the growth of local markets made it profitable. It was, however, the decrease in credit-worthiness and the concomitant desire for detailed information and control over the production process by the providers of capital which made the merchant's existence imperative. Since under the new credit arrangements creditors had to follow the advice of the Cotton Planters' Association by hiring agents to monitor the activities and prospects of borrowers, there were no better candidates for the job than the merchants who were already in residence or arriving by the score to sell the laborers goods.

These changes were a consequence of the loss of wealth collateral embodied in the slave population and were prior to and independent of share tenancy. But the development of share tenancy did increase the scope and necessity of merchant operations.

A correspondent of the *Atlantic Monthly* described the merchant's close observation of his clients in the early 1880s: "During the summer, and all the time the crop is growing, the dealer rides about the country and inspects each man's fields, or sends some competent man to do it, so that he can estimate the probable product." A more colorful description, also explaining

<hr/>

[29] *New York Times*, Nov. 7, 1866. *The Colored Tennessean*, March 31, 1866, p. 1. *BRFAL*, film M869, roll 34, Aug. 22, 1865. John Hammond Moore, ed., *The Juhl Letters to the Charleston Courier* (Athens, Ga., 1974), p. 78. *New York Tribune*, May 23, 1881.

the merchant's source of control over the work effort of borrowers, was given in the early 1870s by one Mr. Solomon, a southwestern Mississippi merchant: "I have three horses riding on saddle—my own one of de best pacers in de country; and when Sunday comes I say to my clerks, 'Go you dis way and dat,' and I go de other, and we see how de work is going on; and if negro is doing nothing we put them all,' with a wave of his hand, 'outside.'" "Outside" referred to a suspension of credit: a tenant or planter whose crops were not looking good would be placed "outside de store."[30]

The early postbellum origins of this method of labor control can be seen in 1867 lien contracts between merchants and share laborers. The practice, from the perspective of the employer, was absolutely necessary. The sharehand who was to receive an uncertain payment at the end of the year had a strong incentive to remain on the plantation, but the strength of the incentive effect upon day-to-day performance was uncertain. Given sharecropping, for both the planter and the merchant the most effective tool for inducing work throughout the season was the withdrawal of rations. The practice was too reminiscent of antebellum techniques.

The share cropping solution ensured that the grand scheme of the radical Republicans was destined to fail from its inception. The structural foundation of the southern economy very quickly showed itself to be too weak to provide a base for the creation of a black bourgeois farmer class. The first aspect of the radical plan to perish was the hope that freed labor would be paid a fixed, certain wage and thereby develop habits of thrift, foresight, and calculation. To those at all acquainted with the conditions of the financial system in the South, it was clear that the money wage path to the Protestant work ethic was for most laborers and planters inaccessible.

Historians generally agree that after the Civil War planters moved to continue the prewar gang system of centralized field labor. This belief is verified by contemporary accounts. For example, Frances Butler Leigh described the organization of work on her father's plantations in the Georgia Sea Islands thus: "In all other ways the work went on just as it did in the old times. The force, of about three hundred, was divided into gangs, each working under a head man, the old Negro drivers, who are now called captains, out of compliment to the changed times." Whitelaw Reid gave a description of a common arrangement in the bistate area about Natchez, Mississippi: "two gangs made up the working force on each plantation; and each was under its own Negro driver." Of the day's work routine, Reid wrote:

> The laborers here went to the fields at daybreak. About eight o'clock all stopped for breakfast, which they had carried with them in their little tin buckets. Half an hour later they were at work again. At twelve they went to the quarters for dinner; at half-past one they resumed work, and at sunset they could be seen filing back.

[30] *Atlantic Monthly*, vol. 51, Jan. 1883, p. 90. Somers, *Southern States since the War*, pp. 241, 242. King, p. 274.

An examination of labor contracts made by freed laborers and employers for the years 1865 and 1866, along with the field reports of local agents of the Bureau of Refugees, Freedmen, and Abandoned Lands, completely validates this consensus.[31]

According to most historical accounts of the period, the great majority of employers sought not only to maintain the gang labor regime but also adopted a fixed money wage payment system.[32] This claim is factually incorrect, and its general invalidity is of paramount importance to a proper understanding of the origins of postbellum labor organization.

The stress upon the planters' supposedly unsuccessful attempt to organize labor upon the basis of fixed money wage payments is inherited from the planter school of historians. That point played an important role in their argument that free black wage laborers constituted an inherently unstable and inferior labor force that brought ruin to southern agriculture.[33] In actuality, taking the South as a whole, no general attempt to perpetuate the centralized plantation system on the basis of money wage payments was ever made! In the first years of Reconstruction a great variety of payment systems were in use, and the evidence shows that money wage contracts were not nearly as prevalent as most historians have implied. In particular, the system stipulating that "half the money wage was paid weekly or monthly, the other half retained by the employer until the end of the contract period "(one year), described as common by one historian,[34] was in fact the least often used.

This issue illustrates the importance of focusing upon the cost and availability of credit to planters. If the problem of inadequate credit was as important as contemporaries claimed, it would be surprising if that problem did not manifest itself in some way during the first years of economic reconstruction. Reports received by the assistant commissioners of the Freedmen's Bureau from local agents do not support the contention that the great majority of labor contracts were for a stipulated monthly money wage. In January 1866, for example, James Davison, the citizens' agent of the Georgia bureau in the northeastern Black Belt counties of Greene and Oglethorpe, reported that

[31] Frances Butler Leigh, *Ten Years on a Georgia Plantation* (London, 1883), p. 56. Reid, pp. 485, 495–96. My statement here and most others made in this book about the functioning of the labor market in the early years of Reconstruction are based upon an analysis and codification of thousands of original labor contracts and thousands of pages of bureau field reports, all on file at the National Archives, Washington, D.C., Record Group (RG) 105, cited hereafter as *BRFAL*.

[32] Ransom and Sutch, p. 57: "so it was that in 1865 and 1866 most of the prewar plantations were reestablished. The freedmen were hired for fixed wage payments, and the work-gang system was reintroduced. . . ." Mandle, *The Roots of Black Poverty*, p. 17: "In the immediate aftermath of emancipation, wages were used in an effort to attract labor to plantation agriculture." Also see Higgs, p. 43. Earlier statements to this effect can be found in Brooks, *The Agrarian Revolution in Georgia*, p. 18; Hammond, *The Cotton Industry*, p. 125; Joel Williamson, *After Slavery: The Negro in South Carolina during Reconstruction, 1861–1877* (Chapel Hill, 1965), p. 130.

[33] Brooks, Hammond, also Walter L. Fleming, *Civil War and Reconstruction in Alabama* (New York, 1904); E. Merton Coulter, *The South during Reconstruction* (Baton Rouge, 1947).

[34] Higgs, p. 45.

"The contracts on large farms, generally, are made for from one quarter (¼)
to one half (½) of the crops." An analysis of "many current accounts of farming operations which were published in the Georgia papers," led historian
C. Mildred Thompson to conclude that "in South Georgia" in 1865 and 1866
"a large majority of laborers worked for a share of the crop." In South
Carolina the same trend was prevalent. The agent for the districts of
Orangeburg and Barnwell reported in the spring of 1866 that "the freedmen
are at work quietly, and the most of them contract for a share of the crop."
For the entire state of Texas the assistant commissioner, General Kiddo,
reported for the period May to November 1866 that "there has been no uniformity in the contracts used. At least three fourths of the freedmen work for
a share of the crop, varying from one fourth to one half." In Florida the report
for the year 1866 read: "about one half the laborers on plantations, are
working for a share of the crop, from one half to one third."

In the heart of the tobacco region of Virginia, a typical enumeration was
made by the bureau agent for Halifax County when he reported in the fall
of 1866 that since "the system of working for part of the crop" had not
proven satisfactory, next year "the greater portion will be wages."[35] The
use of share-of-the-crop payment systems was pervasive and immediate. It
is clear, however, that the great majority of laborers working for shares
were not sharecropping on the family tenancy plan that would appear later.
In the bureau reports sharecropping on the family tenancy plan is mentioned
only a few times, and always as an untested and idiosyncratic experiment.

There were two compelling reasons why many planters did not adopt
weekly or monthly moneywage payment systems:

(1) Both planters and bureau officials were concerned that a newly emanci-
    pated body of agricultural slaves might be a very unsteady labor force.
    The workers might take time off every payday until their cash ran out
    and the abdominal pangs of utter necessity drove them back to the
    fields.
(2) The planter's need for financial credit could be greatly reduced if he did
    not have to make a complete settlement with his laborers until after the
    crop had been sold.

The planters, with the reluctant sanction of the bureau, attempted to solve
both of these problems by the use of postharvest labor payment schemes.
The labor-stabilizing aspect of postharvest payments has often been stressed
by historians. Few planters expected their black workers to be so impious as
to fail to devote sufficient amounts of the planters' time to the celebration
of Saint Payday.

A variant of the second function of postharvest labor payments has also
often been stressed. Many historians have argued or at least suggested that

---

[35] BRFAL, synopsis of Letters and Reports, Jan. 1866–March 1869, p. 30. C. Mildred
Thompson, Reconstruction in Georgia (New York, 1915), pp. 76, 70. BRFAL, film M869, roll
34, p. 519. BRFAL, Synopsis, pp. 70, 23. BRFAL, film M1048, roll 45, p. 721. See Appendix C
for more evidence on this point.

the planters' decision to remunerate labor with a share of the crop was dictated by a shortage of "cash." Few economists have made this argument. The reason for this is that shortage of currency had little to do with the situation. It was not cash but *credit* to obtain goods that producers required. A shortage of cash was not by itself sufficient to explain a resort to postharvest payments of any kind. If the employer could have obtained credit from the businesses from which the laborer would buy, a deficit of cash could have been easily overcome. Planters with sufficient access to credit from readily recognized solvent financial institutions could have issued their workers weekly or monthly notes or banks drafts which would have been readily accepted by merchants selling goods to laborers at no risk and therefore little or no discount. In the antebellum period the planters themselves had transacted many of their purchases in just this way by obtaining drafts of credit from factors and banks that were well known to merchants. In the postwar South the severe credit situation forced many lenders to ration credit so that all would-be borrowers were not able to obtain all the credit they required to begin and carry on their planting operations. This rationing of credit created the conditions that made the second function of postharvest payments attractive to planters.

The credit-reducing function of postharvest payments can be explained in simple terms. To carry on planting operations an employer needed to finance the cost of capital and labor inputs. Postharvest labor payments affected only the later. The laborer's earned income may be divided into two parts. The first part is simply the minimum income necessary to provide food and clothing for the laborer's family so that they would be able to subsist throughout the year in a condition capable of providing the labor. This part of the worker's wages could not be avoided. Employers provided subsistence by obtaining credit in kind and making in-kind payments throughout the year. The second part of the worker's income was the excess above minimum subsistence requirements. From this excess the worker could save or purchase consumption items above minimum requirements. But since this second part of labor income was not a physical necessity the employer could reduce credit obligations by deferring payment of part or all of the worker's wages until after the crop was sold. The worker was forced to defer part of consumption—in effect, to save.[36]

In Chapter 9 we shall see that a major problem with this forced saving was that even for wagehands the implicit rate of interest on such savings could be negative. Note also that the magnitude of the forced savings or, equivalently, the minimum level of subsistance provided throughout the year by the planter was a variable determined by local credit conditions and market competition.

---

[36] Hammond, p. 124. In rejecting the possibility of credit rationing the economists have missed the basic correctness of the shortage-of-cash argument. Insofar as the laborer could obtain credit of her own from a merchant on the basis of her promised wage or crop share, this forced savings could be reduced, at the cost of paying interest. This possibility introduces important complications that are best left to Chapter 9.

In 1865, before the credit system could be reorganized at all, recourse to a postharvest payment system was so pervasive throughout the southern states that it seems incredible that historians could have endorsed the view that money-wages prevailed. Faced with a lack of money and credit, planters were virtually forced to make payments in kind or with a share of the crops. Chaplain Samuel S. Gardner, reporting as bureau agent in the midst of the Alabama Black Belt in the midsummer of 1865, covered each of the salient details:

> General Wilson stripped almost every plantation in this vicinity of its stock, so that, as an average, not more than half a crop can be raised. Under these circumstances, connected with the utter collapse of the confederate currency, I have approved many contracts in which labor was exchanged for subsistence, clothes, home, and medical attendance. Many more, however, have given a share of the crop, usually a fifth or fourth.

When planters entertained the idea of paying money wages for the year 1866, they generally agreed with Colonel J. L. Haynes of the First Texas Cavalry that "it would be difficult for the employers to pay their laborers quarterly, as required by present orders." Since "money can only be realized yearly on a cotton crop," it would be better if the laborer received "his pay at the end of the year."[37]

Officials of the Freedmen's Bureau assessed the problem in terms similar to those used by the planters. At the beginning of the 1866 season General Eliphalet Whittlesey explained why so many North Carolina planters were adopting postharvest payments: "It should be said that landowners are much embarrassed for money to reward labor." Whittlesey had expended some effort in encouraging northern capitalists to furnish funds by loans but apparently had been less than successful, for he emphasized that "a great loan and trust company with millions of capital to be invested in farming would do much to develop the industry and supply the wants of all classes of people in the South." His lack of success is confirmed by the colonel's later explanation that, in 1866, "the great difficulty so far has been that they [planters] have not had money to pay them and have been obliged to make some other arrangement which left a settlement to be made for their work at some future time. Even for this year they are obliged to resort to some arrangement for a division of crops." On a more local level, Captain E. L. Clark explained that the resort to shares by a majority of the planters of Catahoula Parish, Louisiana, in 1866 was "rendered unavoidable on the account of the impoverished condition of the planters," placing "them in a state of comparative destitution, rendering it necessary that they should cultivate their crops this year entirely upon credit."[38]

[37] "Final Report of T. W. Conway, July 1, 1865," *Bureau of Free Labor Department of the Gulf*, in Civil War Pamphlets, Yale University Library, p. 33. *Senate Executive Document*, 39th Cong., 1st sess., no. 2, 1866, p. 75. Also *Southern Cultivator* (Athens, Georgia), Dec. 1865, Editorial. BRFAL, film M869, roll 34, South Carolina, p. 348.

[38] BRFAL, film M752, roll 23, p. 203. *Report of the Joint Committee on Reconstruction*, North Carolina, p. 182. BRFAL, film M1027, roll 28, p. 304, also film M869, roll 34, South Carolina, p. 321.

We see that the credit aspect of postharvest or delayed payment systems was well understood by those involved in agricultural production. The importance of finding sufficient and inexpensive creditors was a task of the highest importance in an economy where regular channels of finance were disrupted and characterized by credit rationing at expensive rates. The Southern Land Agency at Vicksburg, Mississippi, in a written statement estimating the costs and profits of cotton operations, explained the matter concisely:

> But as men's financial abilities differ materially, we think it quite possible to cultivate land with smaller capital. Many are hiring men, agreeing to pay but a *small portion* of their wages monthly, and the *balance at the end of the year*; while others *save the use of capital* by procuring supplies on a short credit, or by allowing a portion of the crop for rent.[39]

Recourse to similar solutions of the problems of credit and payment of labor would be a continuing problem for planters during the Reconstruction period. At the end of the disastrous season of 1876, South Carolina rice planter Gabriel Manigault advised his brother Louis to adopt a labor system many planters had been using throughout the Reconstruction era. Under this plan the employer offered laborers the use of land in exchange for two days' labor on the landlord's crops. Additional labor could then be obtained only as it was needed. Manigault, whose brother was at the time undergoing a severe financial crisis, recommended the plan because the system reduced the planter's "cash expenses from $5,000 to $3,000 a year" and, most importantly, allowed him to avoid "the paying of wages at every step."[40]

The examination of the extent of and reasons for the adoption of post-harvest payments takes us a long way toward an understanding of the decentralization of postbellum southern agriculture, if we recognize the cogency of W. Arthur Lewis's dictum that decentralized "production has the advantage, compared with factory production, that it economizes two scarce factors, capital and supervisory skill."[41] This explanation and the data support the somewhat cautious conclusions of the English historian T. S. Ashton, who in reference to the decentralized organization of mid-nineteenth-century British mining believed that "it may well be that the group contract, and the truck system that came to be associated with it, were methods of overcoming the difficulties arising from lack of capital on the part of the mine-owner: if so, it was natural that they should appear in the midlands, where the employers were relatively poor, and that they should be absent in Northumberland and Durham, where the employers were men of rank and property."[42]

---

[39] Trowbridge, p. 381 (my emphasis).   [40] Williamson, p. 136.

[41] W. Arthur Lewis, *The Theory of Economic Growth* (Homewood, Ill., 1955), p. 137. The relative importance of supervisory skill will be addressed in chapters 10 and 11.

[42] Thomas Southcliffe Ashton, *The Coal Industry of the Eighteenth Century* (Manchester, Eng., 1929), p. 111. Truck, the payment of wages in kind or goods, often involved the laborer in a long period of waiting for monetary pay, as in the postharvest contract. The "group contract" was a manifestation of decentralized production. Much more will be said about these institutions in later chapters.

Lewis's point about the scarcity of capital and supervisory skill provides another dimension that shall be of signal importance to this study. Up to this point nothing has been said which distinguishes between the employer's ability to defer the laborer's payment and, given that, the factors which determine the form of payment. Both postharvest money wages and shares in the crop economize on credit needs and stabilize the labor force. But share payments provide an additional bonus. Insofar as the laborer's level of work effort is stimulated by the immediacy and amount of the pay, a fractional share of the crop—since the value of payment increases as effort increases—has distinct advantages over the fully deferred postharvest wage. With the time of payment constant, workers with this incentive might be expected to do more work with a given level of supervision than those simply waiting for a fixed payment at year's end.

Alternatively, workers who know they will receive payments proportional to their daily work efforts at fixed short intervals of time also have incentives to perform with some alacrity. The employer forced to reduce capital expenditures would therefore perceive an advantage in paying shares over fully deferred wages. What we would expect is that regions facing the most severe credit constraints would have a high incidence of share payments. Planters not especially constrained by credit rationing would therefore never adopt fully deferred money wage contracts and would use partially deferred wage payments solely for the purpose of stabilizing their work force. This thesis strongly implies that the incidence of full postharvest and specifically share payments should have been highest in regions where the devastation of property was greatest.

The southeastern seaboard states, and especially the southeastern districts of South Carolina, where planting operations had to be abandoned early in the war because of the Union Army's occupation, must place high upon the list of devastated regions. An enumeration of the 1866 labor contracts in the subdistrict of Kingstree in Williamsburg County, South Carolina, embracing a total of 2,211 laborers, shows that a great majority of laborers (96 percent worked for a share of the crop. It is no doubt possible that this local example represents quite an extreme case. But the extent of our belief in this possibility must be tempered by the information that a large majority in favor of the share payment system was typical for the state of South Carolina, as is evident from the extracts of Freedmen's Bureau reports by county and districts compiled in Appendix C. As we move west, data collected for the state of Louisiana provides detailed information about the nature of extreme cases.

The great Black Belt parishes of Concordia, Carroll, and Tensas, situated upon the alluvial bank of the Mississippi River, do conform to the view that the vast majority of planters paid their hands money wages for the year 1866 (see table 3.2). But the atypicality of this region may well make it the proverbial exception which proves the rule. These parishes and their neighboring counties across the river in the state of Mississippi, being on the alluvial flood plain of that river, contained the most productive cotton

Table 3.2 The Incidence of Share Payments in 1866, Louisiana

Region	No. of plantations	% paying shares	% paying wages	Confidence interval for % employers paying shares (95%)		No. of laborers	% Shares	% Wages	Confidence interval for % laborers on shares (95%)		Average no. of laborers per plantation	
				Minimum	Maximum				Minimum	Maximum	Shares	Wages
1	253	22	78	17,	27	8,270	20	80	15,	25	27.51	35.34
2	465	69	31	66,	72	7,576	69	31	56,	82	13.65	21.41
3	193	54	46	49,	59	2,287	50	50	45,	55	11.56	13.37
4	169	58	42	53,	63	3,397	53	47	48,	58	16.00	20.89
5	436	60	40	57,	63	6,332	57	43	54,	60	15.26	15.32
6	154	62	38	56,	68	3,144	56	44	50,	62	18.46	23.56
7	—	—	—	—		2,495	60	40	56,	64	—	—

SOURCE: *BRFAL*, film M1027, rolls 28–32; film T142, rolls 66–72. A full description of the data source and explanation of estimation methods appears in Appendix A.

NOTE: Regions 1–6 are all in Louisiana, as follows:
1. Parishes of Tensas, Concordia, Madison, and Carroll
2. Parishes of East Feliciana, West Feliciana, Point Coupee, West Baton Rouge, and East Baton Rouge
3. Parishes of Caldwell, Franklin, and Catahoula
4. Parishes of Avoyelles, Rapides, and St. Landry
5. Parishes of Bossier, DeSoto, Caddo, and Natchitoches
6. Parishes of Morehouse, Union, Ouachita, and Claiborne
7. Region 7 is Shelby County, Tennessee.

lands in the United States. It would not be surprising that the owners of the finest plantation lands would have the least difficulties in obtaining credit. In addition, this region, which was opened for the leasing of plantations during the war, was the recipient of capital from thousands of northern cotton speculators acting as financiers and lessees or buyers of plantations. Although estimates of the precise magnitude of this immigration vary, most studies agree with Michael Stuart Wayne's assessment that "by 1866 most of the great cotton estates bordering the Mississippi" were operated by northern "lessees." Extracts of Freedmen's Bureau reports verify that plantations in the alluvial counties offered a high incidence of partially deferred wage contracts.[43]

As one moves away from the Mississippi River and toward regions with smaller plantations or farms, the dominance of wages over shares recedes sharply, until in less favorably situated parishes share payments dominate. Although data gathered for the riverbank counties of Mississippi is not statistically reliable enough to allow confident inferences, research shows it is highly unlikely that these counties differed much from their neighbors in Louisiana in this regard. Again, as one recedes into the interior of the state, research indicates, the same trend shows up as in Louisiana. Especially noteworthy is the major incidence of share payments in Shelby County, Tennessee (see table 3.2, region 7), one of the greatest cotton-producing counties in the South and second in the country in 1870.

Since the average size of plantation labor forces was uniformly greater where wages were paid, the Louisiana data are consistent with the thesis that large planters paid money wages. But any extrapolation to the entire South must be viewed with extreme caution. South Carolina, for example, shows little respect for this thesis. It is most likely that Major General Carl Schurz's report that northern men "almost universally pay wages in money" while "Southern planters," who "had no available pecuniary means," have "almost uniformly contracted with their laborers for a share of the crop" is the best explanation of the incidence of money wages between Natchez and Vicksburg, Mississippi. A close survey of the incidence of share payments in the state of Alabama indicates that in counties outside the plantation belt the majority of laborers were engaged in 1866 for a share of the crop. Even in the heart of the plantation region, the Freedmen's Bureau agent for the district about Demopolis, after discussing the large variety of contracts, could not be sure that wages were most common but ventured the cautious "estimate that rather more than one half the freedmen in this district are employed in this way," with "the balance [receiving] certain shares in the crop."[44]

[43] The itemized enumeration of labor contracts for the Kingstree district can be found in *BRFAL*, RG105, entry 3351, vol. 1, no. 272, pp. 1–30. Michael Stuart Wayne, "Ante-Bellum Planters in the Post-Bellum South: The Natchez District, 1860–1880" Ph.D. diss., Yale University, 1979), p. 89. John W. Garner, *Reconstruction in Mississippi* (New York, 1901). Lawrence Powell, *New Masters* (New Haven, 1980), discusses northern lessees. *BRFAL*, film 1027, rolls 28–32. See also Reid and Trowbridge.

[44] Carl Schurz, *Report to the President, Senate Executive Document no. 2*, 39th Cong., 1st

Extracts from bureau reports (see Appendix C) make it clear that with the possible exception of a very small number of localities blessed with high rates of northern immigration, payment by a share of the crop was dominant throughout the cotton, tobacco, and rice regions of the South in 1866. The heavy incidence of share payments throughout the southern states in that year makes it clear that at the very inception of Reconstruction, Southern agriculture displayed a basic weakness. With respect to the development of southern institutions, by far the most important and ironic aspect of post-harvest payments was the fact that they made the laborer a creditor of the planter! It has apparently not occurred to those making the argument that sharecropping, or any other delayed payment contract, allowed the planter to economize on financial requirements that since the planter was, by doing so, a borrower, somewhere there had to be a lender.

For these reasons, few planters remunerated their workforces on a weekly or monthly basis. Many planters who elected to pay money wages adopted some variant of what came to be known as the "standing wage system." Thomas J. Edwards, a school supervisor in the Alabama Black Belt during the early twentieth century, had this to say about that system of payment: ". . . the standing-wage system which was originated immediately after the Civil War is not now in vogue. The method of work got its name . . . because 'hands' worked for a period of six months or a year, before a complete settlement was made."[45]

Few planters in 1866 were in a position to choose that system of payment which they considered optimal. Their choice was greatly constrained by conditions in the credit market over which no planter had any control. This problem was to haunt the southern economy for years to come.

To assess the extent to which these problems might have been avoided had the federal government adopted a different agrarian policy requires an intensive preliminary analysis of the social forces at work in the South during Reconstruction. How far a purely economic analysis can take us will be seen to be restricted by the revolutionary context of the period. The principal actors performing in the drama produced by the actual federal reconstruction policy were labor, employers, and the Freedmen's Bureau. Any historical analysis must appreciate the fact that after emancipation, ex-slaveholders and ex-slaves found themselves in altered social relations of production, and just as important, that the policymakers of the Freedmen's Bureau were cognizant of what this meant.

---

sess. For Alabama see *BRFAL*, Record Group 105. The quotation comes from *BRFAL*, film M809, roll 18, p. 11. Oct. 24, 1866. This is consistent with collected data for the neighboring counties of Lowndes and Noxubee, Mississippi, which indicate a heavy incidence of shares, with perhaps a majority even on large plantations.

[45] Thomas J. Edwards, "The Tenant System and Some Changes since Emancipation," *Annals of the American Academy of Political and Social Science*, vol. 49 (Sept. 1913), pp. 38–39.

Part II
Two Kinds
of Paternalism

White men from the North are already crowding that way; buying the embarrassed estates, and preparing by free labour to raise the coloured population of the South from its deep depression.

Reverend H. M. Storrs of Cincinatti, 1865

Making money there is a simple question of being able to make the darkies work.

General Francis Barlow of Massachusetts

4

Why should one part of the country derive such great benefits from compulsory labor under contract—for while a contract lasts labor is compulsory—when the farmer of the North and West has to eke out an existence, by everlasting application and toil—and without a contract.

1st Lt. J. E. Quinn

Work-Leisure and the Meaning of Freedom

Will the free Negro work? This, we are often told, was one of the great questions of the day. In a preceding chapter the question was phrased somewhat differently. If there were any differences in opinion about the Negro's proclivity to toil, they were over how the word *work* was to be defined. The dispute was not over whether the blacks would work in the everyday meaning of the word. Nearly everyone agreed that the Negro, like all men and women, would choose to work when the only alternative was starvation. The disputed question was not even, Would the free Negro continue to work like a slave?' A successful Louisiana overseer explained to journalist J. T. Trowbridge that "no Nigger, nor anybody else, will work like a slave works with the whip behind him." The disputed question was whether or not the freed black would soon respond to market incentives with the work discipline of an acquisitive agricultural proletary striving to gain entry into the landowning class.

In the North, the radicals seemed to be arguing that in a fairly short time, given the proper guidance from disciplined white philanthropists and the help of the Freedmen's Bureau, each freedman would soon rise to his or her natural station in society. Some southerners agreed that the proper course

---

¹ Trowbridge, p.401. Dennett, *South as It Is*, p. 84. George Fitzhugh, "What's To Be Done with the Negroes?" *DeBow's Review*, vol. 1, June 1866, pp. 577–81.

of economic adjustment and development called for the elevation and fair treatment of the Negroes. What seems more likely to be a good characterization of the "representative" southerner's position was given by journalist John Richard Dennett. Free labor, a Virginia planter told him, could only end in "the woods full of outlaws; idleness followed by want, and the country impoverished for lack of laborers—all because fanatics were determined," by experimentation, "to lift up the nigger to a level which by himself he could never attain." The same opinion was stated much more eloquently by George Fitzhugh in his classic argument justifying the need for a "black code."[1] Fitzhugh understood perfectly that the real issue involving the Negroes and the work-leisure choice concerned their lack of an appropriate accumulative work ethic. He hastened to assert that such an ethic was not something that could be learned, but was racially inbred: "The White race is naturally provident and accumulative. . . . they have many wants, and to supply those wants, generally labor assiduously and *continually*."

"The planters," wrote U. B. Phillips, "had a saying, always of course with an implicit reservation as to limits, that a Negro was what a White man made him."[2] The planters had many sayings, but this is one we may accept as being at least marginally true. If the black slave had been trained to be satisfied with few wants, it was either because superior training was impossible or because the planter's perceived self-interest was best served by maintaining slaves at a low level of human existence. Having inherited its own legacy, the problem for the southern ruling class was how a civilized society, in which gentlemen and ladies of the "respectable" leisure class were devoted to politics and the higher arts, could be maintained by a laboring class *free* to make its own choice between work and leisure? If the blacks would be satisfied to work barely enough to maintain their own subsistence, from where would the agricultural surplus needed to support a class-stratified society come? The radical Republicans had answered that the surplus would be forthcoming, and more equitably distributed, once the blacks had been stamped with the acquisitive work ethic. Nonradical northerners were not so sure. Could the blacks be so elevated? And if so, how long would the process take, and at what social cost? This disagreement in the North had ended in something of a compromise: the experiment in black elevation would have to be conducted under a *supervised* free labor market, without benefit of black land ownership.

To Fitzhugh, the threat of a truly free labor system posed a dilemma which appeared to have no solution. If the blacks were as inherently inferior and degenerate as he claimed, so that without the close guardianship of whites they were destined to "perish and disappear from the face of the earth," then the South would be left with no labor at all. Alternatively, if the inferiority of blacks did prove to be a consequence of slavery, the impli-

---

[2] Ulrich Bonnell Phillips, *American Negro Slavery* (1918, reprinted Baton Rouge, 1966), p. 291.

cations, pointing to a mongrelized society, were too awful to contemplate. The one acceptable solution to this dilemma was its dissolution. The free labor experiment could not be attempted. The only conclusion Fitzhugh and any similar thinker could make was simple: "A great deal of severe legislation will be required to compel Negroes to labor as much as they should . . . we must have a Black Code."

Since George Fitzhugh is a somewhat notorious historical personality, it might be thought that his is an unlikely opinion to portray as representative. But any perusal of contemporary testimony should convince the most obstinate researcher that this is not so. Even as honest and ostensibly benign a racist as Frances Butler Leigh was fundamentally in agreement with Fitzhugh. The "pure Negro," she thought, was hopelessly inferior and thus "incapable of advancement to any degree that would enable him to cope with the White race, intellectually, morally, or even physically." Emancipation could only mean "that our properties will soon be utterly worthless" since "no Negro will work if he can help it, and is quite satisfied just to scrape along doing an odd job . . . to earn money enough to buy a little food." Finally, she capitulated to the lowest common denominator of the planter mentality. "I confess I am utterly unable to understand them, and what God's will is concerning them, unless he intended they should be slaves."[3]

Eugene D. Genovese calls attention to the fact that Mrs. Leigh's traditional white southern attitude towards blacks, so different from that of her mother and so reminiscent of that of her father, did not blind her to certain realities. She observed that the blacks "performed no worse and often better than English laborers . . . under similar conditions."[4] This is a crucial point which provides insight into the psychological makeup not only of nineteenth-century southerners but of members of the master class in caste-dominated societies in general. Mrs. Leigh made no attempt to reconcile these cross-racial observations with her race-specific beliefs about blacks. How could she? If the Negroes were not inherently inferior, if it were possible to raise them from the depths of their ignorance, what did that imply about her father, her grandfather, and the entire slaveholding class? Mrs. Leigh was not free to take the viewpoint of her mother, Fanny Kemble, an English woman with no stake in southern institutions.

Halfway across the world, similar rationalizations were being made by European imperialists who, in the South Pacific and Asia, were confronted with the problem of mobilizing the recalcitrant labor of indigenous populations for the production of plantation staples. From the start of commercial operations, the colonial powers encountered severe difficulties inducing adequate voluntary labor supplies. The natives, it was usually argued, would not respond favorably to attempts to raise their standard of living. The majority of the people were content to maintain the standard of living

[3] Leigh, pp. 25, 93, 147.
[4] Genovese, *Roll Jordan Roll*, p. 310.

to which they had always been accustomed. In the twentieth century it was still possible for a friendly observer to write about India's Bengal peasants in the following manner:

> The needs of the Bengal peasantry are very modest and limited. A little food, some scanty clothing, a few crude utensils, a humble shelter, a few lean animals to plough with and the simplest instruments for tillage. . . . he lives on the land of his ancestors and he would be happy to die in the same open yard where all those who have gone before him breathed their last.[5]

To offer such people high wages, the argument ran, instead of lifting them to a better way of life would simply allow them to maintain their habitual standard of living while working less and ruining the employer. Gunnar Myrdal states that the colonialists seldom experimented with actual offers of high wages in any systematic fashion. It was much simpler to develop an explanation of the natives' backward-bending labor supply curve on the basis of racial doctrines and "innate characteristics."[6] The darker races were naturally improvident, wanton, and lazy. It was the white man's burden to compel them to work for their own as well as the world's benefit. Neville Chamberlain put forth the European view when he argued that the advance of civilization required that the native be "convinced of the necessity and dignity of labour." "I think it is a good thing," he continued, "for the native to be industrious, and by every means in our power we have to teach them to work."[7]

The southern states, in creating the "black codes," were expounding a conventional position. The northern Republicans excoriated the "black codes" and then, through the instrumentation of the Freedmen's Bureau, set out to enforce compulsory labor, one-year contracts, and the obedience of labor to management. Were the Republicans mere hypocrites? Did their behavior simply follow the established pattern of European colonists toward darker peoples?

[5] Huque Azizul, *The Man behind the Plough*, p. vi, quoted in Ramkrishna Mukherjee, *The Dynamics of a Rural Society* (Berlin, 1957), p. 57. Mukherjee calls Azizul a "sympathetic and patriotic" leading political figure in regard to the problems of the Bengal peasantry during the 1930s. Mukherjee, however, did not believe that the agrarian problems in Bengal were due in any way to the backwardness of the peasants, but to the colonial system imposed on the country and the role of the parasitic landowning class as an appendage to that system." He seems to argue further that given incentives, the peasants would have made capital improvements and accumulated.

[6] Gunnar Myrdal, *Asian Drama* (New York, 1968), vol. 2, ch. 22. It was also, from the European point of view, more profitable. The economic doctrines upon which colonialism was based gave little priority to the creation of diversified economies in colonies filled with well-paid, intelligent natives. After all, the similarities between the master-slave and mother country–colony relationships are hardly subtle. European countries which professed that their occupation of militarily weaker states was as much for the benefit of the weak as the strong had to face the same kinds of inconsistencies between their behavior and their words as did slave owners.

[7] Quoted in J. S. Furnivall, *Colonial Policy and Practice* (Cambridge, 1948), p. 342. Van den Bosch, the governor-general of Java, used the same argument when introducing the culture system in Java during the 1830s.

The complexities of the psychological makeup of nineteenth-century Republicans reveal similarities to their counterparts in nineteenth-century Britain. David Brion Davis has pointed out the "striking discrepancy" between Jeremy Bentham's "competitive model, based on individual self-interest, and his obsessive concern for social discipline." This same discrepancy was evinced by the "early industrialists whose ideal of free market conditions gave no justification for attempts to control the behavior and leisure time of their workers."[8] The roots of this discrepancy lie in the very foundations of laissez-faire philosophy. Adam Smith, while expounding the virtues of individualistic competition and self-interest, failed to give sufficient study to the fact that, to operate as he claimed it would, his system required not simply free workers but modern proletarians—workers so adjusted to the requirements of industrial life that they, too, adopted the habit of accumulation and strove for advancement in the socio-economic hierarchy. Without such workers, how could production be continuous and orderly? What sense could be made of a social order based upon freedom of contract if the majority were irresponsible and owned no property which could be attached for malfeasance? Bentham provided both the model of and the mechanism for reaching the "true" state of a free society. If the working poor's ideas about the meaning of freedom were incorrect, then they must be taught the truth by their masters. There was no contradiction involved. If the lesson was painful to a given generation of the poor, that would matter little once the lesson was learned. In the resulting system of laissez-faire, workers could not only enjoy true freedom but also a higher material standard of life because of, not in spite of, past sacrifices. What more could be asked of a utilitarian?

The gyrations of the mind required in this process are perhaps best seen in British abolitionist William Wilberforce. He had opposed any attempt to improve the living conditions of West Indian slaves through legislation. This was not only because it would be impractical but because he thought it an atrocity to intervene in the personal lives of the masters:

> How impossible would it be found to enter into the interior of every family, and with more than inquisitorial power to ascertain the observance or the breach of the rules . . . but supposing the means of enforcing the regulations to be found, how odious, how utterly intolerable.

Improvement in the lot of the slave must be effected through the self-interest of his master, Wilberforce argued. The abolition of the slave trade, with its consequent shutting off of cheap recruits for the slave gang, would accomplish what legislation could not. "Abolition would give the death-blow to this system. The opposite system with all its charities, would force itself upon the dullest intellects. . . . ruin would stare a man in the face if he did not conform to it."[9] The West India Islands would, by necessity, become dotted with self-regulating panopticons inherently based upon the

[8] Davis, *The Problem of Slavery in the Age of Revolution*, p. 456.
[9] Ibid., pp. 414, 415.

principle of contract management. With respect to the English working class, Wilberforce apparently attached no odium to a system which sanctioned inquisitorial infringements upon the personal lives of the workers. He repeatedly served the interests of business and gave support to the industrial discipline meted out by paternal capitalists. In Benthamite fashion, there was no hypocrisy involved in this apparent shift in regard to the sanctity of individual freedom. The masters, whether their servants be slaves or "freemen," possessed both property and, *a fortiori*, an accumulative habit. They could be trusted to react in an orderly and predictable manner to market incentives, while their servants could not. This philosophy lay at the bottom of both the Union Army's and the Freedmen's Bureau's basic labor policies.

To argue as some writers have done, that there was very little in the labor policy of the bureau that should have been displeasing to the planters is completely to overlook the radical Republican view of Reconstruction and the role assigned to the bureau in that view. A previous chapter argued that the radical and moderate elements of the Republican Party envisioned Reconstruction as an opportunity to create a new South that would, in their eyes, be congruent with northern society. The success of the radical plan depended upon using the Negro as a primary instrument of reconstruction. A complete reconstruction of the South would be impossible unless the blacks could be raised to a level at least approximating social and civil equality with the whites. To obtain this objective, it was obvious to all concerned that the total "plane of living" of the black population must be raised to a level far above that allowed by the peculiar institution. To state the issue concisely, few southerners desired such an elevation for their black residents. As George Fitzhugh put the issue:

> We are not perfectionists, like the northern people, and should not expect or try to make Solomons, nor even Fred Douglasses of the Negroes. We should be satisfied to compel them to engage in coarse, common manual labor, and to punish them for dereliction of duty or nonfulfillment of their contracts.[10]

For radical and moderate Republicans the Freedmen's Bureau was far more than simply a guardian of the free labor system ensuring that social order was maintained and the South's valuable staples produced in large quantities. The bureau was in fact the right arm of the Republican Reconstruction plan. The attack upon the Bureau from the Democratic Party and the conservative elements of the Republican Party must be seen in this light. General Howard recorded the cause of some of the first signs of opposition

[10] Two who make the Freedmen's Bureau and the planters bedfellows are William McFeely, *Yankee Stepfather: O. O. Howard and the Freedmen* (New Haven, 1968), and Wiener. Fitzhugh, p. 578. Fitzhugh in making these arguments was attacking the entire libertarian school, whose devotion to rationalism led them to the conclusion that even Negroes could be elevated to a plane of equality with the rest of mankind. In this same article he opened with the following barrage: "Rationalism is the appalling evil of the day. . . . there is hardly a reformer in New England who would not undertake to make to order a far better social, moral, political and religious world. . . ."

from President Andrew Johnson: "At first President Johnson was apparently very friendly to me, yet, while Mr. Stanton favored our strong educational proclivities, the President declared that the true relief was only in work. One member of his cabinet, Secretary William Dennison, said about the time I took charge: 'General, it is feared that the Freedmen's Bureau will do more harm than good'. . . . it was hard for them to realize that the training of the mind and hand, particularly with negroes, could go together."[11] Oliver O. Howard was perhaps too generous to the opponents of bureau policies. Nevertheless, he was aware that much of the underlying criticism stemmed from the bureau's activities in the areas of education and civil justice for the freedmen. Honing in directly upon its labor policies, modern criticism of the bureau has hit much harder. This criticism misses much, because it discusses those policies without addressing their relationship to the bureau's overall program and scope of operations.

From the inception of economic reconstruction, government officials were acutely aware of the many problems which confronted the free labor system. Given that the federal government chose not to deliver land to the freedmen, the only avenue open for them was to enter some form of contractual arrangement with the reinstated owners of the land. In agriculture, landless laborers are restricted to accepting some kind of wage labor or rental contract. But for most freedmen rental contracts were also unavailable, due to their lack of capital and the associated inaccessibility of credit. These circumstances meant that for the great majority of blacks the choice was wage labor or starvation.

Agents of the Freedmen's Bureau had gained experience with the free labor experiments conducted during the war, and many with top-level positions entered the postwar era with firm convictions in regard to the work arrangements that would best suit the new regime. In Mississippi, General Wood assessed the conditions facing the freedmen and made his recommendation: "as a great majority of the freedpeople are too poor to lease or buy lands and provide the means of cultivating them, it is recommended that they be advised, though not required, to continue the system of labor under written contract." In regard to the form of payment under a wage contract, Commissioner Howard indicated his agreement and approval of Colonel Whittlesey's judgment that "the freedmen, as a rule, worked more faithfully for money than for a share of the crops" by commenting that "the best outlook was on the plantations where employers paid cash at short intervals."[12]

Above all else, the extent of the change in economic relations for black and white labor would ultimately be defined by the degree of mobility which the black worker would be able to exercise. The right to invoke the option to

[11] Oliver Otis Howard, *Autobiography of Oliver Otis Howard* (New York, 1907), vol. 2, p. 227. Edwin M. Stanton, secretary of war in Lincoln's cabinet, became the lone radical in Andrew Johnson's postwar cabinet. Johnson's vigorous efforts to remove Stanton from the cabinet were successful when Stanton resigned following the president's acquittal at his impeachment trial: Kenneth M. Stampp, *The Era of Reconstruction*, pp. 147–54.

[12] *BRFAL, Synopsis of Reports*, 1866, p. 76. Howard, pp. 233, 250.

leave one's present employer at any time is the essential requirement of a free labor system, and this point was quickly appreciated by the freedpeople. "After the coming of freedom," wrote Booker T. Washington, "there were two points upon which practically all the people on our place were agreed, and I find that this was generally true throughout the South: that they must change their names, and that they must leave the old plantation for at least a few days or weeks in order that they might really feel that they were free."[13]

Great numbers of ex-slaves expressed their new freedom by going into the cities and towns; thousands, unable to find regular employment, led a precarious existence. The freedpeople who had come into the cities and towns had done so for a variety of reasons. To many whites it was obvious that this migration was an expression of the Negro's desire for a life of idleness and city pleasures. There is little doubt that this was the motivation of some of the migrants, but the problem was more complex. In fact, the blacks were coming into the towns for the same reasons that they had flocked to the contraband camps of advancing Union armies. Some were driven by destitution and fear of violence in the interior, and others came simply because it was thought that information about better employment conditions could be secured in the towns occupied by government troops. Whatever the individual motivations, the living conditions in the Negro areas that sprang up in these towns became appalling. The city streets and the outskirts of towns were full of unemployed men and destitute women and children. Hunger and disease were so common that many whites were predicting that the entire Negro race would soon disappear.[14]

Under these conditions, it did not take long for those concerned to realize that for the vast majority of blacks, skilled in the production of southern crops, the only demand for their labor was on the farms and plantations. In the early fall of 1865 the black editor of the *Colored Tennessean*, a Memphis newspaper, implored the unemployed freedmen who were crowding the city streets to "go to the country where employment exists for manual laborers" and "become independent of the government and work."[15]

Verbal remonstrations alone failed to provide an adequate stimulus to move agricultural labor into the uncertain environment of the rural interior. Freedpeople who had been forcibly driven off the plantations without pay were understandably loath to return. For many others the rumors of impend-

[13] Booker T. Washington, *Up from Slavery*, in *The Booker T. Washington Papers*, Vol. 1, ed. Louis R. Harlan and John W. Blassingame (Urbana, Ill., 1972). Large numbers of freedpeople, probably the majority, simply stayed on the old plantation or quickly returned: Susan Dabney Smedes, *Memorials of a Southern Planter* (1889, reprinted New York, 1965), p. 217, is a case in point.

[14] John Preston McConnell, *Negroes and Their Treatment in Virginia from 1865–1867* (Pulaski, Va., 1910), pp. 19–23. Sidney Andrews, *The South since the War* (Boston, 1868), pp. 24–25 (South Carolina); Reid, p. 31 (North Carolina), pp. 325–7 (Virginia). Vernon L. Wharton, *Negro in Mississippi during Reconstruction*, pp. 53–56. Kolchin, *First Freedom*, pp. 7–9. J. D. B. DeBow, *Report of the Joint Committee on Reconstruction*, pt. 4, 1866, p. 135.

[15] *Colored Tennessean*, Oct. 14, 1865, p.2.

ing land distributions to freedmen were enough to recommend the policy of refusing to contract and waiting to see exactly what would occur.[16] Government officials and interested private citizens sought a method of solving the unemployment problem in the nonrural areas of the South. The solution agreed to was a variant of what might accurately, although anachronistically, be termed the Tanzanian plan.[17]

The *Colored Tennessean* had isolated the crucial point. If those blacks who could find little or no employment were to remain in the cities, they would have to be supported by government rations. This course was practically inconceivable. Even if General Howard and his assistant commissioners had believed such a policy was desirable, they would have found it infeasible. Oliver Howard was quite sensitive to charges from various groups that "the bureau would," quoting its detractors, "feed niggers in idleness." Howard's first published order as bureau commissioner expressed what would be a major orientation of the bureau's labor policy:

> While it shall be my object to secure as much uniformity as possible in the matter of employment and instruction of freedmen, I earnestly solicit cooperation from all officers and agents whose position or duty renders it possible for them to aid me. The Negro should understand that he is really free but on no account, if able to work, should he harbor the thought that the government will support him in idleness.[18]

Fifteen days later, on May 30, 1865, the bureau drafted Circular 5, its *Rules and Regulations for Assistant Commissioners*. A joint product of General Howard and his assistant commissioners, it was written by John Eaton and approved by President Johnson. With respect to labor policy the circular ordered that "relief establishments" should be "discontinued as speedily as the cessation of hostilities and the return of industrial pursuits" would permit, and that "every effort will be made to render the people self-supporting." If these conditions had the effect of making labor compulsory, in the sense of work or starve, the conditions of that labor were to be radically altered. Article 7 of Circular 5 required that "Negroes must be free to choose their own employers, and be paid for their labor." Agreements were to "be free, *bona fide* acts, approved by proper officers, and their inviolability enforced on both parties." Finally, "The old system of overseers, tending to compulsory unpaid labor and acts of cruelty and oppression, is prohibited."[19]

[16] McConnell, p. 38. "Statement of the North Carolina Freedmen's Convention," Raleigh, Oct. 3, 1865, reprinted in Andrews, p. 129. General Lorenzo Thomas, *Report of the Joint Committee on Reconstruction*, pt.4, p. 141. Captain T. J. Mackey, ibid., p. 158. Trowbridge, p. 455. Wharton, p. 51.

[17] In the 1960s, the government of Tanzania, reacting to an immense volume of rural-to-urban migration and a rising level of urban unemployment, adopted the policy of forcibly removing the urban unemployed back to rural areas. This phenomenon and the official reaction to it is a historical commonplace in underdeveloped countries.

[18] Howard, *Autobiography*, vol. 2, pp. 214, 213.

[19] Eaton, *Grant, Lincoln and the Freedmen*, pp. 238, 239–40.

Having participated in the process of drafting these regulations, most assistant commissioners agreed with both the letter and spirit of the rules. Samuel Thomas, assistant commissioner for the state of Mississippi, admonished Negroes for their reluctance to sign contracts. Thomas informed the freedpeople that the government would neither give them land nor support them. His men had his approval in compelling both employers and employees to fulfill their contracts. "The state cannot and ought not to let any man lie about idle, without property, doing mischief," he ordered. "A vagrant law is right in principle."[20] Here was the ultimate point. Few people, white or black, believed that penniless, uneducated, and healthy men and women should be allowed to misuse their liberty by refusing to work, thereby signaling their intention to live by begging or stealing.

On this the black press and blacks of stature and position were in agreement with the bureau code. In speech after speech black leaders exhorted the freedpeople to return to the plantations and work hard. Black residents of Savannah, Georgia, went far beyond the *Colored Tennessean* and other black voices when they appointed a committee of nine "to assist the officers of the Freedmen's Bureau in freeing the city of vagrants."[21] The adoption of such an activist position by Savannah's black residents suggests the possibility that they were concerned with more than the well-being of immigrant freedpeople from the country. The arrival of a large population of ignorant country folk willing to accept meager wages could not have forecast prosperity for those black laborers already living in the city. In Georgia, General Davis Tillson offered a solution to these problems that was apparently satisfactory to General Fisk, assistant commissioner of freedmen for Tennessee and Kentucky. An article in the *Nashville Dispatch* reported:

> The census shows that in and around Memphis there are seventeen thousand freedmen. The city will only afford legitimate labor and support for a small portion of this large number during the coming winter. There is a large demand for labor in the country. Persons are daily calling upon General Tillson applying for labor, and are willing to pay the highest wages. The Freedmen as a class, seem to prefer a life of precarious subsistence and idleness in town, to good wages and comfortable homes in the country. . . . General Tillson has determined that he will compel them to leave the city, even if it required military force, and has already conferred with the Assistant Commissioner of Freedmen in Kentucky and Tennessee, informing them of his intention in this respect. He intends soon to send a patrol through this city and suburbs, whose duty it shall be to make a thorough tour of the city, and learn who have not employment, or any means or prospect of support. These will be notified that within a certain time they must leave the city and find employment; at the expiration of which time if they do not comply with the order, they will be arrested, and labor furnished for them by the bureau and they will be forwarded

---

[20] Ibid., pp. 243–44.

[21] Charles Wesley, *Negro Labor in the United States, 1850 to 1925* (New York, 1927, reissued 1967), p. 147. For less aggressive positions see *Resolution of the North Carolina Freedmen's Convention* in Andrews, p. 126.

under guard to the place assigned them. . . . no Negro will thus be permitted to break his contract, but will be arrested and compelled to work at the prices and time agreed to.[22]

The bureau, with the aid of the friends of the Negro, could push the blacks into the rural areas, but it was up to the planters to keep them there by offering attractive employment arrangements. With respect to this, the bureau offered one of its most useful and successful services by filling the role of a huge labor market information center and clearinghouse. Outside the world of the Candidean economist, rural labor markets, especially those with illiterate working populations and poor transportation systems, are renowned for the imperfect mobility of labor they often foster. Under competitive conditions, wages for similarly productive workers tend to equality. In a spatial setting this tendency is more or less realized depending upon whether and how fast labor in areas paying relatively low wages receive that information and how fast they are able to traverse the distance to higher-paying areas. In particular, when the laboring population is very poor, the costs of migrating are often a serious impediment to competitive tendencies. This is why migrations of poor populations are usually financed through loans, often leading to a form of indentured servitude.

The Freedmen's Bureau ameliorated these conditions far beyond what would have prevailed in its absence. Bureau offices, receiving daily inquiries from prospective employees and employers as well as information from bureau headquarters, were well informed of the market value of employment opportunities in their own localities and other states. Davis Tillson, after directing that blacks lacking the capital for self-support be forced to enter contracts as laborers, set out to ensure that these contracts gave the laborer a competitive payment. Evidently General Tillson's insistence upon enforced contracts was predicated upon this assumption. Upon assuming his post as assistant commissioner, he objected strongly to the fact that the Georgia Senate and House desired to make labor contracts binding and enforceable, stating that "it would entail the grossest injustice upon the freedpeople, many of whom have contracted at inadequate prices." Tillson voided labor contracts, many of which had been entered into for wages between two and seven dollars a month. Planters were advised that since men from other states were willing to pay fifteen dollars a month, the bureau, if it permitted freedpeople to accept much less, would be permitting them "to be swindled." The tactic Tillson and other state bureaus used to see that free labor received as pay what it was worth on the "open market"

[22] *Nashville Dispatch*, Sept. 1, 1865, quoting the *Memphis Argus* of Aug. 24, 1865, cited in Taylor, *Negro in Tennessee*, p. 126. Orders of a similar but less forceful nature were resorted to by General Scott in South Carolina: Joel Williamson, *After Slavery*, p. 92. Colonel Orlando Brown's "Instructions to Agents in Virginia" is reprinted in Walter L. Fleming, *Documents Relating to Reconstruction* (Morgantown, W.V., 1904), no. 6, p. 28. See also McConnell, p. 48, for military orders of Virginia commanders similar to General Tillson's. A. Taylor, *The Negro in Tennessee 1865–1880* (Washington D.C., 1941), p. 126.

was the removal of labor from regions of low pay to highly remunerative areas.[23]

This policy of the bureau, coupled with the fact that more fertile and productive land enabled planters of the southwestern states to outbid their competitors to the east, resulted in a westward migration of black labor that continued throughout the Reconstruction period. John William DeForest described the mechanics of the emigration of poverty-laden laborers from South Carolina, where he served as a local bureau agent. "As a result of this wretched remuneration there was an exodus. During the fall of 1866 probably a thousand freedpeople left my two districts of Pickens and Greenville to settle in Florida, Louisiana, Arkansas, and Tennessee. Only a few had the enterprise or capital to go by themselves; the great majority were carried off by planters and emigration agents."[24]

Given the bureau's disposition to favor money wage contracts, the planter's inclination to restore the plantation gang labor system was, in principle, consistent with bureau guidelines and plans. It was not the work gang per se that was objectionable; after all, laborers in the North and most of the world worked in gangs. It was the day-to-day use of physical intimidation to compel labor, the overseer's lash, that epitomized that which was reprehensible in the plantation gang. To the true believer in the powers of democracy and free labor there was no doubt that once these methods were replaced by the "usual" relations between employee and employer in the North, the term "work gang" would lose its connotation of exploitation.

The shaping of the Negro freedpeople into free Americans laborers required a total change in their relationship not only to whites but to themselves and their familial obligations. General Nathaniel P. Banks outlined the lifestyle of the model to be copied:

> The laboring man, the mechanic, and the farmer of the North requires not merely a home, but that his wife and children shall remain at home. They attend to domestic duties, he labors to support them. This is a new idea to the Negro. He has had a faint conception of family relations before.[25]

Organization of family life on the lines of the most desired contemporary model was not something that all freedpeople had difficulty learning, however. Colonel J. Sprague, assistant commissioner of the bureau in Florida,

---

[23] BRFAL, Synopsis of Reports, Jan. 1866, p. 35. Trowbridge, p. 495.

[24] John William DeForest, A Union Officer in the Reconstruction, ed. James H. Croushore and David Morris Potter. (New Haven, 1948). Under Assistant Commissioner Robert Scott, the bureau in South Carolina strongly endorsed the policy of aiding planters and freedmen in making the best possible contracts for both, which often required emigration of Negro labor. See BRFAL, Synopsis of Reports, 1866, p. 149, for Scott's statement that to effect this "he is instructed to furnish the freedpeople with rations, to any point." See also The American Freedman, vol. 1, Jan. 1867, p. 151.

[25] N. P. Banks, Emancipated Labor in Louisiana (address delivered to YMCA, Boston, Nov. 1, 1864), Civil War Pamphlets, Yale. Also William F. Messner, Freedmen and the Ideology of Free Labor: Louisiana, 1862–1865 (Lafayette, La., 1978), p. 74. BRFAL, film M1048, roll 46, Virginia, p. 38.

reported one of the earliest social changes emerging from emancipation and its important economic consequence: "The men are averse to their women and children going into the field as common laborers, desiring them rather to attend to housework, as they express it, like *White* folks. If this prevails it will reduce the working force in the cotton fields one half."

General N. P. Banks and a certain faction of the Republican Party may have been pleased with this development, but the planters had little interest in encouraging family and social arrangements based upon the "White folks plan." The blacks had been brought to the Americas for the sole purpose of providing labor for the growing of New World staples. To that end, the social organization of their lives was to be dominated by the planter's requirement of maximum production.

Early in the free labor regime, planters reaffirmed this objective. A group of planters, presenting their views of what constituted a free labor system, expressed their displeasure with certain changes under the new dispensation by demanding "that the laborers should be required to have their meals cooked in a common kitchen by the plantation cooks, as heretofore. At present, each family cooks for themselves. . . . *The extravagance in wood and the loss of time* by this mode must be apparent to all."[26]

There is no better area in which to compare the social relationships supported by the two modes of production, chattel slavery and free labor. Under advanced capitalism, employers seek as far as possible to control the social existence of employees while they are on the job. What the employees do with their leisure time is largely beyond the control of the employers, although often of interest to them.[27] Under New World plantation slavery there existed a blurred distinction between work and leisure. Since all of the slaves' time was owned by the master, both leisure and the kinds of work that held a secondary place in the master's profit calculation were subservient to the schedule that best allocated the slaves' time toward the maximization of someone else's profit. For example, no matter how a planter may have desired to maintain stable "family relations" in the quarter, he could not help but find that a contradiction existed between the two objectives. John Blassingame captures this aspect of the slave mode of production: "After marriage, the slave faced almost insurmountable odds in his efforts to build a strong stable family." Since "the master determined when both he and his wife would go to work, when or whether his wife cooked meals, and was often the final arbiter in family disputes," it often happened that "the routine of the plantations prevented the lavishing of care upon the infant. On many plantations women did not have enough time to prepare breakfast."[28] An important point to digest from this is that if a stable family

<hr/>

[26] *BRFAL, Synopsis,* Jan. 1866, p. 48. Schurz, *Report to the President.*

[27] For an economy in transition to advanced capitalism this maxim need not hold true in general. If the work force is not sufficiently socialized to the needs of industrial capitalism and is politically weak, the employer may, for reasons of profit, be avidly influential in employee activities away from the job. This point is essential to the argument presented in this book.

[28] John Blassingame, *The Slave Community* (New York, 1972), pp. 88, 94.

were to be built and maintained, it would be the master's doing as much as the slaves'. With emancipation, there could be little doubt that the freed-people would not be willing to submit to the planter's work-leisure schedule. To the emancipated slaves leisure became not something stolen from the master but a commodity purchased for themselves.

Though the mode of production seemed to be shifting rapidly in a way disadvantageous to the planter class, there was one aspect of the slave regime that the planters tenaciously insisted upon. It was, argued the planters, absolutely necessary for agricultural production that they have stable labor forces throughout the entire season. This insistence on the need for a labor contract stipulating that the laborer remain under employer control for a specified length of time, usually one year, and fixing a monetary payment before either party could know what the actual value of labor would finally be, was nothing less than a desire for indentured servitude. This was a point well understood by Julius J. Fleming, Sumter correspondent to the *Charleston Courier*, when he assessed the bureau's free labor system. "Sometimes the indentured labor deserts his employer or throws other embarrassments in the way of the planter," he wrote, "but upon the whole the contract system may be considered successful thus far." Many planters, like J. B. and Peter Barker of Franklin County, Alabama, who began their labor contracts with the prescription "this indenture," made quite clear their understanding of the nature of a labor contract.[29]

If the planters' ideal labor contract was founded upon the principle of bound labor, their ideological inclinations may well have been reinforced by a practical understanding of the relationship between slavery and the flow of credit to agricultural enterprise. In a letter to General O. O. Howard, Robert V. Richardson, treasurer of the American Cotton Planters' Association, delivered a not too subtle lecture on the subject. The association's request for loans, Richardson reported, had been "favourably received" by "capitalists in France," who while considering the proposition, requested "information upon the reorganization of free colored labor." The capitalists desired general information concerning "the powers, and duties" of the "Bureau, upon the subject of freedmen." In addition to concern over the protection of freedmen's rights, the financiers were anxious to know if the "general system of management of Freedmen" would ensure that they were "required to seek employment" and required "to fulfill their contracts."[30]

It would be naive not to believe that one reason for bureau compliance in enforcing contracts was that its officers were also committed to the realization of a good crop and therefore sought to secure the available labor. There were also other reasons for the bureaus' actions. From the perspective of those interested in the making of a free black laboring class, long-term fixed contracts seemed to have some exceptionally attractive

---

[29] Moore, *The Juhl Letters*, p. 78. *BRFAL*, RG 105, entry 220, vol. 1, no. 160, p. 170, 1866. Also ibid., film M826, roll 50, p. 505, Mississippi, film M843, pp. 41, 535, 545, N.C.; T142, nos. 66–72, Tennessee.

[30] *BRFAL*, film M752, roll 23, p. 632, Nov. 8, 1865. See Chapter 3 for the text of this letter.

benefits. From the bureau in the State of Tennessee, the institution was praised by General J. R. Lewis: "Results have demonstrated the advantages of long contracts. They serve to fix the freedmen in permanent homes, keep the families together, enable them to calculate and make plans for the future, and increase the amount saved at the end of the year."

If long-term contracting had the advantage of creating a school for embryonic free workers, it also had the concomitant disadvantage of providing a class of teachers who perceived little personal gain in setting up curricula which would allow for matriculation. From the beginning of federal labor operations it was recognized that planter and bureau objectives were not suited for a harmonious marriage. Chaplain Thomas W. Conway, then superintendent of the Bureau of Free Labor, Department of the Gulf, reported to General Banks in 1864 that "the experiences of the past year furnished an excellent conception of the disposition of those who must employ them, namely the planters," who "would have been pleased with the chance of having all their old habits regarding the slaves revived, and applied to men whom they are now forced to employ as free men." This obstacle, too, was provided a solution in the grand Republican scheme of reconstructing the South.

General Nathaniel Banks explained the northern view as well as anyone:

> But when a new class of men occupies the southern states, and undertakes the cultivation of the land, the transaction of business, and the organization of government upon different principles—when northern habits of industry are transferred to the southern soil, and men are up at five o'clock in the morning, never ceasing in their labors—when they are seen to grow powerful and rich . . . it is quite possible, if not probable, that the Negro will imitate the wiser and better example.[31]

The implicit pact was sealed. Northern immigration was to provide a class of free labor managers capable of inculcating the freedmen with discipline, thrift, intelligence, and that strong sense of self-rightousness that can only be purchased with the sober morality of the Protestant work ethic. For its part of the bargain, the bureau was to provide the minimal external force required to stabilize a preindustrial population so that their education might have enough continuity to succeed. In addition, the bureau was to protect the freedpeople from their own licentiousness, unscrupulous northern profiteers, and southern planters uncommitted to the gospel of free labor.

If the revolution were to be complete, staunch Republicans with high-ranking positions in the bureau understood that the presidential position

---

[31] *BRFAL, Synopsis of Reports*, 1866. T.W. Conway, "Report of the Condition of the Freedmen of the Department of the Gulf" (New Orleans, 1864), p. 5. Banks, p. 23. Also Bliss Perry, *Life and Letters of Henry Lee Higginson* (Boston, 1921), p. 255; Charles Stearns, *The Black Man of the South and the Rebels* (Boston, 1872), p. 155; Reid, p. 578; Lewis C. Chartock, "A History and Analysis of Labor Contracts Administered by the Bureau of Refugees, Freedmen, and Abandoned Lands in Edgefield, Abbeville, and Anderson Counties in South Carolina, 1865–1868" (Ph.D. diss., Bryn Mawr College, 1974), p. 61.

—that the bureau's job was merely to insure that the Negroes were at work —was insufficient. Colonel John Eaton, assistant commissioner for the District of Columbia, wrote that while he served as general superintendent of freedmen in Tennessee during the war his officers were committed to the position that "employment and protection were necessities preceding instructions in order only,—not in importance." In a letter to Governor James L. Orr of South Carolina, Major General Robert K. Scott, replying to a complaint about the bureau's labor policies, delivered the following unsolicited lecture:

> The prejudice of class, caste, or color, can in no way enter into questions of justice between the parties at variance. . . . When the system of free labor shall have become regulated by time and experience it will be found that the fixed principles which govern it all over the world, cannot be ignored in South Carolina. To the establishment of these principles the bureau is committed.

There were subtle inconsistencies in the bureau's stated position that did not go undetected. If "class, caste, or color" could in no way enter into questions of justice, how could the bureau, whose very existence obviously contradicted this statement, be justified? With the bureau caught in such contradictions, there was cause for criticism from all parties concerned. Possibilities of "reverse discrimination" were obvious—indeed, this was behind one of Governor Orr's complaints to General Scott. In Arkansas, General J. W. Sprague instituted a policy which sought to use in cases involving freedpeople "the same means to enforce compliance with contracts that it was lawful to use where both parties were white," "except that it was made a finable offense . . . to entice a *freedman* to violate a contract." How could differential treatment, such as the policy of returning all freedmen who left their employers "without just cause," be reconciled with the principles of free labor?[32] The inconsistencies were understood by members of the bureau itself. In Alabama, General Wager Swayne explained that the contract system was merely a temporary expediant until the law of "supply and demand" could provide "the true security of labor," wherein a laborer who "finds himself ill-treated, or his wages insufficient or unsafe," could "quit and without having to account to anybody."

In the North, free labor advocates were always ready to attack any policy that they perceived to be a continuance of slavery. The bureau had to tread softly if it was to do the impossible job of satisfying both free labor advocates and the demands of employers. In the fall of 1865 Colonel Whittlesey, moving to enact the Tanzanian plan in North Carolina, communicated a discussion he had had with one of his agents over the refusal of freedpeople to accept contracts on Roanoke Island: "'I judge,' says Captain James, 'that the rations on the Island should be reduced *one fourth* within thirty days; as soon as it is done, however, loud complaints will go out to General

---

[32] *BRFAL, Synopsis of Reports*, 1866, p. 150; General J. W. Sprague, ibid., p. 22.

Howards office as was the case in June last when an attempt was made to induce the people to seek self-supporting labor.'"[33]

The most active criticism came from the black population. What was the intent of the contract system? Were all whites exempt from deportation to the countryside? Did all blacks have to contract? Melton R. Linton, an educated black soldier, outlined the issues involved in a letter to the editor of the *South Carolina Leader*, a black-operated newspaper:

> I hope soon to be called a citizen of the U.S. and have the rights of a citizen. I am opposed myself to working under a contract. I am as much at liberty to hire a White man to work as he to hire me. I expect to stay in the South after I am mustered out of service, but not to hire myself to a planter.

A leading voice in the struggle for black rights was the black-owned and black-operated *New Orleans Tribune*. In December 1867 its editor, with a touch of irony that must have had a bitter ring for some Republicans, critiqued the bureau's labor policy in much the same way as the Republicans had critiqued the slave economy. The bureau's labor policy under Generals Banks and Hurlbutt was a "bastard regime which is not slavery" nor "the civilized regime of freedom of labor." The *Tribune* argued that the principles of political economy recognized that it was essential to a free labor system that "the laborer be free at any time to leave his employer." This was necessary so that labor could go wherever wages were highest, securing labor its best reward and maximizing output by allocating that labor where it was most valued. The *Tribune* called for an end to the contract system and institution of a true free labor system fair to both parties: "no contract, but on the contrary, freedom to both parties, is the motto of free labor all over the civilized world." "Is slavery abolished? If abolished, let us follow the rules of free labor. If not abolished, let us know it."

The *New Orleans Tribune* represented what was perhaps the most sophisticated black voice in the South. In spite of this it may well have expressed the desires of the black masses, at least in regard to the meaning of freedom. In Alabama, Whitelaw Reid received a far less eloquent but nonetheless equally expressive statement of the meaning of freedom from a freedman who had left the plantation to take up habitation in a tent shanty outside Mobile: "I's want to be free man, cum when I please, and nobody say nuffin to me, nor order me roun."[34]

Throughout the parishes of Louisiana during early 1866, agents of the bureau were gaining practical experience with the freedpeople's own critique of the contract system. "All or nearly all dread a contract," complained Lieutenant J. D. Rich, "as it seems too much to them like slavery, as they are bound for a certain time." In particular, announced another agent, the "freedmen recently discharged from the army" infected the laborers with

---

[33] Wager Swayne, "Report to General Howard, 1866," *BRFAL*, film M809, roll 2, pp. 396–97; E. Whittlesey, ''Report for 1865," ibid., film M752, roll 23, p. 112. Also ibid., film M752, roll 20, p. 739, for northern criticism of Davis Tillson in Georgia.

[34] *South Carolina Leader*, March 31, 1866, p. 2. *New Orleans Tribune*, vol. 7, Dec. 31, 1867, p. 1. Reid, p. 389.

"enormous and incongruous notions of liberty," which caused them to "decline to make an agreement by which they are to be in any manner restrained for the period of a year." The reasoning behind the distaste for contracts was recalled by Willis, a freedman of Burke County, Georgia; who had refused to sign the contract offered by his ex-master, saying, "What you want me to sign for, I is free." Willis understood the essentials of free labor. When told by his ex-master that the contract would "hold" both of them to their "word," Willis responded, "If I is already free, I don't need to sign no paper. If I was workin for you and doin for you before I got free, I can do it still." When Willis's parents asked, "How you know Marster givine pay," he answered with a precision that would have made the finest political economist of his day proud: "den I can go somewhere else."[35]

To many whites, regardless of their other beliefs, such utterances were expressions of undisciplined wantonness and ignorance. If the rural freedman's idea of freedom was a life of leisure and idleness, then he had to be protected from his own ignorance. After all, idleness of any kind was clearly a public vice, since it served to diminish the social output. T. W. Conway had spoken well for the federal authorities' point of view in his defense of the Bureau of Free Labor's activities in the Gulf states during the War. Bureau regulations had "been strict and even severe" but overall "has benefitted them." Idleness was "license not liberty." The duty of the bureau was not only to employ and protect the freedmen and govern the planter but to "protect property, promote industry, supply markets." Eric Foner has demonstrated that the phrase "free labor" meant many things to the mid-nineteenth-century Republican.[36] But one thing it did not mean was freedom not to labor. Just whose freedom was being discussed anyway? Chaplain Conway and George Fitzhugh may have had many points of disagreement, but with respect to at least one issue they had little quarrel. What good was a free labor economy if emancipation of the underclass resulted in the enslavement of those above?

[35] *BRFAL*, film M1028, roll 28, pp. 159–60; ibid., p. 128. Norman R. Yetman, *Voices from Slavery* (New York, 1970), p. 320. Also Trowbridge, p. 424. *BRFAL*, film M1048, pp. 40–41, Virginia. *Report of the Joint Commission on Reconstruction*, General Edward Hatch, p. 6, Mississippi.

[36] Conway, pp. 4, 5. Foner, *Free Soil, Free Labor, Free Men.*

5

*You white folks spose cause you white, and we all black, that us
dunno noffin, and you knows eberyting; Now missus, youse one bery
good white woman, come down from de great North, to teach poor we
to read, and sich as that; but we done claned dishes all our days, long
before ye Yankees hearn tell of us, and now does ye suppose I gwine to
give up all my rights to ye, just cause youse a Yankee white woman?
Does ye know missus that we's free now? Yas, free we is, and us ant
givine to get down to ye, any more than to them ar rebs.*

Margaret, a freedwoman, Georgia, 1866

Conflict and Irresolution

Carl Schurz's investigative journey through the southern states in 1865 had
left little doubt in his mind that the planter class in particular, and whites in
general, not only had little faith in the new labor system but harbored a
desire to see it fail. "In at least nineteen cases of twenty the reply I received"
was "you cannot make the Negro work without physical compulsion," he
reported. To Schurz, such beliefs, although sincere, were hopelessly
grounded in the prejudices of "the late slaveholder still clinging to the traditions of the old system." The effect of such attitudes upon labor relations in
the South was somewhat exaggerated by Schurz: "a northern man knows
from actual experience what free labor is, and understands its management." "When a northern man discovers among his laboring force an individual that does not do his duty," he continued "his first impulse is to
discharge him, and he acts accordingly. When a late slaveholder discovers
such an individual among his laborers, his first impulse is to whip him, and
he is very apt to suit the act to the impulse."[1]

If the radical Republicans were to create a black laboring class in their
own image, the primary instrument in this process was to be the northern
capitalist, manager of free labor par excellence. Who were these northerners

---

[1] Schurz, *Report to the President.*

in whom Carl Schurz could express such boundless faith?[2] Most conspicuous were the prototype carpetbaggers who came with the objective of exploiting the South for a quick profit and then returning North. Perhaps the more typical northerner had mixed motives. There was surely nothing wrong with the desire to make money. If during the process much-needed capital came to a devastated economy, the bearer of this capital could simultaneously do God's work by contributing to the raising of the black masses from the poverty and ignorance that had been their fate in the house of bondage. Many of these men and women were destined to receive a miserable lesson. To be successful the planter, whether an immigrant from the North or an experienced southerner, had to learn to handle recently created free black labor. To the experienced planter, there seemed to be plenty of good evidence that this task would not be simple.

Eugene Genovese's provocative analysis of the "black work ethic" provides the starting point for any contemporary discussion of the black worker during the years of Reconstruction. In many ways I find Professor Genovese's treatment something of an enigma. His luminous discussion, it seems to me, touches upon each of the cardinal points of the matter, but it focuses upon so many that the reader cannot help but feel somewhat like the proverbial child turned loose in a candy store. In particular, one can appreciate the inherent differences that must exist between an agricultural and urban industrial lifestyle and still believe that Professor Genovese places too much emphasis upon the time rhythm of agricultural seasons as a determinant of the slave work ethic. Peasant producers in agricultural societies often adopted irregular work rhythms in time with the length of the day and crop seasons. In societies where the producers were at least in principle free, the work-leisure decision and therefore the rhythm of work was adapted to nature by the worker's *choice*. E. P. Thompson, upon whose work Genovese draws, was careful to point out that the task-oriented approach to labor and time typical of rural workers presupposes "the independent peasant or craftsman as referent." The important distinction is that in societies where the direct producers are not free, the work-leisure decision is largely made by someone other than the worker.[3]

Although it may be an empirical fact that most agricultural societies have adopted an irregular work pattern bound to the time frame of nature, there is no logical connection between rural work patterns and nature. The seasons and the weather condition not so much the rhythm of work as the nature of the work performed. If the free peasant chose an irregular work rhythm, it was because he valued leisure during the rainy season more highly than the output which could be produced inside his hut. If the peasant was poor, the output he could produce in his hut was likely to be of very low value to him. Similarly, the value of work to the slave during the

[2] For a most detailed and comprehensive answer to this question consult Powell, *New Masters*.

[3] Genovese, *Roll Jordan Roll*, pp; 285–309. E. P. Thompson, "Time, Work-Discipline, and Industrial Capitalism," *Past and Present*, no. 38, Dec. 1967, pp. 56–97.

off-season was likely to be low, but the point at issue is whether that work had sufficient value to offset its costs to his master.

The master's valuation of the optimal use of the slave's time was studied by Charles Sydnor, who reported that "when bad weather interrupted work in the field, negro women were frequently required to spin or weave." Ex-slave Adrianna Kerns affirmed the point: "When we had to come out of the field on account of rain, we would go to the corn crib and shuck corn if we didn't have some weaving to do." The men, too, were at their master's disposal. Texan Andrew Goodman recalled his ex-master's policy in regard to the effect of nature upon the slave's work rhythm: "He didn't never put his niggers out in bad weather. He gave us something to do that we could do in out of the weather, like shelling corn and the womens could spin and knit." From the master's point of view, "order and system must be the aim of everyone on this estate, and the maxim strictly pursued of a time for everything and everything done in its time." Freedwoman Sarah Ford had no trouble making the analogy: "Massa Charles run dat plantation jus like a factory. Uncle Cip was sugar man, my papa Tanner and Uncle John Austin, what have a wooden leg, am shoemaker." The rhythm of work was consistent with the division of labor: "when its fallin weather de hands is put to work fixin dis and dat."[4]

The planters' objective was to maximize production to accumulate. To this end, they sought to construct a factory system. How else can we interpret the organization of work, the discipline, the division of labour, the function of drivers and overseers? If the planters failed to achieve this paradigm of factory efficiency and if, as Genovese writes, "the slaveholders presided over a plantation system that constituted a halfway house between peasant and factory cultures," it was not because the plantation was in a rural setting, and it was not because the slaves were descendants of Africans from a traditional society.

The original planters, much like the early English industrialists, possessed work forces composed of men and women from a nonaccumulative tradition. The masters in both cases were accumulative and their production objectives required that workers be made accommodative. Genovese quotes E. P. Thompson: "the process of cultural transformation had to rest on economic and extra-economic compulsion and ultimately on violence." The industrialist, faced with free workers who had the choice of quitting, no matter how unpleasant the alternative employment, could allow employees a wider range of behavior than could the slaveholder, who had to own up to the ideology justifying slavery. The creation of a self-disciplined, intelligent, and accumulative slave force might have undermined the entire

[4] Charles Sydnor, *Slavery in Mississippi* (New York, 1933), p. 24. George P. Rawick, *The American Slave* (Westport, Conn., 1971), *Arkansas Narratives*, vol. 9, part 4, p. 193; *Texas Narratives*, vol. 5, part 4, p. 1523. Phillips, p. 262. Joseph A. S. Acklen, "Rules in the Management of a Southern Estate," *DeBow's Review*, vol. 21, Dec. 1856, pp. 617–20, vol. 22, April 1857, pp. 376–81, offers a detailed system of rules for overseers. Rawick, *Texas Narratives*, vol. 4, part 2, pp. 44, 43. Also ibid., pp. 1, 18, 205, and vol. 5, p. 20. Walter L. Fleming, *Civil War and Reconstruction in Alabama* (New York, 1905, reprint 1945), p. 170.

system. The result was that the industrialist, in the end, stressed economic incentives while the slaver stressed compulsion, violence, and a work force that remained preaccumulative. Under slavery, the average field hand was not just accustomed to a low standard of living but kept ignorant of what a better life might be. What the Negro slave and the English preindustrial worker had in common was what W. Arthur Lewis calls "the most important limitation on men's desire for goods, namely their limited horizons." George Fitzhugh understood this legacy of slavery, although he chose to see it as a race trait: "How make him work ten hours a day, whose every present want can be supplied by laboring three hours? To give the Negro emulation, ambition, to make him aspiring" and desirous "to keep up with the times and fashions will be as knotty a subject as the solution of the Riddle of the Sphinx."[5]

Paternalistic capitalism became a basic tenet of upper-class southern aspirations for the creation of a "New South." With respect to the Negro, this was a straightforward attempt to continue the social ordering which had been maintained in the antebellum period. C. Vann Woodward has instructed us that "it was upon the tradition of paternalism that the Redeemer regimes [of the postbellum period] claimed authority to settle the race problem and deal with the Negro."[6] The appeal to the tradition of planter paternalism was, of course, very different from the ideas espoused by free labor advocates. The white southern ethic had no initial plans for voluntarily elevating blacks above their previous existence. Planter paternalism for Negroes and industrial paternalism for whites in the emerging cotton mills were seen as solutions to poverty and aids to order and commercial growth.[7] Events would dictate otherwise, and planters who witnessed the decline of their brand of paternalism would be forced to seek a new vision.

The planter class, trapped in its own ideology of paternalism and black inferiority, had with a few exceptions failed to take full advantage of the most efficient institution ever created for the purpose of molding efficient laborers. With the termination of slavery, northerners, determined to teach slave and master the benefits of free labor, flooded the South, bringing with them a new kind of paternalism. Some southerners were already versed in Benthamite labor management techniques, and we might do well to introduce the subject through the medium of a South Carolinian writing under the pseudonymn A. N. Ideal. During the antebellum period, Mr. Ideal had amassed a substantial fortune as a merchant. Traveling through Europe after retiring from the mercantile trade, he observed that the custom of those countries "exists, to a great extent, of forcing from the laborer the utmost work for the lowest possible price." Such methods struck him as both exploitive and non–profit-maximizing.

[5] Genovese, *Roll Jordan Roll*, p. 291. W. Arthur Lewis, *The Theory of Economic Growth* (Homewood, Ill., 1955), p. 29. George Fitzhugh, "Freedmen and Freemen," *DeBow's Review*, vol. 1, April 1866, pp. 416–20.

[6] C. Vann Woodward, *Origins of the New South* (Baton Rouge, 1951), p. 215.

[7] Ibid., pp. 117, 134, 222–38.

In postwar South Carolina, Ideal invested $200,000 in a large tract of land for the purpose of growing rice and cotton. On each of three separate 250-acre tracts he built one hundred four-room houses of uniform design with one-acre garden plots. The residence areas were lined with attractive streets which "intersected at right angles." Two thousand employees—laborers for the cotton and rice fields and workers in all necessary mechanical specialties—were to live in the nearly self-sufficient town. Employees were paid cash wages every Saturday and no accounts were kept at the stores. Ideal's motives were explained in detail: he "wished to invest [his] capital profitably," to "improve the physical condition of the people employed, and determine whether they—the negro population—cannot be made reliable and profitable as laborers. To attain this object, and to obtain an influence over them so that their moral condition could also be benefited, it became necessary to bring into action the power of sympathy." The laborers all needed "training, watching, and teaching." Each aspect of this prescription was carried out with meticulous care:

> The physical comfort of my laborers is cared for. They have good food, com-
> fortable clothing and regular medical attention by a resident physician. They
> have religious services on the sabbath, and are taught to dress themselves neatly
> on that day. There is a school for children from 8 to 12 o'clock each day
> except Saturday. In the evening they are permitted and requested to meet
> together, to visit each other, to enjoy the music of the violin and banjo, to
> dance and be cheerful, making themselves happy as they can.[8]

This is the paternalism of incipient industrializing capitalism. Early industrial Britain had received an especially large dosage of this social medicine from Robert Owen. Owen's famous mill town, New Lanark, was, as E. P. Thompson puts it, "instituted to meet the same difficulties of labour discipline, [in this case] the adaptation of the unruly Scottish labourers to new industrial work-patterns," that had to be overcome by other industrialists. The American paradigm is represented by George Pullman's model factory town, Pullman. The paternalism of industrializing capitalism requires that the employers, in order to remold the workers into inferior images of themselves, buy both the laborer's services and loyalty through material incentives that exceed the going market rate. The quotation from A. N. Ideal contains all the key ideas of the philosophy. For Robert Owen, who, Professor Thompson argues, could not comprehend the "notion of working-class advance by its own self-activity towards its own goals," a key question was "what are the best arrangements under which these men and their families can be well and economically *lodged, fed, clothed, trained, educated, employed*, and *governed*?" Pullman was expected to attract a superior type of workman who would be "elevated and refined" by his new environment, which would exclude "all baneful influences." George Pullman practiced the panopticon principle with a

---

[8] *Rural Carolinian*, "My Plantation," March 1870, pp. 392, 393–95. Ideal's mercantile background should not be ignored.

vengeance. There were ten censuses in the town's first four years. His methods approached those of Mr. Ideal: "We have provided a theatre, a reading room, billiard room, and all sorts of outdoor sports. . . . our people soon forget all about drink, they find they are better off without it, and our work is being done with greater accuracy and skill."[9]

Robert Owen and his followers may have understood more than Professor Thompson supposed. The problem was that early working-class goals, and especially the time frame of self-advancement, failed to conform to the desires of the manufacturers. Consider the comments of a Mr. Carr of Leeds, who, in describing the paternalistic management of the cotton mills at Rothsay in the Isle of Bute, noted that "much, however, among *free* laborers, must depend upon the people themselves. In this respect there are great difficulties: too many of the parents do not see the advantages, which arise from cleanliness (particularly in their houses), from proper methods of preparing their food, and from having their children well educated."[10] The resulting conflict is responsible for much of that which is common to the two kinds of paternalism. In the South Carolina Sea Islands, Edward Philbrick, whom Willie Lee Rose describes as "the one evangel who thought hard about economics," encountered the problem early and recognized an effective but infeasible solution: "We find it very difficult to reach any motive that will promote cleanliness as a habit. It requires more authority than our position gives us as employers to make any police regulations very effectual in their quarters." The "people" being "too independent and too ignorant to see the advantages" could not be organized in the most productive way.[11]

Philbrick's engineering business background and his unwavering commitment to market capitalism—"combining a fine humanity with honest sagacity and close calculation"—led Edward Pierce to make the endorsement "that no man is so well fitted to try the experiment" of elevating slaves to free laborers. Skeptical of being able to protect the Negroes through the vehicle of "disinterested philanthropy," Philbrick believed that "Negro labor has got to be employed, if at all, because it is profitable." The world must be shown that free black labor is "profitable, and then it will take care of itself." The argument was a blueprint for a system of contract management, and Philbrick set about his task with alacrity.[12] He organized a joint-stock company with twelve northern investors. With the funds he purchased eleven plantations and leased two others. The engineer whose "orderly bones ached" when he thought of petite agriculture found scope for his own managerial abilities. The vast holdings required the hiring of a group of young northern free labor advocates to superintend individual plantations

---

[9] Thompson, *Making of the English Working Class*, pp. 780, 781. Stanley Buder, *Pullman* (New York, 1967), pp. 44, 42, 69.

[10] *Reports of the Society for Bettering the Condition and Increasing the Comforts of the Poor* (London, 1805), p. 55 (my emphasis).

[11] Pearson, *Letters*, pp. 15, 109.

[12] Willie Lee Rose *Rehearsal for Reconstruction* (New York, 1965) p. 216. Pearson, p. 221.

under his guidance. Philbrick's system of plantation management was both successful and innovative enough to influence the practices of many planters in South Carolina and Georgia. But Edward Philbrick was far too atypical a man to dwell upon in isolation.

Perhaps most representative of the northern immigrant to the postwar South was George Benham. A druggist by trade, Benham entered a partnership with a physician to find "a short cut to a fortune" growing cotton on a Louisiana plantation. The lure of profit was Benham's first consideration. After convincing himself of the soundness of his partner's estimate of a profit totaling $343,511.57 in four years, his first thoughts were "what a puny, sickly thing our drug store looked to me now." The partners deluded themselves into believing that their venture could not fail. They would introduce new techniques of farming and labor-saving machinery. In the same spirit Benham later reasoned, "were we not then while serving ourselves so well, also acting a good part toward the poor black people." He and his partner "were simply pioneers in the enterprise of starting this tide of immigration, which was just as essential to the prosperity of the southern country as was capital itself." This optimism was to be short-lived. One serious error in the profit estimate involved the productivity of emancipated relative to that of slave labor. Shortly after the cotton grew large enough to commence cultivation, heavy rains stimulated the rapid growth of cotton-killing weeds. Benham cajoled and reasoned with his employees to hoe the weeds, but the freedpeople would have none of it; "this was what they had to do in the days of slavery, but now "dey was free, and dey wouldn't work in de mud an' de water for nobody." George Benham quickly grasped the implications: "here, manifestly, was a serious hitch in cotton-raising under the new dispensation."[13]

In the summer of 1866 the Freedmen's Bureau office in Augusta, Georgia, received a letter from a Mr. C. Stearns, who was cultivating a plantation of about 1,500 acres near Augusta. Mr. Stearns wished to contradict the report of General Steedman "that the system of contracts enforced by the Bureau in the South is simply slavery in a new form."[14] Stearns wrote:

> One who would make such a statement must be either grossly ignorant of business North as well as South, or else an intentional falsifier of the truth. . . .
> A system of contracts for work by the year is absolutely essential to the working of cotton plantations by free labor.

This statement appears to be a class-biased defense of a system of indentured servitude offered by a typical southern planter. Yet Charles Stearns

---

[13] George Benham, *A Year of Wreck* (New York, 1880), pp. 12, 16, 198.

[14] *BRFAL, Synopsis,* 1866, p. 437. Generals J. B. Steedman and J. S. Fullerton were appointed by President Johnson to conduct an investigation of the local operations of the Freedmen's Bureau. Their report, which caused something of a sensation in the press, resulted in many charges of corruption but only a few convictions. See *House Executive Document No. 120* 39th Cong., 1st sess. Steedman, a Democrat, was later appointed collector of internal revenue for the first district of Louisiana. The *New Orleans Tribune* attacked this appointment as a political payoff: *New Orleans Tribune,* April 9, 1867.

was neither a southerner nor a typical planter. Born in Massachusetts, he had devoted much of his life to the abolition of slavery and had come to Georgia with the expressed objective of laboring for "the perfect development of the colored race." He hoped to accomplish three things: "their *education* and moral improvement; secondly, their *right to vote*; and thirdly, making them the *owners of the land* they cultivate." Charles Stearns's "Hope on Hope Ever" plantation came close to being that personification of Jeremy Bentham's panopticon most consistent with an ideal conception of free labor.

The elevation of the freedmen to the status of efficient free laborers required that their education include "lessons of industry, neatness, punctuality, obedience, honesty and benevolence." To achieve this, Hope on Hope Ever plantation was to be established as "a model farm, or self-supporting industrial school where all that related to man's external welfare should be taught and practiced."[15] As discussed earlier, the free labor ideal which would give the laborer complete freedom to recontract with other employers on a daily basis, an option which need seldom be taken in an advanced society, is inconsistent with the idea of the panopticon. How could the employer become master of his employees' destinies through behaviour modification if they had the option of leaving his employ at any time it suited them? "What hold," Bentham had asked, "can any other manufacturer have upon his workmen, equal to what my manufacturer would have upon his? What other master is there that can reduce his workmen, if idle, to a situation next to starving, without suffering them to go elsewhere? What other master is there, whose men can never get drunk unless he chooses they should do so?"

Few northern planters seemed to grasp the full implications of the nature of these contracts. George Benham provides a clear picture of the ideological apparatus which shielded the northern planter from any possible psychological damage resulting from the obvious similarities between his own position and that of the slaveholder. One evening Benham and his partner discovered one of their employees "with a bundle suspended from a stick over his shoulder, looking just as we had seen pictures at the head of advertisements for runaway slaves—but one of our own laborers, running away!" They arrested the would-be runaway, shackling him to the floor of their house until he could be taken to the Freedmen's Bureau for punishment. Although the runaway black was to Benham an obvious symbol of the corrupt slave regime, he made no mention of the next logical step, which would have cast Benham himself as a symbol of the slave catcher. His own role, as seen by himself, was not as owner of unfree labor but as employer of undisciplined "babes" needing guidance. If, because they were "so long deprived of their freedom and having now the crudest ideas of its true

---

[15] Charles Stearns, *The Black Man of the South and the Rebels* (Boston, 1872, reprint New York, 1969), pp. 28, 155. Edward Philbrick in Pearson, pp. 1, 11; William Channing Gannett, ibid., p. 137.

meaning, they were mistaking discipline for an attempt to rob them of their priceless treasure," it was obviously one part of his own duty to instruct them in the "true meaning of freedom."[16]

Over one hundred years earlier, a few English industrialists pioneering in capitalist management techniques had endeavored to mold what we would now term a modern industrial work force out of crews of undisciplined "men who were nonaccumulative, non-acquisitive, [and] accustomed to work for subsistence" by making them "obedient to the cash stimulus." They had, as Josiah Wedgwood put the issue, "to 'make *Artists* . . . [of] mere *men*' and second to 'make such *machines* of the *men* as cannot err.'"[17] Post-emancipation planters inherited an agricultural work force that for the most part had been accustomed but not acculturated to the habit of steady work. Josiah Wedgwood inherited an industrial work force that had not even become accustomed to steady work. Who had a harder task is perhaps an unanswerable question, but the amazing similarities in the behavior patterns of the two groups, thousands of miles and over a century apart, should not only provide an answer to the question of whether or not the slaves had imbibed a Protestant work ethic, but also enable us to gain a much deeper understanding of labor relations and economic development in the southern economy.[18]

Earlier it was shown that after the war, planters, believing it most efficient, moved to reinstitute the centralized gang system of work. Southern planters were not alone in this assessment. After close observation, even Republican Whitelaw Reid concluded:

I was convinced that when they were employed in gangs, under the supervision of an overseer who had the judgement to handle them to advantage, they did as well as any laborers. . . . Put one at some task by himself, and there was every probability that he would go to sleep or go fishing. Even in gangs, not half of them could be depended on for steady work, except under the eye of the overseer or driver.

Here was a clear recommendation for that principle considered to be the fundamental advantage of the panopticon: "To be incessantly under the eyes of the inspector is to lose in effect the power to do evil and almost the thought of wanting to do it."[19]

Few experienced planters needed to be instructed in the productive virtues of a factory organization of their labor forces. The antebellum labor system

[16] Bowring, vol. 4, p. 56. Benham, pp. 218, 222. Bureau agent, St. John Parish, Louisiana, *BRFAL*, film M1027, roll 28, p. 170, Jan. 1866. Stearns, pp. 107–8.

[17] Sidney Pollard, "Factory Discipline in the Industrial Revolution," *Economic History Review*, 2nd ser., vol. 16, 1963, p. 254. Neil McKendrick, "Josiah Wedgwood and Factory Discipline," *Historical Journal*, vol. 4, 1961, p. 34.

[18] Robert Fogel and Stanley Engerman, in *Time on the Cross* (New York, 1974), argue that southern slaves had inculcated a Prostestant work ethic at the behest of their masters. The claim aroused vehement criticism.

[19] Reid, p. 503. Halevy, p. 83. Thomas W. Knox, *Campfire and Cotton Field* (New York, 1865), p. 371. Pearson, pp. 113, 109.

had been predicated upon the principle of the omnipresent eye of authority. Bentham himself, who advocated "that the persons to be inspected should always feel themselves as if under inspection" by inspectors who make "contrivances for seeing without being seen," might have learned a few tricks if he had had the privilege of consulting with Mr. Edward Covey, the "Negro breaker" made famous in the autobiographies of Frederick Douglass. Of Covey's training techniques Douglass wrote that

> it was, however, scarcely necessary for Mr. Covey to be really present in the field, to have his work go on industriously. He had the faculty of making us feel that he was always present. By a series of adroitly managed surprises, which he practiced, I was prepared to expect him at any moment. . . . He would creep and crawl, in ditches and gullies; hide behind stumps and bushes, and practice so much of the cunning of the serpent, that Bill Smith and I—between ourselves—never called him by any other name than "*the snake.*" . . . One half of his proficiency in the art of Negro breaking, consisted, I should think, in this species of cunning.

For Douglass this was the essence of the slaveholder's system of inducing continuous labor.

Few planters were willing to allow emancipation to force drastic changes in their methods of close supervision. A contributing expert to an agricultural journal, noting that "the lax discipline of free labor" obviated the previous advantages of locating the laborers centrally "in a quarter," stressed that on a well-run plantation "the manager's cottage is central and gives opportunity of close and constant supervision."[20]

What many ex-masters required was instruction in the management of free labor. In many localities, the excess of demand for labor over the available supply meant that most laborers would not be forced to contract with an unsatisfactory employer to avoid starvation. The competition for workers among planters allowed freedpeople to be selective about whom they worked for, where they worked, and what they were to receive. It quickly became clear that employers with bad reputations were going to have a difficult time recruiting labor. In 1866, in Mississippi, a southerner complained that "they won't hire to southern men anyhow, if they can help it. . . . They've no confidence in us." This was an understatement of some magnitude. Many freedpeople had been cheated of their pay in the two previous years, and their ignorance of the value of their labor and the market system made the situation all the more difficult. Whitelaw Reid illustrated this point with one of his brief encounters with a freedman. Upon being asked "what wages he received" the elderly servant answered, "twenty dollars a month sah but I'se gwine to quit. Tain't enuff is it?" After being assured he was receiving fair wages, the servant explained his position: "I was afeared of dese men cheatin me, because I knowed dey would if dey could."

[20] Bowring, vol. 4, p. 44. Frederick Douglass, *My Bondage and My Freedom* (New York, 1969), pp. 215–16. "Notes on Planting," *Rural Carolinian*, vol. 4, no. 5, Feb. 1873, p. 234.

If it was tough hiringing hands who knew one's personal reputation, Mississippians had the additional burden of overcoming the reputation of an entire state. A Mississippi planter in Louisiana was refused his offer of five dollars a head for all the workers a black labor agent could supply. "I wouldn't send you a man ef you gave me a hundred dollars a head," the agent replied. Upon inquiring why, the planter was informed that the agent "didn't like our Mississippi laws." In Florida, the Freedmen's Bureau's assistant commissioner reported in 1866 that the "Negroes have a kind of telegraph by which they know all about the treatment" of laborers on the plantations in a wide area. The commissioner believed that this information was used "in the choice of their employers," and retribution was so swift that "some large planters were unable to employ a single laborer," while others were able "to obtain all the help and more than they can employ."[21]

All southerners were by no means slow or unwilling to learn. If the lash and other old methods would not hold free laborers on the plantation, employers would have to alter their methods. Agricultural journals of the period were filled with advice from and to planters about the best methods of remunerating and supervising free labor. In 1869 the *Southern Cultivator* printed a communication from the transactions of the Pomological and Farmers' Club of Society Hill, S.C., on "The Labor Question." This article is not only representative of the key issues found in most others, but the best I have seen. The only contractual forms it thought merited consideration were *annual* hiring for either a share of the crop or a stipulated wage.

The theoretical advantages of the share contract were its stimulus to the laborer's exertions, due to the laborer's own interest in the crop; the easier recruitment of workers, most of whom considered it a higher form of contract; and the fact that sharing shifted some of the risks of production to the laborer. The planters felt that in practice the incentive advantage only applied to a small portion of the best laborers. The disadvantages of the share contract were the difficulty of discharging inefficient or refractory hands, because doing so under a share contract often led to a civil or Freedmen's Bureau court; the difficulty of carrying on general work of improvement on the farm because share laborers refused to do any work not connected to their current crop; and, finally, the annoyance and perplexity of dividing the crop.

A wage contract was advantageous because it facilitated the planters' control over laborers who, not having a claim to the crop, could be discharged when inefficient. It also stimulated the farmer to give close attention to operations and invest more, since he did not share profits, and it enabled a general system of improvement to be carried out. The disadvantages of the wage contract were reputed to be the following: it might lead to ruinously

---

[21] Trowbridge, pp. 365–66. Reid, pp. 362, 447. Joe M. Richardson, "A Northerner Reports in Florida: 1866," *Florida Historical Quarterly*, vol. 40, p. 383.

high wages due to competition, and it involved far greater labor in supervising the crop. The Farmers' Club correspondent decided in favor of the wage system primarily because it allowed the employer close control over a class of "ignorant and indolent" laborers. Irrespective of their opinions as regards choice of contract, most planters agreed that black workers were lazy and inefficient and that what was needed was more control over them.[22]

Much of this assessment was to be expected. Since for many planters the measure of comparison for free labor was the amount of work accomplished by slaves, few freedpeople would be willing to provide a satisfactory performance. Some employers, like Virginia tobacco planter Robert T. Hubard, invoked the old standard but reluctantly admitted that the requirement was unfair:

> My crop hands are behaving tolerably well and doing about 2/3 rds of the work they did ante-bellum. They are working and behaving *infinitely better* than they did last summer and fall and give me less trouble than I had anticipated. The Darkies should have *their due*, since even the *devil* is said to be entitled *to his*.

"Niggers are working well," complained an Arkansas planter, "but you can't get only about two-thirds as much out of 'em now as you could when they were slaves."[23]

If we set aside judgments based upon the work performance of the slave, the planters' general set of complaints about the efficiency of free black labor is consistent with the reports of the fairest observers available. The theme of these complaints corroborates the conclusions, if not the causal arguments, of Eugene Genovese. A North Carolina farmer, whose mistreatment of one hired black was no doubt part of the problem, provides a convenient starting point: "He took the shovel and worked right well for five or six days. Ye know how it is with 'em—for about three days its work as if they'd break everything to pieces; but after that its go out late and come in soon." Missionary Elizabeth Hyde Botume explained why employers of all persuasions would demand the close supervision recommended by Whitelaw Reid: "They were not idle or lazy. On the contrary, they seemed to have a passion for work. But they knew nothing of order, or system, or economy of time. So nine-tenths of their labor went for nothing. In this particularly they needed so much help, and got so little. They wanted a clear, judicious head to plan for them." Miss Botume believed that few "planters or teachers would see this or take the trouble." For those who did try the task was difficult. "Now they work and then they stop,—and some stop before they begin," complained William Channing Gannett of the "very wayward" South Carolina laborers who were giving him such a "Hard time."[24]

[22] "The Labor Question," *Southern Cultivator*, Feb. 1869, pp. 54–55. Ibid., 1867, pp. 1, 111. Brooks, pp. 19–23. *American Farmer*, 1866, pp. 15, 138–39.

[23] P. M. Thomas, "Plantations in Transition: A Study of Four Virgina Plantations" (Ph.D. diss., University of Virginia, 1979), p. 401. Trowbridge, p. 391.

[24] John Richard Dennett, *The South as It Is* (New York, 1866), p. 114. Elizabeth Hyde Botume, *First Days amongst the Contrabands* (reprint, New York, 1968), p. 220. Pearson, p. 138.

Francis Butler Leigh lamented the absence of the work ethic among the emancipated slaves. "From the first, the fixed notion in their minds has been that liberty meant idleness," she charged. It seemed hopeless to get them "to settle down to steady work." In particular, under share contracts she found "unless I stayed on the spot all the time, the instant I disappeared they disappeared as well." In August 1870 Charles Stearns wrote into his diary, "I can never cultivate a farm successfully with the blacks as laborers. Nothing can be done without incessant and minute supervision." Harriet Beecher Stowe found that black women serving as housekeepers demonstrated leisurely and haphazard habits and complained of "the continuous nature" of their tasks. Ida Higginson, who had come from Massachusetts to join her husband, Henry, on a Georgia cotton plantation for "the making of money" and "the work of great importance to be done for these Blacks," lacked the patience Mrs. Stowe's description conveyed, and was somewhat harsher: "Of course there are exceptions, but they seem to need supervision, and spurring on and urging and system to guide them; however, time will show, whether this is merely the result of slavery and dependence, or whether they can ever be wholly independent." Within three weeks, Mrs. Higginson apparently had the answer to her question: "curious creations these darkies are. I don't believe they could work, entirely left to themselves." A group of northerners, some ex-missionaries, in Port Royal, South Carolina, made similar complaints. "They (blacks) worked when they pleased" and "at what they pleased and only so long as they pleased." A black supervisor on a Louisiana sugar plantation, reputed to be an intelligent but hard taskmaster, had much the same to say about his brethren:

> I sposed, now we's all free, dey'd jump into de work keen, to make all de money dey could. But it was juss no work at all. . . . Why sah, after I ring de bell in the mornin' twould be hour, or hour'n half for a man'd get into de fiel. Den dey'd work along maybe an hour, maybe half hour more . . . and next thing I'd know, dey'd be poking off to de quarters. When I scold and swear at em dey say we's free now, and we's not work unless we pleases.[25]

The blacks were everywhere asserting their demand for freedom and self-determination of the work-leisure choice. It is a commonplace that the great majority of freedmen were exceedingly reluctant to contract as wage laborers if there existed any alternative. A Charleston freedman, who had relocated with his family to James Island, stated it all in one simple sentence: "I heard there was a chance of we being our own *driver* here; that's why we come." Then, in order to emphasize his position to the two white strangers he was addressing, he added, "If I can't own de land, I'll hire or lease land, but I won't contract." To the planter, seeking to employ laborers for the express purpose of producing for the market and a monetary profit, the prospect that blacks might obtain small plots of land by which a near-subsistence could be grown brought to planters' minds ominous images of

25 Leigh, pp. 124, 83, 56–57. Stearns, p. 279. Harriet Beecher Stowe, *Palmetto-Leaves* (Boston, 1873), p. 309. Perry, *Henry Lee Higginson*, p. 257. Dennett, p. 212. Reid, p. 462.

Thomas Carlyle and Demerara. Nowhere was this more evident than in the Sea Islands, where the freedpeople actually possessed the land, believing it to be theirs. A group of northern planters there had the common complaint that the large employer had no "control" over his laborers. A man with land of his own "seldom labored more than five hours in the day, and half of that time was spent cultivating his own crop of corn, potatoes, or ground nuts."

As far as planters were concerned, the problem was far from being local to the Sea Islands. The complaint was being made throughout the South. In 1869, the Boston cotton brokerage and agency Loring and Atkinson conducted a survey of the cotton states to ascertain "opinions relative to the labor, the methods of cotton culture, and the general condition and capacities of the South." From the responses it seemed that a major cause of the "unsatisfactory condition of Black labor" was the desire of the freedpeople to be "entirely independent of white men," which led to "squatting" on the piny lands of the uplands. From the perspective of northern businessmen like Loring and Atkinson, "nothing could be more encouraging than this if experience proved that their labor in such cases was productive; but we regret to say that it seldom amounts to much." What the cotton brokers lamented was that the squatters "seldom raise much to sell." Here was a potential conflict between the northern merchant and the southern planter. The merchant cared not who and how the raw staples were grown, as long as they were supplied in large quantities. If a profitable alternative to the planter's organization could be developed, northern merchants would accept it. The planters were in agreement with the merchants that staples had to be produced in ample quantities for commercial purposes, but they had an obvious incentive to retain production under their control.

This posed a dilemma for the planters that was difficult to ignore. Perhaps the radicals up north were at least half correct about the upward mobility of free labor in a competitive economy. More than one planter must have worried about the potential problem posed by the prospect of a good crop and high prices. A prosperous year would also enrich the Negroes. In that case "his wife will decline to become a field hand," his children will enter the "schools instead of the fields, *and the producing population will decrease fifty percent.*" And if this weren't bad enough, those Negroes most efficient as laborers will take their families and "settle down upon some poor tract" just to "eke out their existence" so they could enjoy more independence than that "in the relation of master and servant."[26]

The planters were at times both angered and perplexed by the behavior of freedpeople seeking to escape from the control and discipline of plantation work. Such behavior seemed noneconomic and irrational when remunerative offers were refused for the sake of a little independence. In 1866

---

[26] Trowbridge, p. 545. Dennett, p. 213. F. W. Loring and C. F. Atkinson, *Cotton Culture and the South Considered with Reference to Immigration* (Boston, 1869), pp. 22–23. "Mississippi Sentinel," printed in *DeBow's Review*, June 1867, p. 586.

Georgia planter and ex-confederate general Howell Cobb, wrote that he was "thoroughly disgusted with free Negro labor." Blacks who were offered "some little thing of no real benefit to them but which looks like a little more freedom" invariably "would catch at it with avidity," willing to "sacrifice their best friend without hesitation and without regret." The price of freedom in the determination of the work-leisure choice was high.

As late as 1905 historian-planter Alfred Holt Stone was frustrated by this same phenomenon. In 1898 Stone and his associate, Julian H. Fort, began a "plantation experiment" by offering what they considered better than fair market conditions to their tenants to encourage a more stable class of laborers with enhanced productivity on their Mississippi plantation. After five years of this experiment, Stone was sure that he and Fort had "demonstrated their ability to make independent property-owning families out of poverty-stricken material." The benefited Negroes reciprocated by showing "their independence by severing relations" with their landlords "almost as promptly as we put them on their feet." Stone reported that after three years he and Fort began "to feel reasonably certain that even the most practical appeal . . . to radically improved material welfare would be generally overcome by an apparently instinctive desire to move."[27]

To A. H. Stone and Howell Cobb this behavior was easily explained as a race trait. But in an earlier time and another place a different dominant group was, as T. S. Ashton suggests, distressed by the "irrationality of the poor" who "took no thought of the morrow" and were "indolent, improvident and self-indulgent." The change from the domestic or "putting-out" system to factory organization of the textile industry during the English industrial revolution provides a useful comparison. The craftsmen, the master weavers, their journeymen and apprentices were loath to give up the freedom and loose supervision of work in the home for the rigid discipline demanded by factory employment. The extreme disinclination of the weavers in Coventry to subject themselves to factory conditions was described by a government commissioner:

> With all its usual distress and degradation, the trade of single hand weaving (requiring a minimum of strength and skill) offers half the liberty of savage life, for which the uninstructed man is almost tempted to sacrifice half the enjoyments of the civilized. Thus, there is a well known feeling among the farm laborers, the brick layers, and other ordinary artizans in this district, that it is very hard on them to be turned out at early hours every day instead of being able to take what hours they please like the ribbon weaver, and like him, take *saint* Monday, and *saint* Tuesday too if they choose.[28]

[27] Ulrich Bonnell Phillips, *The Correspondence of Robert Toombs, Alexander Stephens, and Howell Cobb* (Washington, 1913), p. 684. Alfred Holt Stone, "A Plantation Experiment," *Quarterly Journal of Economics*, Feb. 1905, pp. 270–87.

[28] T. S. Ashton, *An Economic History of England: The Eighteenth Century* (London, 1955), p. 201. Government commissioner quoted in Abbot Payson Usher, *An Introduction to the Industrial History of England* (Boston, 1920), pp. 349–50. See also Mantoux, and Thompson, *Making of the English Working Class*.

The inclination of precapitalist workers to show their disdain for continuous labor has manifested itself in a variety of ways that have been a source of employer complaints. One eighteen-century English hosier complained that his men had the "utmost distastes" for any regular hours and habits. Such distaste led not only to the individualized absenteeism that could significantly slow production—as in the case of an eighteenth-century London businessman who repined, "I have not half my people come to work today"—but, as Josiah Wedgwood discovered, could also actually precipitate wholesale stoppages at most inopportune times: "Our men have been at play 4 days this week, it being Burslem Wakes. I have rough'd and smoothed them over and promised them a long xmass, but I know it is all in vain, for wakes must be observed though the world was to end with them."

A passion for four-day wakes is quite possibly an inherent characteristic of the human species. It may well have seemed that way to George Benham as he unsuccessfully tried to prevent the work stoppages for funerals that occurred sporadically in the area of his plantation and would often interrupt operations on "adjoining plantations" for "four solid days." And Charles Stearns, when his entire work force spontaneously decided to have a holiday during the height of the cotton planting, might well have saved himself the anxiety and effort expended in the attempt to bargain with them by offering two holidays *after* the cotton was planted. Stearns did not know that the master of discipline himself had failed using the exact tactic ninety-one years earlier. Stearns argued in vain, for "they wanted it then, and have it they would."

In addition, the Americans were not disposed to be bested by the English in demonstrating expressions of their individuality. Sidney Andrews reported from South Carolina that railroad companies complained that black laborers would only supply "three or four days work per week." However, in a face-to-face encounter with a group of English agricultural workers brought to Georgia by Mrs. Leigh, there seems to have been little or no contest. Mrs. Leigh reported that the whites "were constantly drunk and shirked their work so abominably" that the Negro foreman wanted them to work in separate fields from the blacks so as not to "set so bad an example."[29]

The proof of the pudding is in the comparisons. How contemporary observers gauged the work performance of the freedpeople depended heavily upon the observers' previous experience with wage workers. Lieutenant J. D. Rich, a bureau agent in Louisiana's sugar parishes, stressed the right points of comparison when he proclaimed that "as a general rule one whiteman from the North is worth two freedmen and will perform as much labor in a day, [because] he loses a great deal less time." But in view of the contemporary evaluations of northern laborers, one wonders if Rich was referring to northern laborers or to their employers. Perhaps Lieutenant

[29] Pollard, p. 225. Ashton, *Economic History*, p. 211. McKendrick, p. 46. Benham, p. 262. Stearns, p. 171. Andrews, p. 100. Leigh, p. 206.

George Rollins, who explicitly limited his comparison to "the same number of whites *similarly situated* or as laborers," made a more objective test when he asserted that "no set of men, be they white or black, work more diligently, cheerfully, and with more energy than the freedmen of this parish."[30]

What might best be inferred from quotations of this variety is illustrated by the fact that General Edward Hatch, within the span of six months, could make the seemingly inconsistent testimonies that the best labor organization was "to compel by some intimate close fitting system of prescriptions every able bodied negro to work," and that Negro men "worked as well as any men."[31]

After the end of his first year of southern residence, Charles Stearns assessed the prospects of the freedpeople becoming efficient and disciplined laborers: "I had sounded the depths of their degradation, and ascertained that they were greater than I had previously believed." Actually, Stearns and other planters had rediscovered a law of human nature that had perplexed Josiah Wedgwood to no end: "It is very strange how long workmen are in quitting a habit they have been long accustomed to, though I have told them, scolded them, and shewn them so often." But Stearns was perhaps even more surprised at this than Wedgwood had been. "Another trait of theirs, is the last one we should suppose they would possess, and that is *conservatism*," he noted. "Old ideas, old customs, old ways of performing labor, and especially old sins seem almost sacred in their estimation." Harriet Beecher Stowe, making similar observations of the freedpeople's behavior on a Florida plantation, evinced no surprise in finding this characteristic among them, remarking that "like all uneducated people, the Negroes are great conservatives."[32]

To profit-maximizing capitalists, the precapitalist workers' wasteful and inefficient habits, their playing at work, were just as odious as their propensity to maintain irregular hours. Such behavior was no doubt most vexatious to the most able and exacting employers. If Josiah Wedgwood desired machines that could not err, it soon became apparent that such workmen could not be hired but must be made. "We have stepped forward beyond the other manufactors and we must be content to train up hands to suit our purpose," he declared. In antebellum times, slave owners possessed an advantage over managers of free labor; having full control of their laborers, often from birth, they could more easily fit them to given purposes. After emancipation, the more talented managers received lessons in the working

30 *BRFAL*, film M1027, roll 28, pp. 160, 406 (my emphasis). Rollins referred to Carroll Parish, the location of George Benham's plantation. Benham, too, found the freedmen superior to a group of white northern laborers: Benham, pp. 98–102, 121. Also see Pearson, pp. 11, 18; Reid, pp. 503–4. On nineteenth-century northern wage laborers consult Herbert G. Gutman, *Work, Culture and Society in Industrializing America* (New York, 1977), pp. 3–78, and Foner, chap. 7.

31 *BRFAL*, film M752, roll 15, p. 472. *Report of the Joint Committee on Reconstruction*, p. 5.

32 Stearns, p. 153. McKendrick, p. 41. Stowe, *Palmetto-Leaves*, p. 290. (Mrs. Stowe had gone to Florida with her son, who had entered a partnership with two former Union officers to grow cotton and make their fortunes.) Also see Henry Lee Higginson in Perry, pp. 257, 265. Pearson, pp. 110, 315, 317.

of competitive labor markets that went beyond those received by their less gifted brethren. David Dickson, who owned a 15,000-acre plantation near Sparta, Georgia, and was reputed to be one of the South's most successful and efficient cotton planters, learned quickly:

> During the last year [1866], I learnt some valuable new lessons: one was the training of hands to do double the amount of work, with more ease and less waste of sweat and muscle. My former hands [slaves] being better trained than others, had better offers than I could give and nine tenths of them left me. I then employed hands from as many as forty plantations, and got none that knew how to work to any advantage.

That the problem would continue is clear from the remark of a Virginia tobacco planter in 1878: "the only drawback is that formerly you had the comfort of servants whom you could bring up to your ways and be sure of keeping, but now they do as they like."[33]

It is a commonplace that the training of hands to industrial routines is a simpler task when the laborers are young or inexperienced so that old habits need not be broken. The point is consistent with the experience of an English manufacturer who disclosed that his men were "considerably dissatisfied, because they could not go on just as they had been used to do." Again, it is consistent with the problems of a South Carolina saw mill proprietor who, after offering his freed workers twenty-five dollars a month plus rations in an attempt to change the hours of work, was informed that the fact that they were being offered higher wages and shorter hours than northern laborers received made no difference: "we want to work just as we have always worked." The famous story of Wedgwood inspecting the work of his employees and shattering the inferior pieces upon the floor with a scornful "This will not do for Josiah Wedgwood" is another example. Whether it is literally true may, as Neil McKendrick suggests, be subject to some doubt. We have it from no less an authority than the pen of Edward Philbrick himself that upon a surprise visit to the cotton house during the ginning process, two bales "which I ripped open to inspect and found, as I had feared, that it wasn't half cleaned" caused him "to set the women at work picking over the whole of it again" for the third time.

The presumed retrogression from the standards of the antebellum period for both Virginia tobacco workers and Sea Island cotton hands should make us somewhat skeptical about the importance of habits or experience relative to the independent decisions of free workpeople to do their tasks in the manner least costly or most profitable—to themselves. This point was insightfully recognized by another superintendent, who wrote that "there is

---

[33] McKendrick, p. 34. *Southern Cultivator*, 1867, pp. 58, 34. Campbell, *White and Black*, p. 286. Philip A. Bruce, *The Plantation Negro as a Freeman* (New York, 1889), pp. 182–85. During the antebellum period, Dickson's slaves had worked without the spur of either an overseer or driver, producing ten to fifteen bales and working thirty-three acres per hand. According to Loring and Atkinson (p. 22), the best planters in South Carolina and Georgia made from five to eight bales per hand and Dickson claims his worked twenty acres per hand.

much lazyness to be overcome in them," for "they sometimes neglect well-known precautions because they cost too much trouble."[34]

The profit-maximizing capitalist requires laborers with a work discipline compatible with the demands of continuous and efficient production. The owner of slaves can more or less enforce this discipline, but free workers must be trained to a point where they have internalized as their own the fundamental virtues exhibited by the capitalist. Above all, capitalist workers must economize on time, perceiving it as a commodity that can be alienated from themselves in distinct efficient units. Of all the work habits demoralizing to employers, the waste inherent in nonproductive units of time was perhaps most distressing. Planters heavily ingrained with beliefs about the inferiority of Negroes were obviously at a disadvantage as free labor managers, since their ideology might stifle their creativity. Those with less of a stake in racist ideology were free to experiment with various attempts at behavior modification.

If, for the capitalist, time was a valuable commodity, the most obvious way to instruct workwomen that this was so was to use wage incentives to give quick and expensive lessons in the dictum that time is money. Thomas W. Knox, a Civil War correspondent for the *New York Herald*, leased a Louisiana plantation near Vicksburg, Mississippi, in 1864. It used a variant of the standing wage system, whereby the laborers were to receive their wages "as soon as the first installment of the cotton crop was sent to market." But some of the Negroes, "having no fear of the lash," began to shirk labor by feigning sickness. The solution to this problem was the plantation version of the clocking-in system:

> We procured some printed tickets, which the overseer was to issue at the close of each day. There were three colors, red, yellow, and white. The first was for a full days work, the second for a half day, and the last for a quarter day. . . . These tickets were given each day to such as deserved them. They were collected every Saturday, and proper credit given for the amount of labor performed each week.

Knox reported that the "effect was magical": just one day after the adoption of this system the number of sick workers fell by one-half. Whitelaw Reid reported similar systems and results on the Mississippi side of the river near Natchez.[35]

Once the workers were assembled on the job, the planter was concerned with work habits and obedience. What was the profit-maximizing amount of labor to require of a workman? Just how much work could a man do? The modern industrial engineer spends a great deal of time in school learning the mechanics of time-work studies. Planters in the Reconstruction South anticipated many of these methods. At Natchez an old plantation

---

[34] Pollard, p. 255. *BRFAL*, film M869, roll 34, p. 323. McKendrick. Pearson, pp. 116, 122, 112. The issue, which has an importance all its own, is discussed further in Chapter 11.

[35] Knox, *Campfire and Cotton-Field*, pp. 373, 426–27. Reid, p. 489. *BRFAL*, film M1027, roll 27, p. 128. Easterby, *South Carolina Rice Plantation*, p. 231.

overseer explained a procedure that clearly grew out of his experiences under the old institutions: "I got up a race, and give a few dollars to the men that picked the most cotton, till I found out the extent of what each man could pick; then I required that of him every day, or I docked his wages." Another planter and his laborers submitted their disagreement over work and pay to a Freedmen's Bureau agent for arbitration. The agent "decided the matter in the only way that it could be decided by any other than an experienced planter. Viz by putting three or four hands at work and timing them on one or two rows and calculating from that, I found they could easily perform three tasks per day which I concluded would secure good wages."[36]

In this area no system of free labor management was more indicative of efficient capitalist management techniques than those of W. Miles Hazzard of Santee, South Carolina. Captain Hazzard and his relatives operated a number of plantations. Three of the captain's plantations employed 175 laborers, producing rice as the major crop. In 1869 Hazzard paid weekly wages, with those rated full or first-class hands receiving fifty cents per day and three-quarter or half hands proportionately less. All work that could be so organized was done under the task system. The captain's incentive system was ingenious enough to rival the creativity of modern-day economists' best theoretical work on incentives and wage payment systems. Each contracting laborer was allowed to choose his own rating, under the provision that no wage would be paid unless the task were entirely completed. Those doing more than a task were paid proportionately more, so that one-and-a-half task work brought one-and-a-half day's wages. It was found that this system greatly decreased disputes about work, and that the extra pay for extra work and the total loss of pay for less than a full day's work provided "adequate incentives to diligence."

For every twenty-five laborers there was a foreman, who received double the pay of the laborers and was responsible for their work. Deductions from pay or fines for improper work were taken from the foreman. A similar system was operated by Colonel Lockett of Georgia. The colonel classified three grades of labor and paid wages accordingly. Wages were contracted yearly but computed on a daily basis and paid weekly in tickets that were accepted at noncredit prices at the plantation store, just as was done on Captain Hazzard's places. Workers were "docked by the manager for every hour and even half hour lost." The colonel was reported to be a rigid but scrupulous employer whose system of rules worked to "preserve discipline, encourage industry," and promoted "contentment and happiness."[37]

The basic model of Captain Hazzard's operation had been an innovation Edward Philbrick brought to the culture of cotton. Both historians and contemporaries who criticized this adaptation of the antebellum planters' task

---

[36] Trowbridge, p. 386. *BRFAL*, film M752, roll 31, p. 113.

[37] Knox, *Campfire and Cotton-Field*, p. 373, 426–27. *Southern Cultivator*, May 1869, p. 206. Loring and Atkinson, pp. 27–28. For comparable systems in plantation management on Louisiana cotton and sugar plantations see *BRFAL*, film M1027, roll 27, p. 128. Reid, pp. 270–78.

system as merely a minor innovation which paid wages that were too low were wide of the mark. The basic wage, which varied from twenty-five cents to seventy-five cents per task after the war, could be earned with half a day's labor. This, as in Captain Hazzard's system, allowed individual workers to self-select full-time or part-time work on a daily basis, and allowed households to arrange their own work schedules, inside and outside the home, to suit individual needs—a choice rarely open to wage workers anywhere. As for the putatively low rate of pay, Philbrick answered his critics by saying that "the smart hands earn more than this in a day, for they do one and one-half times or twice as much per day as they used to" under "the old master's day's work or task." This meant that "whenever the amount of work done in a day approaches the standard of a day's work in the North, the wages also approach the limit of northern wages, under similar conditions." These points were independently corroborated by bureau agents after the war.[38] Whatever the thoughts of northern critics, joined by Generals Steedman and Fullerton during their 1866 inspection tour of the Freedmen's Bureau for President Johnson, northern and southern employers found the system a good one.[39]

Other than the argument of racial inferiority, there were three related theories which contemporary observers used to explain the work behavior of the blacks. The most straightforward was the arguments that the freedmen had developed slovenly work habits and expensive notions of time because as slaves these work habits had been expensive to their masters but not themselves. A proper, that is capitalistic, valuation of time could be instilled if the freedpeople were forced to bear part of the costs of their own behavior. Those who really accepted this argument also saw that an effective training program required that punishment for divergence from acceptable behavior must be immediate. The worker who took off half a day must pay the cost as soon as possible. Because of this, postharvest payments, especially when they were a share of the crop, were not conducive to the development of market notions of time and work discipline.

Thomas W. Knox expounded this theory as clearly as anyone in describing an experiment with payment systems on three plantations near Vicksburg. On the first plantation, the laborers were promised a share of the crop and provided with food and clothing until the crop was made. On the second, wages were ten dollars a month plus food, with half the wages each month held back until the end of the year. The last plantation paid daily wages of one dollar at the end of the day. Knox described the behavior of the respective workforces:

> On the first plantation, the Negroes are wasteful of their supplies, as they are not liable for any part of their cost. They are inclined to be idle, as their share in the division will not be materially affected by the loss of a few days' labor.

[38] Powell, p. 83; Rose, p. 225. Edward Philbrick in Pearson, pp. 246, 266.

[39] *BRFAL*, film M752, roll 31, pp. 112–14; ibid., film M869, roll 35, p. 138. Perry, *Henry Lee Higginson*, pp. 252–53. James B. Easterby, *South Carolina Rice Plantation*, letter from Benjamin Allston to Charles Richardson Miles on Captain Hazzard's system, p. 231.

On the second they are less wasteful and more industrious, but the *distance* of the day of payment is not calculated to *develop* notions of strict economy. On the third they generally display great frugality, and are far more inclined to labor than on the other plantations.

The reason is apparent. On the first plantation their condition is not greatly changed from that of slavery, except in the promise of compensation and the absence of compulsory control. In the last case they are made responsible both for their labor and expenses, and are learning how to care for themselves as freemen.

William Channing Gannett elaborated upon the theory in detail. Every vestige of the paternalism of slavery and dependence must be rooted out, he argued, for "to receive has been their natural condition. They are constantly comparing the time when they used to obtain shoes, dresses, coats, flannels, food, etc., from their masters, with the present when little or nothing is *given* them." "They still do not understand the value of work and wages," declared Henry Lee Higginson. "Think they ought to get all their living and have wages besides, all extra." The introduction of the "natural laws" of "competition," he thought, would develop "habits of responsibility, industry, self-dependence, and manliness." In the spirit of Lord Howick, there could be no "gradual system of preparation and training. Strike the fetters off at a blow and let them jump, or lie down, as they please, in the first impulse of freedom, and let them at once see the natural effects of jumping and lying down."[40]

Such language might easily be misconstrued as an overly simplistic advocation of nineteenth-century laissez-faire liberalism. "They must work or die," wrote Henry Higginson to his wife. That these men were laissez-faire liberals is unmistakable, but to call them simplistic is to overlook their conviction that "outside of directly spiritual labor, this emancipation work is the noblest and holiest," requiring "the best heads and hearts in the country." Higginson, at the end of a financially disastrous year, comforted his wife and himself by writing, "please remember that one great reason for our coming here was the work of great importance to be done for these blacks. Money is less valuable than time and thought and labor, which you have given and will give freely."[41]

Cold, calculating, and compassionate profit-seeking missionaries of the Reconstruction era tended to ignore their own paternalistic impulses toward the freedpeople. Their Benthamite philosophy may be criticized upon many grounds, but not for being too simple or for attempting to prepare black slaves for a life in a laissez-faire and racist environment.

A closely related theory was the idea that industrious, steady laborers could be created by inducing an increase in the variety of goods desired by the laborer. Frances Butler Leigh saw this as a major problem: "they all raise a little corn and sweet potatoes, and with their facilities for catching

---

[40] Knox, pp. 353–54. Pearson, pp. 147–48, 179. Perry, p. 256. Also Reid, pp. 527–31.
[41] Perry, p. 255. Pearson, pp. 149, 179. Perry, p. 261.

fish and oysters, and shooting wild game, they have as much to eat as they want, . . . not yet having learned to want things that money alone can give." To alter this condition she was "having them educated," in an attempt to "increase their desire for comforts, and excite their ambition to furnish their houses and make them neat and pretty."

Bishop George Berkeley, in a noncharacteristic burst of materialism, had suggested this Veblensque means of improving the work habits of the eighteenth-century English working class:

> Whether the creating of wants be not the likeliest way to produce industry in a people? And whether if our peasants were accustomed to eat beef and wear shoes, they would not be the more industrious?
> Whether comfortable living doth not produce wants and wants, industry and industry, wealth? Whether the way to make men industrious be not to let them taste the fruits of their industry? And whether the laboring ox should be muzzled?

Just as Bishop Berkeley and David Hume found this approach preferable to the traditional suggestion that men would work hard and steady only if driven by starvation wages, some planters sought to stimulate Negro workers. Captain W. Miles Hazzard, in addition to his adoption of tight discipline and payment of weekly wages to dissuade the laborer from idleness and the "pursuit of pleasure" (an "indulgence purchased at his own cost"), also maintained a store stocked with "abundant and diversified" goods sold at the "lowest cash prices." The captain stocked everything asked for, since "the greater the variety the more their wants are multiplied and their industry stimulated in order to procure the means of satisfying those wants." To too many planters the possibility of stimulating the plantation Negroes' wants presented itself not so much as a means of promoting industry and a new way of life but as a means of extracting profit by selling cheap goods of dubious usefulness at expensive credit prices. The profligacy of the rural laborer was to become legendary.[42]

Harriet Beecher Stowe sought to understand the work ethic of the freedpeople as a response to "the influence of climate and constitution, and the past benumbing influences of slavery." To Mrs. Stowe, the habits of southern laborers differed from those of northern men because the latter were "accustomed by the shortness of summer and the length of winter to set the utmost *value* on their *working-time*." In the South, where growth goes on all the year round, "there really is no need of that intense driving energy and diligence in the *use of time* that are needed in the short summers of the North: an equal amount can be done with less labor." Mrs. Stowe failed to recognize that much of the conflict between northern capitalist planters and their southern laborers was concerned precisely with the fact

[42] Leigh, pp. 124, 123. Berkeley quoted in Edgar Furniss, *The Position of Labor in a System of Nationalism* (Boston, 1920), p. 179. David Hume joined with Berkeley in this opinion, but we cannot be sure he would have considered the argument valid for Negroes, given his racial beliefs. *Southern Cultivator*, 1869, p. 207. Reid, pp. 343, 363. Powell, pp. 87–91.

that the capitalists were not satisfied with an equal amount done with less labor, but wanted a greater amount done with equal labor.

To Mrs. Stowe's credit, she did understand that the black work ethic was not inherent in the people but an integral aspect of a whole approach to life. If this is Professor Genovese's essential point, he might agree with Mrs. Stowe that the freedpeople's general "style of attending to things" could be understood no better than through their religion, a religion so oblivious to the capitalist's ideal of the value of time that the same God worshipped by that capitalist could, according to the freedpeople, instruct Gabriel thus on the final day of reckoning:

> And he will say, "Gabriel, Gabriel,
> Blow your trump;
> Take it cool and easy
> Cool and easy, Gabriel:
> Dey's all bound for to come."[43]

To a man as devout as Charles Stearns, it was transparent that the undesirable behavior of the black worker was a legacy of the degradation of slavery and, more fundamentally, of the "failure of the black people's religion to make them obedient laborers." The blacks were "saucy and impudent to their employers," paying no attention to orders unless it suited them. This same complaint was made by Sea Island planters. Here are the problems one of them had trying to get his cotton picked:

> He told the people to go out and pick it. They were busy at the time ginning the cotton already gathered, and at the end of a week he found that not one person had been into the field. He was determined to get the cotton picked so he locked the gin-house door. That exactly suited them, . . . it was just time then to dig the slip potatoes.

"It is a long, long struggle against ignorance, predjudice and laziness," Henry Lee Higginson groaned in despair. "The blacks will advance, if they are led, and if they will trust anyone. Now they cannot be induced to talk, to ask questions. They will listen, but not heed much from a white man."

Southern whites were having the same problems. General Howell Cobb's overseer wrote from Cobb's Georgia plantation:

> Your proposition to hier them has no effect on them at tall tha say and contend that onley three of them agreed to stay that was the three that spoke Sam Alleck, and Johnson the rest claim tha made no agreement whatever an you had as well sing Sams to a ded horse as to tri to instruct a fool negrow some of them go out to work verry well others stay at their houses untell and hour by sun others go to their houses and stay two and three days. Say enny: thing to them the reply is I am sick but tha air drying fruit all the time tha take all day evry Saturday without my lief I gave orders last Saturday morning for them to go to work when they got the order eight went out I ordered Tom to go to mill he said he would not doo so. The air stealing the green corn verry rapped som of them go when they pleas and wher tha pleas an pay no attention to your orders or mine.

<hr>

[43] Stowe, pp. 285–86, 293 (my emphasis).

After a few months with free labor, Virginian Robert Hubbard decided that perhaps the devil wasn't entitled to his due after all: "Some Negroes are very lazy, and others have become impudent. Being free and not used to freedom, they are inclined to give themselves airs." "One great trouble," explained an agent of the Freedmen's Bureau, "is that they have been preached to about *'their rights'* until they are persuaded that nobody else had any rights." "The planters" in St. James Parish, Louisiana, were "generally very well disposed and very honorable men," reported Lieutenant J. D. Rich, "but the freedman does not as yet know what freedom is."[44]

To combat nonpliant workers, Stearns, as had many others before, "relied on fines for non-performance of duties." But fines were not enough to bring about the desired result. To Stearns, if the recently emancipated blacks existed on a "lower plane" laboring "for a mere subsistence," it was not because of any *"native deficiency of intellect,"* and therefore the problem could be overcome with education and their acceptance of true Christianity as a means "to regulate their daily lives." In this regard, Charles Stearns and other devouts and missionaries of the Reconstruction period were demonstrating the cogency of Sidney Pollard's insight that "the preoccuption with the character and morals of the working classes" is a marked feature of the early stages of industrialization. The English working class had also exhibited disrespect for middle-class authority. Speaking of the rural English peasant of the nineteenth century, one outraged employer echoed the words of future Reconstruction planters: "If you offer them work, they will tell you that they must go to look up their sheep, cut furzes, get their cow out of the pound, or, perhaps say they must take their horse to be shod, that he may carry them to a horse race or cricket match."

To the eighteenth-century political economist Josiah Tucker, such license taken by the common classes could and should have been prevented or at least modified by the conjoining of a "plan of Labour and Industry" with "that of a virtuous and religious education" in the charity schools for children. Tucker, who observed that "the children of savages, without any formal plan of tuition, grow up to be the same savage beings with their parents and the offspring of *Hottentots* become *Hottentots* themselves," would have agreed with Charles Stearns that it was essential to the success of "manual labor" schools that young children be taken away from the influences of "rough and boorish" parents in homes that were little more than "Nurseries of vice" and "ignorance."

Josiah Tucker's enlightened rationalism was hardly calculated to win swift approbation from the employing classes. He, like Charles Stearns later, recognized the motive power in George Fitzhugh's dilemma. Sidney Pollard, addressing the problems of labor discipline during the industrial revolution, explains it best:

[44] Stearns, p. 364. Dennett, p. 212. Perry, pp. 265, 212. Thompson, pp. 71–72. *BRFAL*, film M869, roll 34, p. 952, South Carolina. Ibid., film M1027, roll 28, p. 166, Louisiana. Thomas, p. 400. *BRFAL*, film M1048, roll 45, p. 67, Virginia; film M979, roll 44, p. 683, Arkansas. Messner, pp. 49, 45, 86.

Lastly, the inevitable emphasis on reforming the moral character of the worker into a willing machine-minder led to a logical dilemma that contemporaries did not know how to escape. For if the employer had it in his power to reform the workers if he but tried hard enough, whose fault was it that most of them remained immoral, idle, and rebellious? And if the workers could really be taught their employer's virtues would they not save and borrow and become entrepreneurs themselves, and who would man the factories?

Southern planters, even George Fitzhugh, would have agreed that on another occasion Josiah Tucker spoke for the "civilized classes" of all ages: "Such brutality and insolence, such debauchery and extravagance, such idleness, irreligion, cursing and swearing and contempt of all rule and authority . . . our people are drunk with the cup of liberty."[45]

---

[45] Stearns, pp. 363, 396, 367. Pollard, p. 268. J. L. Hammond and B. Hammond, *The Village Labourer*, quoted in Thompson, "Time, Work-Discipline," p. 77. Josiah Tucker, *A Sermon Preached in the Parish Church of Christ-Church* (London, 1766), pp. 11, 4. Stearns, 493–94. Pollard, p. 269. Josiah Tucker, *Six Sermons*, pp. 70–71, quoted in Thompson, "Time, Work-Discipline," p. 31.

6

They say all the colored people's free; they do say it certain; but I'm a-goin on same as I allus has been. . . . My mistress never said anything to me that I was to have wages. . . . I left it to her own honor . . . she han't spoke yet. . . . I don't low to work any longer than to Christmas, and then I'll ask for wages. But I want to leave the ferry. I'm a mighty good farmer, and I'll get a piece of ground and a chunk of a hoss, if I can, and work for myself.

Old North Carolinian

Cain's Mark

It is an irony of history that when the transformation from the slave plantation to a decentralized domestic system was provoking an intense demand for more labor discipline and social order, the inverse movement, from domestic to centralized factory production, was eliciting the same response from northern industrialists and stalwarts of the middle classes. It was an irony only partially understood by many of its self-proclaimed victims. Social critic Nicholas Paine Gilman, referring to the "industrial warfare" between labor and capital in the latter third of the nineteenth century, attributed the problem to the alteration in industrial organization. Prior to the advent of the factory age, "the craftsmen who kept at work any considerable number of journeymen and apprentices were thus comparatively few, and the latter were not numerous enough to create a labor problem." Gilman's adherence to a traditional view of some merit overshadowed another of his observations, which southern whites would have made primary. Under earlier forms of organization difficulties of consumption, production, or distribution "would be adjusted in accordance with the traditional rules of domestic discipline; it would be, not a 'labor trouble,' but a 'family jar.'"

Josiah Tucker, a man of much deeper intellect, had wrestled with similar problems in an eighteenth-century England that seemingly teemed with social indiscipline. Like William Wilberforce, who was concerned that "an increasing evil" was developing in the manufacturing towns where "apprentices used to live with the master and be of the family" but increasingly began to "lodge out and are much less orderly," Tucker would have placed more emphasis upon Gilman's last point. The social functionalism ingrained in a paternalistic hierarchy was clearly appreciated by Tucker. He believed that the mid-eighteenth-century Irish poor were more orderly than their English neighbors because "the common people" of Ireland were "much more under the command of their superiors, than the common people of this kingdom would submit to be." The consequence was that "many Rules and Regulations, many Restraints and Coercions, can be established in that country, which would be deemed here to be high infringements upon liberty."[1] Former slaveholders had always stressed the social benefits inherent in the institution of slavery and had no illusions about the relative importance of large combinations of workers as opposed to the presence of paternal authority. The views of a group of cotton planters were typical: "This whole body of laborers has been demoralized by the removal of the domestic regulations to which they have been accustomed, and the failure to substitute any other discipline or government over them."[2]

Elimination of the domestic superintendence of slavery would invariably lead to changes in behavior and social relations, but the extent of these changes was not predetermined. If the blacks had exhibited the meek and reactive behavior attributed by their Sambo image, freedom could have brought very little change. For this reason an assessment of the alteration in social relations requires study of the initiatives of the emancipated laborers whose activities precipitated the most significant changes. Of foremost priority upon the employer's list of social offenses committed by freedpeople was the latter's refusal to adopt themselves to the employer's labor market requirements. Much Reconstruction historiography teaches that southern agricultural labor markets were dominated by planter attempts to depress labor's earnings to a subsistence level by the formation of cartels. That there is value in this approach cannot be denied, but it is possible to carry this aspect of employee-employer relations too far.

The issue is not whether such cartels were successful but what the social conditions surrounding them were. Even those writers who most forcibly argue that agricultural workers received a "competitive" return on their labor stress the mobility of individual laborers and the profit incentives of employers to ignore their agreements as the primary reason for the failure of such cartels.[3] This one-sided emphasis, no matter how inadvertently,

[1] Nicholas Paine Gilman, *Methods of Industrial Peace* (Boston, 1904), pp. 6, 7. William Wilberforce, *Life of William Wilberforce II* (London, 1838), p. 164. Josiah Tucker, *A Sermon in Christ-Church* (London, May 7, 1776), p. 14. See *New York Times*, Feb. 24, 1869.

[2] *Edgefield* [S.C.] *Advertiser*, July 5, 1865, in Chartock, p. 78. Also see *Senate Executive Document No. 2*, 39th Cong., 1st sess., pp. 83, 85, 86. William King, "Letter to O. O. Howard," May 30, 1865, *BRFAL*, film M752, roll 15, p. 857.

[3] Stephen DeCanio, *Agriculture in the Postbellum South* (Cambridge, Mass., 1974) and also Higgs, (see Chapter 3, note 22). My complaint with their argument that black laborers received

serves too well the Sambo image of the emancipated blacks. Their major recourse is supposed to have been uncoordinated individual action, manifested by the high rate of labor turnover upon plantations. More importantly, this emphasis obstructs the appreciation of an important theme which aids our understanding of developing black-white relations and the emerging social and economic order.

Members of the planting and commercial classes in southern society anticipated that federally imposed emancipation would fall far short of the social revolution expected by abolitionists. In 1864 a Memphis attorney predicted "that a very large number of the negroes will not accept their freedom and that, by one name or another, pretty much of the old relations will be reestablished." An Alabama planter foresaw a more active role for the landlords: "The nigger is going to be made a serf sure as you live. It won't need any law for that. Planters will have an understanding among themselves: 'you won't hire my niggers, and I won't hire yours;'then whats left for them? They're attached to the soil, and we're as much their masters as ever. I'll stake my life, this is the way it will work."

The landlords observed the unaltered economic dependence of the blacks and then proceeded to make an unwarranted extrapolation from their previous experience. Free Negroes had existed in the antebellum period under a system of protective guardianship meted out by benevolent white patrons. To ex-masters, the only difference abolition made was to swell the number of free Negroes desiring and needing guardians. Major General Henry Clayton of Pike County, Alabama, was sure that "by treating them with perfect fairness and justice" southerners would "secure the services of the Negroes, teach them their places, and how to keep them." Clayton assured his colleagues that with such treatment the Negro "torn loose from the protection of a kind master" would be "proud to call you Master yet."[4]

These beliefs cannot be discounted as mere wishful thinking. Given the racial distribution of property, wealth, and education, a master-servant (if not master-slave) relationship would not have been an unnatural outcome. A propertyless population extremely dependent upon the availability of employment was susceptible to some form of voluntary servitude. Masters who preserved some continuity with the old relationship by offering gifts or payments for past service were in an advantageous position to claim the loyalty and obedience of their people. A cynical outlook would no doubt

a "competitive return" in southern agriculture is that the point itself is rather peripheral to the important issues concerning black-white relations and the economy-wide opportunities available to the black worker. Critiques of their position can be found in Mandle, Wiener, and Harold Woodman, "Sequel to Slavery: The New History Views the Postbellum South," *Journal of Southern History*, vol. 43 (Nov. 1977), pp. 523–54.

[4] Genovese, p. 98. Trowbridge, p. 427. Walter L. Fleming, *Documentary History of Reconstruction*, vol. 1, p. 271. See J. D. B. DeBow in *Report of the Joint Committee on Reconstruction*, pt. 4, pp. 134–35. Stephen Powers, *Afoot and Alone* (Hartford, 1868), pp. 69–76.

place the profit motive as the preeminent reason for such payments. The perspective offered here retains this cynicism, but we should not dismiss the fact that some masters must surely have felt it a moral duty to reward their ex-slaves.

Not a few met their obligations. The payment was commonly referred to as getting a "start." Robert Toombs, Georgia's finest advocate of the doctrine of states' rights and an unrepentant slaveholder, was fondly remembered by one of his ex-slaves: "He give eve'ry nigger over twenty-one a mule, some lan' an' a house to start off wid." Alexander Stephens, vice-president of the Confederacy, made arrangements with a trusted former slave to lease rental tenancies to the families of his freed slaves. One less historically conspicuous figure, Florida planter Redding Pamel, made land grants to all his former slaves who wanted to remain with him. A freedman recalled that Willis Menefee of Loachapoka, Alabama, "called us all up an' told us we was jes' free as him. He give us all a suit of clo'es, some money, a mule, a cow, wagon, hog and li'l corn to start off on." The practice was widespread, if not deep. Colonel E. Whittlesey informed the congressional Committee on Reconstruction that "some of the old aristocratic planters are acting splendid towards their former slaves." Whittlesey had specific praise for a large planter near Wilson, North Carolina, who adopted a more egalitarian version of the South Carolina method of renting freedmen small plots of land to keep a supply of labor available. He "divided up a portion of his immense estate of 5,000 acres, and bestowed a certain number of acres to each of his former slaves who are now working for him. They are paid their regular wages besides, for working the planter's farm, and are allowed so much time per week to cultivate their land."[5]

The intended effect of the gratuities was not lost upon the Negroes. Alonza Fantroy Toombs made the point for Robert Toombs: "atter de s'render, nary a slave lef' Marse Bob." Eighty-three-year-old Ezra Adams explained why he remained on his South Carolina home plantation: "De slaves on our plantation didn't stop working for old marster, even when they was told dat they was free. Us knowed too well dat us was well took care of, wid a plenty of vittles to eat and tight log and board houses to live in."[6]

This attempt to solidify a continuing tradition of paternalism was the cause of much controversy then and later, when it was invoked by early twentieth-century historians to demonstrate that the planters were the Negroes' best friends. It was very likely a major factor in the divergent views of the social condition of the South reported to President Johnson by Carl Schurz and Benjamin Truman. Truman saw these actions as demonstrative expressions of the planter class's willingness to treat the blacks fairly. The problem for free labor reformer Schurz was that he perceived that the planters' conception of what was fair treatment for Negroes, while it might

[5] Rawick, vol. 6, p. 384. Thompson, p. 80. Rawick, vol. 17, p. 215; vol. 6, p. 280. *Report of the Joint Committee on Reconstruction*, p. 196. Also Rawick, vol. 17, *Florida*, pp. 246, 260; vol. 13, part 3, *Georgia*, pp. 85, 207, 134, 92;. vol. 5, part 3, *Texas*, pp. 139, 145; part 4, p. 50.

[6] Rawick, vol. 6, p. 381; vol. 2, part 1, p. 5. See also below, notes 12 and 13.

deliver material comfort, was not conducive to eventual social and political equality and the growth of a democratic free labor society. Planter fairness, in the words of planter J. Jenkins Mikell of Edisto Island, South Carolina, envisioned a "laboring class" settled "with their little farms around our broad acres; with their little comforts, their cow, pig and poultry," content and "always at hand to hire you their services in cultivating your fields and attend you to the ballot-box in support of intelligence and honesty." For anyone convinced of Negro inferiority this future was not only fair but wise. Those planters, and some of their historian descendants, could never see the Negro as being fit for anything other than a field hand. Consequently, they could never understand how anyone who argued for more could be anything but a fool or, worse, an abolitionist troublemaker.[7]

Jan Breman's penetrating study of the *hali* system of bonded servitude in the agricultural areas of South Gujarat, India, during the nineteenth and twentieth centuries is a paradigmatic example of the labor system most plantation employers sought to install in the postbellum South. A *hali* was an agricultural laborer who entered a labor contract which bound him to the employ of a landlord, *dhaniamo*, for an indefinite period which usually lasted for life and continued through later generations, although this last aspect was not binding upon the employer or servant. The *hali*'s contract bound to employment not only the contracting head of the household but also his wife and children.[8] Breman writes that "neither the work nor the working-hours of the hali were clearly defined. At all times of the day, and if necessary of the night, he was at his master's beck and call, always ready to carry out orders." The duties of a bound farm servant in Bihar Province early in the nineteenth century as described in detail by a twentieth-century student of the system are worth quoting in full:

> The master might employ slaves in baking, cooking, dyeing and washing clothes; as agents in mercantile transactions; in attending cattle, tillage or cultivation; as carpenters, iron mongers, goldsmiths, weavers, shoe-makers, boatmen, twisters of silk, water-drawers, ferriers, brick-makers and the like. He may hire them out on service in any of the above capacities; he may also employ them himself, or for the use of his family in other duties of a domestic nature, such as in fetching water for washing, anointing his body with oil, rubbing his feet, or attending his person while dressing, and in guarding the door of his house, etc. He may also have connection with his legal female slave, provided she is arrived at the age of maturity and the master has not previously given her in marriage to another.

A more servile and "exploitative" relationship would be difficult to describe. Yet Professor Breman, who by no stretch of the imagination can

[7] J. Jenkins Mikell, Esq., "Labor," *An Address to the Agricultural Society of South Carolina*, Seventy-sixth Annual Meeting, Jan. 12, 1871, p. 29. The reports of Schurz and Truman are respectively printed in *Senate Executive Document No. 2*, 39th Cong., 1st sess., and *Senate Executive Document. No. 43*, 39th Cong., 1st sess.

[8] Jan Breman, *Patronage and Exploitation: Changing Agrarian Relations in South Gujarat, India* (Berkeley, 1974). Breman translates *hali* "literally" as "he who handles the plow" and *dhaniamo* as "he who gives riches,": pp. 40, 44.

be called an apologist for or worshipper of the principle of voluntary contract, rejected the traditional explanation that *halis* were forced into service
by the landowning class, and argued that bonded servants entered this contractual relationship of their own free will. Indeed, landless laborers of the
Dublas caste who bound themselves as *halis* considered themselves to be
more fortunate than members of their caste who were unable to enter such
contracts. The key to this system was that in an extremely poor region
landless laborers with no private means of subsistence were, except in the
most extreme famine conditions, guaranteed subsistence by their masters in
return for loyal service. In an economy where agricultural laborers were
barely above the margin of survival during the best of times, the failure to
secure a master could literally result in starvation.[9]

Professor Breman explores in depth the complex socioeconomic relationship which developed within the *hali* system under the rubric of a patron-
client relationship, which he defines as

> a pattern of relationships in which members of hierarchically arranged groups
> possess mutually recognized, not explicitly stipulated rights and obligations in-
> volving mutual aid and preferential treatment. The bond between patron and
> client is personal, and is contracted and continued by mutual agreement for an
> indeterminate time.

Within such a system the obligations of the servant are unswerving loyalty,
obedience, and social deference, those of the master, a duty to protect and
give care such as medical treatment, subsistence, loans, and a pension during old age. Under conditions where the laborer received an income that
barely exceeded subsistence in the best of times, a poor season could make
many laborers absolutely redundant. Landlords were able to unilaterally
dictate the terms of the labor contract. According to the laissez-faire interpretation of voluntary contract, *halis* became beings whose very existence is
supposed to imply a contradiction—*free slaves*.

Nevertheless, I must disagree with Professor Breman's assessment that
within the patron-client relationship that developed "there was no question
of a *quid quo pro*." While the *dhaniamo* may have found egoistic gratifi-
cation and increased social discipline to be benefits of his insistence that the
compensation received by the servant was a gift made by the master's bene-
ficence, this social facade was an illusion. The one essential difference
between chattel slavery and voluntary bond service is that in the latter the
patron-client relationship which develops must be able to stand the test of
an improving labor market for the client. Whether he understood it or not,
the patron who accepted a voluntary client would have to meet the market-
determined price of loyalty and subservience. Professor Breman implicitly
notes this point himself when he says that the *hali* would desert his master
with "alacrity" if alternative employment suddenly materialized nearby, as

⁹ Ibid., p. 40. Also see James Scott, *The Moral Economy of the Peasant* (New Haven,
1976), especially p. 38. Breman, p. 18.

in the 1860s, when "many farm servants left their masters to labor on the construction of the railroad through South Gujarat."[10]

It is within this framework of *market paternalism* that we must seek to understand the planters' unsuccessful attempt to reestablish slavery under the guise of the contract system overseen by the Freedmen's Bureau. The explanation usually offered is that the planters failed to recognize that they were bargaining under new social relations with free labor.[11] For some planters this was certainly true, but for most it was not their failure to understand the differences between the institutions of free and slave labor that caused trouble but their inability to understand the aspirations of the workers. A comparison of the initial labor policies of two planters who ostensibly failed or refused to recognize free labor is instructive.

In August 1865 an aging John Hartwell Cocke, one of antebellum Virginia's most famous agriculturist-abolitionists, convinced of the divine sanction of slavery as recognized by master and thrall, wrote to a relative:

> My people are still working on satisfactorily under their voluntary preference of bondsmanship to me, to Yankee Emancipation—to the end of this year—. . . when if they find they can do better for themselves—they will be free to make any change they may deem for the better—.

Apparently confident that his people would find voluntary slavery under him the best they could do, Cocke had a contract drawn up for the year 1865 with the intention of convincing the Negroes to enter a long-term relationship upon similar terms beginning in 1866. The laborers were to agree to

> render service to him under the same rules and regulations for food, clothing, labour and discipline as have been customary on this place before the war and that we prefer to live with him subject to that system of management to the end of the current year, [rather than] to accept of the freedom which is given to us as a result of the war.

Not surprisingly, Cocke's laborers later registered complaints with the Freedman's Bureau, which ordered him to discontinue whippings and to pay his workers wages.[12]

Contrast the Cocke plan with that of a Warren County, Mississippi, planter who boasted, "my niggers are all with me yet, and you can't get 'em to leave me."

> I give my boys a heap more money than I should if I just hired 'em. We go right on like we always did, and I pole 'em if they don't do right. This year

---

[10] Breman, pp. 18, 63–64.

[11] Mandle, Wiener (see Chapter 3, note 4).

[12] Martin Boyd Coyner, Jr., "John Hartwell Cocke of Bremo: Agriculture and Slavery in the Ante-Bellum South" (Ph.D., University of Virginia, 1961), pp. 578, 578–79. Cocke, a leader among Virginia proponents of black colonization outside the USA, had undergone a tortuous metamorphosis during the war. For further examples of this behavior see Dennett, pp. 77–95. *BRFAL*, film M826, roll 48, Mississippi, p. 396; film M752, roll 23, North Carolina, p. 79–81; film M869, roll 34, South Carolina, p. 283; film M798, roll 15, *Georgia*, p. 161. Rawick, vol. 5, *Texas*, p. 33; vol. 13, part 3, *Georgia*, pp. 6, 10, 92, 185.

I says to 'em, "Boys, I'm going *to make a bargain* with you. I'll roll out the ploughs and the mules and the feed, and you shall do the work; we'll make a crop of cotton, and you shall have half. I'll provide for ye, give ye quarters, treat ye well, and when ye won't work, pole ye like I always have.["] They agreed to it, and I put it into the contract that I was to whoop 'em when I pleased.[13]

These lines reveal all. A system of market paternalism, to be successful, must be built upon an illusion which conceals the commercially oriented nature of the employment compact. If servants understand that the payments they receive, in cash and kind, are market-determined payments for services rendered, the entire facade of paternal reciprocity is stripped away for both parties. For the master who knows that the services and deference he obtains from his servants could just as easily be given to any other master who paid the market price there can be no illusions of paternalism. Where relations of deference and servilty are recognized market variables the servants are in truth just as independent as any proletarian who has the choice of changing employers with no diminishment of status or material well-being. Under such circumstances deference must clearly be perceived as social role playing. Paternalistic contractual relationships must necessarily be personal: "such a one that no one else would do with them unless similarly situated," explained a Georgia planter in 1866. The master, even while insisting that his servants are materially better cared for than they would be anywhere else, will also insist, like the Warren County planter or Mr. McGee of Lumpkin County, Georgia, "that no one could entice them from me for even high wages."[14]

The labor market under such a system is not a competitive one. The worker's services are not valued identically by all employers, and therefore wages are not determined precisely by an impersonal market. On the other hand, the good master must remunerate his servants above the standard set by the "average" employer. When that standard is remarkably low, such as subsistence for Gujarat *halis* or antebellum slaves, there is a relatively low cost to masters. To pay above that which is required to attract a servant is to purchase deference and enforce the belief that payments are not market-determined but the blessings of a kind master and dependent upon obedient behavior. The question which needs to be answered is, why did the employers have this need for the personal relationship exhibited by market paternalism? Professor Genovese, more than any other, has provided us with the most insightful glimpses into the slaveholders' "collective world view." If one accepts his powerful argument that the planters' ideology of paternalism was a logical, if not necessary, class-based justification of an oppressive and exploitative social system, it is not difficult to believe that this view had to continue as a justification of the past.[15]

---

[13] Trowbridge, p. 391. Rawick, vol. 5, *Texas*, pp. 30, 31; vol. 17, *Florida*, pp. 239, 360; vol. 6, *Alabama*, pp. 314–15. See below, notes 19 and 20.

[14] The statement of Mr. McGee is in *BRFAL*, film M798, roll 13, p. 508.

[15] Genovese, p. 86.

Beyond this psychological need, there existed an economic benefit to be derived from the extreme demonstrations of loyalty by workers involved in a paternalistic employment relationship. For employers concerned with the problem of mustering adequate supplies of labor during peak periods of the year, the stable labor force provided by loyal servants promised to be an efficient solution. In fact, the very nature of the paternalistic labor contract, its generality and extreme ambiguity, promised the employer a great amount of labor services per dollar spent, and this was in spite of a possibly high absolute payment to hired servants. The attempt to institute market paternalism as the free labor system of the South was probably the most crucial step in the planter class's aborted effort to preserve continuity with antebellum conditions.

The sanguine prospects they saw for the preservation of self-respect and social continuity were founded in their understanding that the Emancipation Proclamation did not in and of itself entail the end of slavery. To be more precise, planters understood the nature of voluntary contract better than many abolitionists. The difference between John Cocke and the Warren County planter is that Cocke not only desired a master-slave relationship with his laborers but believed that he could in fact own slaves of their own free will. The Mississippian understood that free labor negated the possibility of his owning slaves, but he believed that by virtue of a quid quo pro he would be able to purchase not slaves but the master-slave relationship.

The distinction lays bare a source of conflict that existed in Anglo-American societies during the nineteenth century. For free white men, emancipation provided a frame of reference within which cherished ideals of individual freedom and democracy could be well defined. It could do so because their view of the slave was from above. But as the distance was not uniform, neither was the frame of reference. Middle- and upper-class abolitionists, free from the vicissitudes of physical necessity, could define freedom (but not free labor) as the antithesis of slavery in the abstract terms of voluntary contract. For the emerging labor movement, on both sides of the Atlantic, slavery provided a much more personal frame of reference. "There is," asserted the Massachusetts machinist Ira Steward, "a closer relation between poverty and slavery, than the average abolitionist ever recognized." David Montgomery, with succinct clarity, isolates the rupture which in the United States prevented a fusion between radical Republicanism and the labor movement: "The criticism to which labor subjected the doctrine of freedom as the absence of external restraints, therefore, was non philosophical—not the substitution of a new theory of freedom—but practical."

Ira Steward's positivism was a direct challenge to the continental rationalism of the middle class:

> All the distinction it is possible to make between poverty and chattel slavery, is the difference between a natural right, and a natural necessity. We call liberty a natural right; and food and shelter, that we may not starve or perish, natural necessities; but whether a natural right is more or less sacred than a natural

necessity, is a question that decides when answered, how close the relation is between slavery and poverty.[16]

The very fact that under conditions of extreme poverty voluntary contract could result in self-inflicted slavery seemed to settle the matter, from the perspective of the empiricists. The labor movement was forced to make paramount a question that seldom occurred to abolitionists: if voluntary contract in a "free" market could result in material and social conditions no better than slavery, what was the difference between free wage labor and slavery? The answer depended upon the definition of slavery. In providing it, free laborers upon both sides of the Atlantic drew from within their own experience an answer that was remarkably consistent with the definition offered by Negro slaves.

Let us begin with the Americans, whose experience has been aptly termed by Montgomery "the hours of labor and the question of class." The General Congress of Labour at Baltimore signaled the movement for total emancipation with the death of chattel slavery: "The first and great necessity of the present, to free the labour of this country from capitalistic slavery, is the passing of a law by which eight hours shall be the normal working day in all states of the American union." In the same year a group of working men in Dunkirk, New York, resolved: "We, the workers of Dunkirk, declare that the length of time of labour required under the present system is too great, and that, far from leaving the worker time for rest and education, it plunges him into a condition of servitude but little better than slavery." Individual laborers reached the same conclusions. A Massachusetts bootmaker, responding to the practical meaning of the hours of labor, was most specific: "my experience as to the hours of labor is, that long hours tend to destroy health, bring on premature old age . . . and above all degrade both men and women, taking all manly feelings and independent thought out of them, making mere machines and creating a feeling akin to slavery."[17]

In Great Britain the discourse was carried on in the same terms. Slavery, as defined by the workers, went far beyond abstract conceptions of self-ownership and freedom of choice. White workers on two continents asked themselves: what is the meaning of self-ownership if the length of the working day, the time that a laborer is controlled by the employer, is so prolonged that the time remaining for self-control must be largely consumed by sleep and regeneration? They, as did the slaves, concluded that ownership without control was a specious concept.

This was the basis for the idea of wage slavery as understood by an English factory inspector who, in 1862, declared that "there is a time when the master's right in his workman's labor ceases, and his time becomes his own." Different members of the middle and upper classes who came to

[16] Ira Steward, "Poverty," *Fourth Annual Report: Massachusetts Bureau of Statistics of Labor* (Boston, 1873), p. 411. Montgomery, *Beyond Equality*, p. 252. Steward, pp. 411–12.

[17] Quoted in Karl Marx, *Capital* (New York, 1967), vol. 1, p. 301. *Massachusetts Bureau of Statistics of Labor*, p. 284.

understand the sociological nature of the workers' complaints offered radically varying critiques. Marx and Engels believed that the greatest boon of the legislation shortening the working day was "the distinction at last made clear between the worker's own time and his master's," considered the legislation to be an initial step in the working class's struggle for independence. In the United States, Republican E. L. Godkin sought a solution that would avoid not only class divisiveness but also class division. Such division was considered harmful to the growth of democracy, and impossible to prevent under the wage system. To Godkin and many Republicans the permanent wage laborer was "a kind of retainer or vassal, who was favored by being allowed to work, and from whom the employer was entitled to exact, not simply the service agreed upon, but deference and obedience with regard to the control of his whole life."

The Republican dilemma was that legislative action to create a maximum working day would violate the sacrosanct principle of individual liberty with freedom of contract. Yet without some changes Godkin believed that the workmen would "never cease agitating and combining until the *regime* of wages, or, as we might perhaps better call it, the servile regime, has passed away as completely as slavery or serfdom, and until in no free country shall any men be found in the condition of mere hirelings." [18]

It was within the context of a revolutionary labor movement that southern planters sought to reinstitute the master-slave relationship. The conflict centred about the length of the working day at its most basic level. Federal imposition of the institution of free labor upon southern markets made it clear that the government would only recognize contractual relationships made in the market. This forced the employers to appreciate that any desire for personal control over the lives, and not merely the labor, of workers could be satisfied only by purchasing for specified periods of time those politically inalienable rights that the freedpeople might voluntarily sell in the market. Employers who understood this—and they were the rule, not the exception—attempted to perpetuate their previous autonomy by making control over their laborers' activities and behavior obligatory conditions of the wage contract.

An obsessive concern for social discipline and personal responsibility for the actions of employees are manifested by Regulation 2 of the 1866 share contract offered by two planter-partners in Shelby County, Tennessee: "Freedmen shall not leave the farm at *any time* without the permission of the employer. But he shall grant them indulgence when he thinks it not predjudicial to the interests of the farm, or the *quiet of the county*." The contract in which this regulation appears was apparently a standard form used by a firm in Memphis which acted as an employment agency for planters and freedmen. [19]

---

[18] Marx, pp. 300, 302. E. L. Godkin, "The Labor Crisis," *North American Review*, vol. 216 (1867), p. 185.

[19] *BRFAL*, film T142, no. 71. This stipulation should be distinguished from contracts some writers have criticized for requiring that laborers not leave the plantation during working

Planters, through the mechanism of the voluntary labor contract, sought to control the very social existence of the freedpeople. Employment contracts not only required laborers to be "obedient and faithful" but specified that there was to be "no drunkenness, gambling or swearing"—in short, no "vice of any kind." Upon penalty of forfeiture of pay, plantation residents were "to have no assemblies of visitors or frolicking" without the employer's consent. If the hours of labor were contractually specified as from "sun-up to sun-down," that was meant to refer only to the time spent in the fields. In practice the laborers had many tasks that they were expected to perform prior to their arrival in the fields and after they had left them. The employer sought also to regulate the domestic affairs of the workers' families. Women were told when and how much time they were to devote to cooking and the washing of clothes.[20]

The minute enumeration of contractual terms could become tedious. Some employers, like Robert Inman of Robeson County, North Carolina, simply explained that all employees were to render "true and faithful service as they had done when they were slaves," upon penalty of forfeiture of contract and pay. In 1867 freedman James Brown signed with two planters of Holmes County, Mississippi, under the stipulation that laborers "do and perform all of the work usually performed by the colored people when they were owned and held as slaves." "In consideration therefore," Brown was to have and receive one-half of the crop of all kinds that may be produced on said farm by one hand during said year. J. J. Thigpen of Wilcox County, Alabama, gave one-fifth of the crop to the workers, provided rations, clothing, and medical attention, and paid taxes, and merely required that laborers "be obedient servants in all respects." Let us end with North Carolinian Riley Pitman, who in 1866 "leased" ten acres and the privilege of two-thirds the crop for a period of five years to freedman Abraham Monsing "on the condition of his being obedient and demeaning himself in an upright manner."[21] The nature of the contracts attest well to the intentions of the planters. Realization of this intent required the cooperation of the Negroes.

The demand for labor in the postbellum South was too high to allow planters to induce black laborers into becoming *halis*. The Memphis *Argus*, no special friend of the Negro, gave the planters an early warning:

> The events of the last five years have produced an entire revolution in the social system of the entire Southern country. The old arrangement of things is broken up. The relation of master and servant is severed. Doubt and uncer-

---

hours. The regulation quoted in the text turns up repeatedly throughout the tristate region surrounding Memphis.

[20] *BRFAL*, film T142, nos. 66, 71, Tennessee-Mississippi; film M843, no. 34, North Carolina, p. 0541, 0549, 0733; film M826, no. 50, Mississippi, pp. 37, 894. *BRFAL*, entry 3286, vol. 1, no. 237, South Carolina, and entry 220, vol. 1, no. 160, Alabama, p. 170, are all fair samples.

[21] *BRFAL*, film M843, roll 34, p. 547; film M826, roll 50, p. 426; *BRFAL*, entry 104, vols. 1 and 2, no. 128–29, Alabama; film M843, roll 34, p. 627.

tainty pervade the mind of both. . . . The transition state of the African from the condition of slaves to that of freeman has placed him in a condition where he cannot avoid being suspicious of his newly acquired privilege of freedom. He looks with a jealous eye upon anything like an encroachment on what he esteems his rights. It is clear therefore, that while matters remain in this condition, no satisfactory arrangements can, for sometime, be made between two such opposing elements. We fear that too many of the former masters of the negro, forced by the events of a mighty revolution to relinquish their rights in the persons of slaves as property, do it with a bitter reluctance, amounting to absolute hatred.

The *Argus* had isolated an important point. An obvious reason for the coming failure of the contractual system to preserve the masters' authority intact was the simple fact that for most blacks some inalienable rights that were now their property were not for sale at the offered prices. The point had already been made and would be made again by free women and men throughout the world. Paternalistic capitalism, essentially a relationship between patron and client, was assessed by the workers in terms of their self-determination. Robert Owen, in his autobiography, failed to grasp why the workers were influenced by "their democratic and much-mistaken leaders [who] taught them that I was their enemy and that I desired to make slaves of them in these villages of unity and mutual cooperation." Half a century later economist Richard Ely would describe the failure of George Pullman's experiment in terms of the "total power of the company and the absence of self-government." In Pullman, "the resident had everything done for him, nothing by him. . . . the idea of Pullman is un-American. . . . it is benevolent well-wishing feudalism."

Southern planters, for their part, would quickly learn that they could not be *dhaniamos* for emancipated *halis*. "Freedmen are very willing to work when well treated," responded a Yankee general to the charge of General Edward Hatch that Negroes required a "close fitting system of prescriptions compelling discipline." But "slaves cannot be had at any wages whatever."[22]

[22] Quote in Armstead Robinson, "In the Aftermath of Slavery: Blacks and Reconstruction in Memphis, 1865–1870," (Senior thesis, Yale University, 1968), p. 96. Also Genovese, p. 110. Thompson, *Making of the English Working Class* p. 782. Buder, *Pullman*, p. xx. *BRFAL*, film M752, roll 15, pp. 480–1.

7

The land is mine you can either leave it or pay 1/3 crop as rent.

Jane Pringle

But small is the relief that History gives to Charles Lowndes for instance, a model of a good master, whose nigs are leaving him by scores.

James L. Petigru

The Curse of Cain

In practice, individual attempts by employers to contractually control the laborers' lives met with minimal success. Exceptionally burdensome stipulations led to plantation strikes or wholesale refusals to recontract on an offending plantation. In individual cases workers might simply leave. Freedmen Green Gray explained how his parents left their Alabama home after the war when the employer's wife attempted to work him without pay: "White woman told me go do somethin, bring in a load er wood I think it was, and my mother told me not to do it. He [the employer] and my father had a fuss an he tied my father to some rails and whooped him. Soon as they done that we all left. They hunted us all night long." George Welch, while working for a tobacco planter near Rocky Mount, Virginia, "being annoyed by the family" of his employer, simply moved off the plantation but continued to work.

The Freedmen's Bureau agent in Talladega County, Alabama, attributed the constant quarrels between employers and laborers to the fact that "many laborers contracted to perform more duty than the circumstances would admit." Greatly overmatched in the art of negotiating written contracts, the laborers sought to give the employers on-the-job training in the principle that time is money. Whitelaw Reid, usually a discerning observer, explained the process from the employer's perspective and for once failed to see the true nature of the phenomenon he observed. "Nothing seemed more characteristic of the negroes than their constant desire to screw a little higher wages out of" the employer, he reported. Reid enumerated a number of cases which he considered comical. A man was hired under contract to feed the plantation mules. After a lapse of time the employer gave orders that the number of feedings was to be increased. "Straightway Morton presented his claim for extra pay for this extra duty." A wagon driver who did double work on Saturdays, "kase I's ligious an wouldn't work on Sundays no how," asked his employer, "doesn't you tink dat for dat extra work on Saturday you ought to low me anoder day's wages?"[1]

A northerner, attempting to plant cotton in the Sea Islands, provides the best description of the freedpeoples' efforts to alter the mode of production and shorten the length of the working day:

> He chops down the weeds with his hoe, lays them in the furrow between the old cotton beds, and pulls the earth of the old bed over the weeds. That's listing, and he gets so much a task for listing. Then he goes out again and pulls up the earth on the listing. That's *banking*, and he has so much a task for banking. Then comes the planting, and so much a task for that, and so much a task for each hoeing and each hauling. For such an operation so many cents. By-and-bye picking comes, and there's so much a pound for all the cotton picked; everything as easy and regular as clockwork. But if he's wanted to do a quarter of an hour's work at any time, he expects pay for that. If he goes to the house for an axe he's to be paid extra for it. It's well enough to pay a man for all he does, but who can carry on a farm in such a way as that? But suppose you want him very much for some piece of work that must be done. You cannot have him. He's working an acre of cotton for you, but his corn, and his rice and potatoes make a little farm that he's working for himself, and he can't do job-work for you when he's got his provisions to make; he needs to have *control of his own time.*

Control of time was exactly what planters did not want freedpeople to have. A Georgian explained the situation to a Yankee military commander:

> General, formerly the slaves were obligated to retire to their cabins before nine o'clock in the evening. After that hour nobody was permitted outside. Now, when their work is done they roam about just as they please, and when I tell them to go to their quarters, they do not mind me. Negroes from neighboring plantations will sometimes come to visit them, and they have a sort of meeting, and then they are cutting up sometimes until ten or eleven.

The general attempted to placate the complainant by pointing out that when northern laborers relaxed by singing and dancing no one thought there was much harm in it. The reply was that "these were negroes who ought to be subordinate, and when I tell them to go to their quarters and they don't do it we can't put up with it."

---

[1] Rawick, vol. 9, part 3, *Arkansas Narratives*, p. 80. BRFAL, entry 4259, "Complaints 1866 and 1868 Rocky Mount, Virginia," p. 32. *BRFAL*, film M809, roll 18, Alabama, p. 842. Reid, pp. 506–9.

Especially revealing is the experience of Louis Manigault who, after an absence of many years, returned to Gowrie, his Georgia rice plantation on the Savannah River in 1876. "All of this free-labor system was perfectly new to me," remarked Manigault, who approached the venture of active planting innocent of the state of plantation life under freedom. Manigault's horror is evident: "One of the first things we had to do upon taking possession of Gowrie, was to free the settlement from boisterous and turbulent negroes, many of whom had accumulated and settled upon the premises. A regular trading boat kept by a negro, (one Riley) had for months, (perhaps years) been plying between Gowrie and Savannah, in regular daily trips." A special concern was that "a large amount of liquor would be flowing about the place, and the most riotous conduct was going on in the settlement each night." This narrative continued with a harrowing account of a vicious knife fight and homicide that "occurred just under [his] window." These conditions were not to be tolerated. James B. Heyward, Jr., Manigault's plantation manager, "step by step, broke up all trading boats, sent off every worthless negro, and caused every house to be filled with quiet orderly people, in lieu of a savage boisterous set of worthless 'human beings.'"[2]

Louis Manigault believed that "the perfect little hell" which he found at Gowrie had been purposely allowed by the last lessee, Daniel Heyward. Manigault believed Heyward had kept Gowrie Plantation "as a kind of Botany Bay" in order to ensure a plentiful supply of labor, so that it "became a rendezvous for all the turbulent negroes in the vicinity." Manigault was probably unfair to Daniel Heyward. Planters throughout the South had lost much of their personal control over the freedpeople. The combination of emancipation and a postwar economy had opened many lines of entrepreneurship, and petty traders could turn a profit on liquor, sex, and cheap goods. The Freedmen's Bureau had found itself unable to prevent the operation of black markets and the concomitant disorder in the very river districts where Manigault's plantation was situated. A bureau agent who recognized the futility of attempting to ban all such trade recommended that the sale of liquor to freedpeople be controlled by requiring that sellers to blacks have a bureau-approved license. He explained the problem to a superior officer: "I have found out that at different places in my district travelers were selling liquor to the freedmen, and the freedmen [sold] to the soldiers."[3]

In lieu of taking direct advantage of their mobility, many freedpeople sought to win better employment terms by engaging in group work slowdowns and strikes. On the plantations "they are constantly striking for higher wages," testified Dr. James B. Hambleton of Atlanta. Most of this activity was undoubtedly confined to individual plantations and farms, and

[2] Dennett, p. 205. Reid, pp. 387–88. Louis Manigault, "The Plantation Journal of Louis Manigault," handwritten and typed copy in the U. B. Phillips Papers, Yale University; pp. 1–4 handwritten section.

[3] Manigault, "Plantation Journal," pp. 1, 4. *BRFAL*, film M869, roll 34, South Carolina, pp. 0374, 0445; also film M1027, roll 28, Louisiana, p. 6.

may well have been a logical extension of the work stoppages and protests against extra-harsh practices that had sometimes occurred during the slave regime.[4] It is doubtful that continuity along these dimensions would have been very disturbing to most employers. Significant problems resulted from the fact that many freedpeople lost little time in demonstrating their ability to grasp the essentials of labor-management relations under market capitalism.

The unknown freedmen and women who were destined to become grass-roots leaders of the people had little trouble understanding that amelioration of labor conditions depended upon group action. One potential organizer in Mississippi, warned by a white northerner in 1865 that "the whites intend to compel you to hire out to them," replied, "What if we should compel them to lease us lands?" The ability of the Negroes to organize labor combinations involving hundreds of workers and large land areas was amazing to contemporary whites. A South Carolina correspondent of the *New York Times* recorded his wonderment for posterity:

> The freedmen, during the past twelve months, have shown a surprising and constantly increasing capacity to adapt themselves to the novel circumstances by which they are surrounded, and they are rapidly acquiring the habits and modes of thought and action which are the characteristics of free labor all the world over. Perhaps the most striking instance of their advancement in this respect is to be found in their tendency to combine for the redress of grievances by lawful means. A movement is now on foot among the blacks, in many of the interior districts, designed to secure, during the coming year several changes in the system of employment, including a very material increase in the compensation of the laborer.

Underestimating the blacks was no doubt the cause of many labor problems that might have been avoided. A case of this kind occurred on a Louisiana cotton plantation when the initial monthly payroll was a few days late. The northern lessee, instead of explaining the minor problem to his work force, took the advice of his overseer, who advised that "probably they would never know it was the beginning of a new month, unless he told them, and that therefore it was best to say nothing about the payment till the money came up from New Orleans." Shortly afterwards the owner found himself in the midst of a strike of his entire work force, which required the aid of the local bureau agent before it was settled.[5]

Strikes were not always of the wildcat variety, or spurred by reaction to a real or perceived mistreatment. In the summer of 1866 an Edisto Island bureau agent was called to arbitrate "a strike for higher wages on the part of the freedmen at a time when the loss of labor for three or four days would prove the loss of the whole crop. A fact which the freedmen claimed would

[4] *Joint Committee on Reconstruction*, p. 167, Georgia. Phillips, *American Negro Slavery*, pp. 303–4.

[5] Trowbridge, p. 362. *New York Times*, Nov. 30, 1866, p. 2. Reid, p. 547. For descriptions of individual plantation strikes see Perry, p. 256. BRFAL, film M752, roll 20, pp. 152, 157; film M798, no. 32, pp. 216, 666–67. Pearson, p. 300. Reid, pp. 550–51.

bring the agent to terms at once." Leon Litwack has noted that one of the most opportune times to strike was during the period between seasons, when contracts were being negotiated for a new year. This strategy allowed the laborers time to sample the payment and working conditions offered by different employers in their region. In addition, as the planting season approached employers short of labor were often induced to make better offers which frequently led to a panic that destroyed the previously made agreements of employers to hold payments to a given level.[6]

If the waiting game could drive some of the less astute planters into a frenzy of labor bidding, it could also work in the other direction. In a group composed almost entirely of poverty-stricken workers, strikes of almost any duration caused great deprivation, which severely weakened the solidarity required for success. If labor cartels were to extract the maximum payment for labor that the market would bear, recalcitrant workers and would-be strikebreakers willing to accept less remunerative positions than were agreed upon by the cartel would have to be disciplined, and some means would have to be found to enforce the agreement. In searching for possible means of enforcement the laborers, like the employers, turned to time-proven methods of group discipline.

In 1866 a South Carolina bureau agent reported to his superiors that at some of the labor meetings "threats of violence have been made by some freedmen against any other freedpeople who might be disposed to contract for the coming year." In some cases the laborers entered a pact to ensure enforcement of the strike agreement. In Cherokee Country, Alabama, a "meeting of negroes" voluntarily "bound themselves together, under penalty of fifty lashes, to be laid on the naked back, not to contract for any white-man during the present harvest, for less than two dollars per day." In this instance the strike was broken by white men who undertook to gather "the harvest at $1.50 per day."[7]

South Carolina's rice region along the Combahee River was disrupted for several weeks in 1876 when large numbers of field laborers quit work and demanded that planters stop paying them in paper IOU's, called checks, which were redeemable only at plantation stores. The strikers demanded regular cash wages for regular work. Throughout the strike, labor meetings were held to maintain camaraderie and support for the strike. It is said that at such meetings the laborers, setting a precedent for civil rights movements in the twentieth century, sang "the greenback song" to strengthen morale, voice their grievances, and chastise various offenders among the planters:

> We are not afraid to work
> We will labor every day.

[6] *BRFAL*, film M752, roll 31, p. 113. Leon Litwack, *Been in the Storm So Long* (New York, 1979), p. 435.

[7] *BRFAL*, film M869, no. 34, Dec. 31, 1866, p. 1017. Kolchin, p. 39. *New York Times*, June 22, 1866, p. 4. Also see Litwack, p. 438; Wharton, pp. 63–64; E. Merton Coulter, *The South during Reconstruction, 1865–1877* (Baton Rouge, 1947), pp. 110–11; Du Bois, *Black Reconstruction*, pp. 416–17, 508; *BRFAL*, film M1027, no. 28, p. 0127.

All that we want is the greenback.
When the day's work is ended,
Come and bring the pay.
All that we want is the greenback.

Greenbacks, forever, come planters come.
Up with the greenback
And down with the check.
We will labor in your fields
From the morning until night.
All that we want is the greenback.

G. G. Martin, don't you know
That we told you at the store
All that we want is the greenback?
Henry Fuller, don't delay.
J. B. Bissell, what you say?
All that we want is the greenback.

As the duration of the strike lengthened, both white workers and some blacks began to accept employment. Those who wished to remain on strike responded by compelling all workers to respect the picket lines. Negro politician Robert B. Eliot reported that "all classes of workmen are compelled to demand higher wages, and women and children are now beaten as well as men." Two years after the strike had ended a visitor was told by planters that strikers had "used much violence towards non-strikers, hunting them about with whips." The strike spread into parts of Georgia, where a contemporary planter noted in his journal that the negroes "marched from plantation to plantation beating and exciting those who continued to work at the old prices." The strike ended when the state militia was sent to the area.[8]

Defensive strikes similar to the Combahee "riot" could generate offensive labor demands. In 1873 the recession and falling sugar prices provoked many Louisiana planters to alter their payment system from wages to shares. Other planters formed an association and agreed to lower monthly wages from $20 or $18 a month to $15 and $13, giving as reasons low prices and an increased supply of labor due to migration from Alabama and Georgia. On January 4, 1874, in a meeting at Zion Church near Houma, Louisiana, two hundred laborers retaliated by forming their own association. They bound "themselves not to work for any planter for less than $20" and rations, stipulating that the wages were "to be made monthly in cash." The laborers then went further, demanding that they be allowed to "form sub-associations and rent lands to work by themselves." Four days

---

[8] *Beaufort Tribune*, Aug. 30, 1876, quoted in Dorothy Sterling, *The Trouble They Seen* (New York, 1976), pp. 288–89. Williamson, *After Slavery*, p. 104; Campbell, *White and Black*, p. 145; Sterling, p. 289. "Journal of Charles Manigault," in Phillips Papers, Yale University, p. 38. The militia handled this problem without resorting to violence, probably, as Dorothy Sterling believed, because the South Carolina militia that was sent was composed of black troops commanded by Negro congressman Robert Smalls.

later the strikers were called to another meeting, where local leaders W. H. Keys and T. P. Shurbem advised them that if the lands were not given, they should seize them by force and not allow anyone to work for wages below those stipulated. The strike spread throughout the state. On January 15 the *New Orleans Daily Picayune* reported that "large sections of the state are overrun by lawless bands of negroes, who visit plantations, stop all work, threaten the lives of the peaceful and contented laborers, and fill the country with terror."[9] Here as in South Carolina, the strike—or "Terrebone riot," as the white press labelled it—was quelled by state militia.

The presence of this kind of labor market conflict indicates strongly that upper-class concerns with social discipline during Reconstruction went far beyond the usual portrayal of planter concerns with roving, lazy black workers or white concerns with proper social etiquette between the races. Leon Litwack and Marxist historians such as Jay Mandle and Jonathan Wiener have emphasized the importance of an incipient labor movement among black workers during Reconstruction.[10] One element that makes this issue important is the implication that the decision by many black laborers to form coalitions and confront groups of planters as a united front represented a clear and immediate repudiation of the paternal relationship that had existed between an individual master and his slaves. The occurrence of such a marked abandonment was unexpected by the slave-owning class, and it is difficult to explain even if one rejects the thesis that the blacks were willing participants in the paternalist slave compact.[11]

Max Weber, writing about the agricultural laborer in Germany, made a similar point that provides a clue to the answer:

> Except in times of extreme political turmoil, a class consciousness among the rural proletariat directed against the masters could only develop purely individually with relation to one master alone insofar as he failed to show the average combination of naive (sic) brutishness and personal kindness. That, on the other hand, the agricultural laborers were not normally exposed to the pressure of a purely commercial exploitation was in complete accordance with this. For the man opposite them was not an "entrepreneur" but a territorial lord in minature.[12]

Even during slavery, the fact that the planter produced for a market economy often impinged upon his relationship with the slaves. Weber's use of the qualifier "normally" takes on great significance when we recall that

---

[9] *New Orleans Daily Times Picayune*, Jan. 15, 20, 1874, p. 1. See Rogers, *Agrarianism in Alabama*, for an account of the Perry County Labor Union (1871), p. 18.

[10] Mandle, *Roots of Black Poverty*; Wiener, *Social Origins of the New South* (see Chapter 3, note 4). The classic reference for this interpretation of labor relations is W. E. B. Du Bois's *Black Reconstruction*.

[11] Genovese, in "The Moment of Truth," *Roll Jordan Roll*, pp. 97–112, documents the trauma experienced by slaveholders when "their people" foresook them for freedom. My interpretation differs, but draws upon Genovese's insightful perspective.

[12] Max Weber, "Die Ländliche Arbeitsverfassung," in *Gesammelte Aufsätze zur Sozial- und Wirtschaftsgeschichte* (Tübingen, 1924), p. 474, quoted in Breman, *Patronage and Exploitation*, p. 66.

financially constrained slaveholders often had to sell slaves and reduce the amount of gifts made to their laborers. His phrase may be paraphrased with profit. Antebellum planters may very well have been schizophrenic in just the sense that they were "normally" capitalists when confronting all commodities *but* their labor. The fundamental flaw in the planter's ideology was not, as Professor Genovese correctly points out, that it was "self-serving and radically false." This condition is shared with "every other ruling-class ideology."[13] The ideology failed at a point where Genovese himself seems ambiguous. The failure of planter ideology to gain hegemony over the slaves is a condition not shared by certain ruling-class ideologies. The black response to the political turmoil of Reconstruction demonstrates the point.

Slaveholders had convinced themselves that theirs was a benevolent institution. An effect of emancipation was the severing of the connection between economic incentives and acts of benevolence. If the planters' version of paternalism had been more than a politically expedient ideology serving as self-prescribed psychotherapy to rationalize the social and individual guilt that would have been forced upon a society that recognized profit as the basic motive for its exploitation of the slaves, there should have been a very slow repudiation of the old authority. But too many masters were just as anxious as the blacks to discard those aspects of paternalism that had costs, while retaining those with benefits. The point was made well by a bureau agent, who had the opportunity to base his comments upon observation:

> The race of kind human Masters of whom we have heard so much seems to have become extinct and with callous indifference to the condition or sufferings of the freepeople save where their own interest is concerned seems the principle characteristic of the white population of this District.[14]

If Major Williams overstated the case, that travesty would have been forgiven by the many freedpeople who were turned out of their homes to make their way unaided, without food or support. Newspaper correspondent J. T. Trowbridge related the planters' conceptualization of the role of the state: "We can make more out of them than ever. The government must take care of the old and worthless niggers it has set free, and we will push through the able-bodied ones." In 1866 Commissioner Howard reported that "with few exceptions, the state authorities have refused to provide for old and infirm freedmen, or in fact to do anything for the relief of the class of persons supported by the government." Groups of planters throughout the South agreed in their meetings that former slaves should bear their own necessary expenditures.

A Louisiana planter explained the new facts of life by saying, "When I owned niggers, I used to pay medical bills. I do not think I shall trouble

13 Genovese, p. 86.
14 *BRFAL*, film M869, roll 34, 1866, South Carolina, p. 1017.

myself." Virginia planter Robert Hubbard instructed his son that "formerly, negroes were clothed by the hirer, but now they have to clothe themselves." In a few special cases some old Negroes would be allowed to remain on the plantation, "but they (or their children) must clothe them, and even these old negroes must not be allowed any improper privileges." "As far as practicable, all of the non-producing negroes ought to be sent off from Tye River."[15] The turning out of the old, sick, and dependent must be interpreted as a mass repudiation by the slaveholding class of their paternalistic obligations. Booker T. Washington described a scene on his Virginia plantation that must have occurred throughout the South during the early days after emancipation:

> To some it seemed that, now that they were in actual possession of it, freedom was a more serious thing then they had expected to find it. Some of the slaves were seventy or eighty years old; their best days were gone. They had no strength with which to earn a living in a strange place and among strange people, even if they had been sure where to find a new place of abode. To this class the problem seemed especially hard. Besides, deep down in their hearts there was a strange and peculiar attachment to "old Marster" and "old Missus," and their children, which they found it hard to think of breaking off. With these they had spent in some cases nearly a half-century, and it was no light thing to think of parting. Gradually, one by one, stealthly at first, the older slaves began to wander from the slave quarters back to the "big house" to have a whispered conversation with former owners as to the future.[16]

We may suspect with some confidence that the content, if not the assertiveness, of the old people's interviews is well represented by an occurrence described by Miss Caroline R. Ravenel on the occasion of her uncle informing a body of Negroes that "they were now free and could stay on the place if they [chose] to sign the contract or go if they did not." To some, such terms were a clear violation of their earned rights. "The oldest negro on the place, a very old man, was exceedingly indignant, and said, 'Missus belonged to him, and he belonged to Missus, and he was not going to leave her, that Massa had brought him up here to take care of him, and he had known when Missus' grandmother was born and she was 'bliged to take care of him; he was going to die on this place, and he was not going to do any work either, except make a collar a week." The old man had presented an unimpeachable case. Having worked as a slave for many years, he was entitled to his pension. On the same plantation the old family nurse raised the same question: "Missus seemed to want her to go off the place. That was not the way to treat her; if Missus had taken her aside and told her, she would have given her opinion." Missus, according to her niece Caroline, "did not attempt to coax them to stay, on the contrary, I think she was rather disappointed that they all decided to stay."

[15] Trowbridge, p. 409; "Report of the Commissioner," BRFAL (Washington, 1866), Nov. 1, p. 8. Genovese, p. 111. Thomas, pp. 296-97.

[16] Washington, Up from Slavery, in Three Negro Classics (New York, 1965), p. 40.

Like Mrs. Ravenel, many planters who lacked the calculating selfishness to openly concern themselves with their self-interest alone were forced to confront the reality of their own ambivalence toward the welfare of the freedpeople. In the late summer of 1865, Thomas S. Watson of Louisa County, Virginia, became concerned when his house servant Martha Smith enlarged her family with a new child. Watson felt that Martha's husband-less family would become a financial burden upon him. A profit calculation led to the conclusion that if "she should offer to serve me for food, houserent and firewood, it would be to my interest to decline." Watson was not prepared to succumb to the profit motive:

> But I see no way consistent with humanity for us to be relieved of the burden. True I might complain to the Yankees; but the most to be expected of them seem to me to be that they will give her and children rations. She tells me she can rent a house. Well, suppose her fixed in a rented house; and drawing rations—how long will this last? When the winter comes, how about warm clothes and firewood, etc.?[17]

Conversion to strict market determination of the value of goods and services exchanged was a two-sided affair, and it did not come easy. In late May of 1866, Mrs. Mary Jones wrote her son concerning a strike of her entire work force. Mrs. Jones had contracted to pay them a share of the crop that was less than the "half of the crop" many others in the area were receiving. The blacks felt that they were being cheated and forced arbitration by the bureau, which reluctantly informed all parties that the contract was legally binding. Grounds for the break were cemented. Mrs. Jones confided in her son:

> I have told the people that in doubting my word they offered me the greatest insult I ever received in my life; that I had considered them friends and treated them as such, giving them gallons of clabber every day and syrup once a week, with rice and extra dinners; but that now they were only laborers under contract, and only the law would rule between us, and I would require every one of them to come up to the mark in their duty on the plantation. The effect has been decided, and I am not sorry for the position we hold mutually. They have relieved me of the constant desire and effort to do something to promote their comfort.

A few months earlier she had lambasted the Negroes for their "vileness and falsehood" and had longed "to be delivered from the race." One fellow in particular was singled out for special criticism: "Cato has been to me a most insolent, indolent, and dishonest man; I have not a shadow of confidence in him, and will not wish to retain him on the place."[18]

[17] D. Smith, A. Smith, and Arney R. Childs, *Mason Smith Family Letters, 1860–1868* (Columbia, S.C., 1950), pp. 225, 226; Thomas, p. 204. Martha solved the problem by obtaining employment elsewhere. In some instances the new market relationship cut across the boundaries of the labor market, as when a community of North Carolina whites began to charge its black residents $12.50 for the church building which they were formerly granted free: *Report to the Joint Committee on Reconstruction*, p. 197.

[18] Myers, *Children of Pride*, pp. 1340–41, 1313, 1296.

Mary Jones was suffering from an acute case of psychological trauma commonly referred to as cognitive dissonance. In the twentieth century the caste of Anavil Brahmans, who had long served as *dhaniamos* to the Dubla *halis*, would suffer the same experience. When Dublas began to receive more outside employment opportunities, almost simultaneous to a change in agricultural techniques by landlords that lowered labor requirements, two trends occurred. First, Dublas increasingly began to refuse to accept the master-slave relationship: "The servant no longer wishes to be at his landlord's disposal from morning till night, and tries to limit his obligations to agricultural labor if he can." Second, and often alternatively, the employers, no longer requiring as many laborers, began to see "the wide range of obligations to *halis* and their families" as a burden.

Jan Breman details the developing conflict: "without wishing to keep up the former relationships, both parties appeal to their traditional rights while rejecting the obligations that went with them." Young *halis* began to break their contracts more frequently by running off, and a few *dhaniamos* reneged on the pensions of old servants. In such situations a process of group and self-rationalization is required to legitimize the changing relations. Members of the Dubla caste, especially the young, were increasingly seen as immoral, insolent, and undisciplined laborers. A group of older Anavils explained the new situation to Professor Breman in the 1960s: "Formerly the Dublas were our devoted servants, they listened to what we said, were satisfied with what we gave them, and worked hard for us. Now everything has changed. The agricultural laborers are in the village only during the rainy season. They have abandoned us and no longer honor their superiors."[19]

The implication was that the Dublas were no longer deserving of the benevolent treatment they had formerly been given. From here it is a minimal transition to the espousal of a purely businesslike relationship. For Americans the beginning of the process preceded emancipation. Many slaveholders felt their "peculiar institution" had become a curse. In 1863 the Reverend John Jones confided to his sister:

> I am truly tired of my daily cares; they are without number. To clothe and shoe and properly feed our Negroes and pay our taxes requires more than we make by planting, especially when debts have to be paid. I believe the most pressed people in our confederacy are the owners of the slaves who have no way to support them. Sometimes I think that providence by this cruel war is intruding to make us willing to relinquish slavery by feeling its burdens and cares.

Many planters confronted the ideology which had shaped their culture and solved the dilemma by making a break with the past. In 1865, rice planter Allen S. Izard outlined the course which must be taken by planters who meant to retain their property and position:

---

[19] Breman, pp. 126, 123, 144, 115. The Dublas, of course, viewed the change in an entirely different light, one that reflected their own interest.

Our place is to work; take hold and persevere; get labour of some kind; get possession of the places; stick to it; oust the negroes; and their ideas of proprietorship; secure armed protection close at hand on our exposed river, present a united and determined front; and make as much rice as we can at $2 to $2.50 a bushel. . . . Our plantations will have to be assimilated to the industrial establishment of other parts of the world, where the owner is protected by labor tallies, time tables, checks of all kinds, and constant watchfulness.[20]

The break, as incomplete as it was, made the planters as culpable as any other group that participated in the destruction of the social order they cherished so well. The blacks experienced and remembered the ingratitude of masters who denied their obligations. Many freedpeople believed that their long years of forced servitude entitled them to reparations. A Freedman's Bureau agent writing from the sugar region of Louisiana explained:

Many of them seem to have an idea that because they have been slaves, etc., that now the government is in duty bound to get them started as they term it, that is, to give each one a cabin and piece of land. . . . In short, that to make amends for their former servitude they should now receive from every white, far greater privileges than the whites accord to each other. But these ideas are gradually working out of their minds.

Bitterness that resulted from this unrequited demand has been interpreted as being directed toward the government. A study of slave testimony reveals that this cannot be clearly established. A twenty-one-year-old husband and father believed that the government should give the freedmen "the lands of their rebel masters," but his enmity was not directed at the government. "My master has had me ever since I was seven years old, and never give me nothing. I worked for him twelve years, and I think something is due me." An aged freedman related how his past master, refusing to pay wages to his former slaves, saw all of the "able-bodied men and women" leave the plantation. The planter swore "that the rest should go too," and he drove off the "aged and sick." The old man made the universal complaint, "He said he'd no use fo' old wore-out niggers. I knowed I was old and wore-out, but I growed so in his service. I served him and his father befo' nigh on to sixty year; and he never give me a dollar. He's had my life, and now I'm old and wore-out I must leave. It's right hard, Mahster!" Frank Fikes's memory of his owners was blotted by their final behavior: "Ol' Miss and Massa was not mean to us at all until after surrender and we were freed"; then "they got mad at us because we was free and they let us go without a crumb of anything and without a penny and nothing but what we had on our backs." Perhaps Jane Simpson, who had farmed, slave and free, throughout the Mississippi Valley, expressed the anger felt by all those freed slaves who never got a "start":

---

[20] Meyers, *Children of Pride*, p. 1121. Smith, Smith, and Childs, *Mason Smith Family Letters*, p. 231.

I never even heard of white folks giving niggers nothing. Most of de time dey didn't even give 'em what dey 'spose to give 'em after dey was free. Dey was so mad 'cause dey had to set 'em free, dey just stayed mean as dey would allow 'em to be anyhow, and is yet, most of 'em.[21]

If we ask not how many masters proved benevolent or harsh but why they acted as they did, we discover again one of the central structural problems of the Reconstruction period. Some masters were generous or harsh simply because they were benevolent or grasping beings. Men and women of these two types were atypical. We undoubtedly come closer to understanding the behavior of the group by considering the position of a true paternalist who acted neither harshly nor especially philanthropically.

In March 1867 Louis Manigault visited the family plantation for the first time since the war. Financially ruined by that war, he was at the time supporting himself and his wife and children by working as a clerk in a Charleston counting house. Unable to resume the planting of rice on his own accord, Manigault found himself touring the homestead under the guidance of the man who had leased the property. The remarks entered in Manigault's diary offer us some of the most profound glimpses into the inner pysche of the returning master, who for the first time confronts his people in their new state. In the presence of many of his former slaves Manigault's melancholy thoughts are the eloquent last statement of a vanquished class:

I imagined myself for the moment a planter once more as if followed by overseer and driver. The weather was most beautiful, not a cloud on the horizon, and so clear and pure the atmosphere that the Presbyterian Church steeple in Savannah loomed up as if one half its distance. I wished my horse with me to ride over the entire tract as of yore. But these were only passing momentary thoughts, and soon dispelled by the sad reality of affairs.

This reality drove some planters into a despair that vented itself upon the freedpeople. For Manigault, self-pity provided a source from which he empathized with the worse condition of his former slaves. Women "formerly pleased to meet [him], but now not even lifting the head as [he] passed," added to his melancholy. Jack Savage, a formerly unruly slave who had once been sold for indiscipline, became the unlikely inspiration for a mixture of external pathos and self-pity:

Even now I felt sad in contemplating his condition as in fact was the case with all of them. In my conversation with these Negroes, now free, and in beholding them, my thoughts were turned to other countries, and I almost imagined myself with Chinese, Maylays or even the Indians in the interior of the Philippine Islands. That former mutual and pleasing feeling of Master towards slave and vice versa is now as a dream of the past.

[21] *BRFAL*, film M1028, roll 28, pp. 171–72, St. Martin Parish, Jan. 31, 1866. Trowbridge, pp. 151, 155. Julius Lester, *To Be a Slave*, (New York, 1968) p. 147. Yetman, p. 280. Rawick, vol. 5, *Texas*, pp. 12, 70, 114, 130; vol. 6, *Alabama*, pp. 7, 420; vol. 9, *Arkansas*, p. 264.

To Louis Manigault the breaking of the bonds of slavery meant the Negroes were no longer his people and had become almost as anonymous as any group of a foreign race and color. Almost, for he still would have been moved to do something for them if it had not been for his untenable financial position.

For the historian there is in this a valuable lesson. Louis Manigault would have taken little solace in the knowledge that some years later Frederick Winslow Taylor, often acclaimed the father of "scientific management," when explaining why many manufacturers did not adopt his methods, could just as well have been discussing antebellum paternalism. Scientific management, he declared, requires time, patience, and money. Manigault may, however, have garnered some comfort if he could have believed that his former people had the understanding of a Tennessee Negro, who related his mother's feelings about the "start" she received from her master: "I don't think he give them anything when they were freed. He was a kind of poor fellow. Didn't have but six or seven slaves. He offered to let them stay and make crops. My father had a better job than that."[22]

[22] Manigault, "Plantation Journal," pp. 17, 24. Frederick W. Taylor, *Scientific-Management* (Westport, Conn., 1972), p. 60. Rawick, *Arkansas Narratives*, vol. 9, part 3, p. 75; also ibid., *Georgia Narratives*, vol. 13, part 3, pp. 160, 170; *Florida Narratives*, vol. 17, p. 23; *Arkansas Narratives*, vol. 9, part 3, pp. 319, 183; *Texas Narratives*, vol. 5, part 3, pp. 70, 73.

8

In this part of the country the people are disposed to enact from the laborer all that is implied in their own interpretation of the contract and are disposed to bring them in their industrial relations as near to the condition of slavery as possible and extort from them unremunerative labor which was never contemplated by the freedmen when making the contract.

Colonel Orlando H. Moore

Industrial Relations
and the Turnover of Labor

The desire of freed blacks to escape the wage gang and thereby break from the most objectionable features of the slave regime has been a recurrent theme in explanations of the rise of the sharecropping system. It is often argued that planters were forced to abandon both gang labor and the money wage because the freedpeople, under a system of fixed money wages, offered a much too unstable labor supply; abhorred the work gang, which was reminiscent of slavery; and, because of the labor shortage caused by their behavior, were in a position of power which enabled them to force adoption of family sharecropping.[1] This black power argument has not been subjected to critical analysis.

---

[1] Ransom and Sutch, *One Kind of Freedom*, follow a variant of this line of reasoning. "The facts that laborers were free to contract and in relatively short supply gave them the power in the labor market to insist on alternative arrangements," p. 67. "From the onset, the workers had expressed their displeasure with the wage plantation. The wage system as practiced by the large landowners bore an uneasy resemblance to slavery," p. 65. Jay Mandle, *Roots of Black Poverty*: "in the immediate aftermath of emancipation, wages were used in an effort to attract labor. . . . The degree of labor mobility implied by the operation of a labor market meant that the kind of stable and reliable labor force essential to plantation agriculture simply was not available under this method. . .," pp. 17–18. See also Wiener, pp. 43–46. The evidence for this position has never been examined. It will be argued here that it was falsely interpreted by the original historians who expounded it because it conformed well to their preconceived ideas

It is within the context of the planters' attempt to redefine the master-slave relationship under the rubric of voluntary contract that this argument must be examined. According to the economic version of this labor-market-instability view, wage contracts when initially agreed to by the freedmen were attractive. However, as harvest time approached it often became increasingly clear, because of good weather and abundant crops, that the value of labor was greater than the contracted wage. With that information employers would desire more labor and be willing to pay more. This induced large-scale enticement of workers, hence rising wages.

There is a fatal flaw in this argument. The most salient aspect of the postharvest or standing wage contracts that planters were offering is the fact that a portion, often all, of the cash payment would be forfeited if the laborer failed to fulfill the terms of the contract. Therefore, to be attractive any labor-stealing bid from another employer would not only have to be greater than the worker's future earnings during the rest of the current contract but must also compensate for all the forfeited backpay otherwise due the worker. This is a formidable requirement that gave the original employer a great advantage in retaining labor. For all the employer needed to do to retain the worker was raise the wage paid during harvest time enough to keep the entire payment, accrued back pay plus harvest pay, equal to the value of labor at harvest time. For the original employer, this increase in wages at harvest time would always be less than the harvest value of labor, or what the enticing employer would be willing to pay to steal the worker.[2]

Except for a few employers suffering from an idiosyncratic season (very localized flooding, for example), all factors affecting the value of labor in a large land area would be identical and there would be nothing to stop the planter from meeting the market wage at harvest time. There is, therefore, considerable reason for skepticism about the argument that a money wage system failed because it could not stabilize the planter's labor force. It is far more likely that any wage enticement that occurred was early in the

---

about the innate characteristics of Negroes. See Brooks, *Agrarian Revolution in Georgia*, pp. 18–36; Fleming, *Civil War and Reconstruction in Alabama*, pp. 720–34. Hammond, *Cotton Industry*, pp. 120–40, gives the fairest statement of the evidence, but because of his racial beliefs sees no need to seek an explanation along socioeconomic lines.

  [2] As an illustrative example, suppose that a laborer contracts for $10 a month plus rations, and that the contract stipulates that one-half of each month's cash payment will be retained until the end of the contractual period. If at the end of the ninth month the laborer is offered another job, he will require a payment at least equal to $45, composed of his retained pay of $5 a month for nine months, plus the $10 a month for the remaining three months of his contract. Suppose that the maximum value of a laborer to any employer for the next three months is $100, so that the enticer is quite willing to pay up to $100 if the worker will break his contract. The original employer can always top an enticement offer and make a profit. For example, to top the $100 offer he can make an offer of $56 for the remaining three months which, when added to the retained $45, equals $101—so the worker stays. Note also that the employer earns a three-month profit, because he only pays $56 for labor worth $100. If the contract withheld all pay until the end of the year, the worker who quit would lose $120 to gain $100.

contracting season, because some employers attempted to pay wages well below those available elsewhere from the start of the season. But this kind of instability would also affect a share payment contract. Any planter offering share contracts with a smaller share, less land, or other deficiencies relative to the contracts offered by his neighbors could expect to lose his workers fast if the information became known. To give credence to the instability-of-labor argument, we must be prepared to believe that a significant percentage of the laborers were naturally unstable, hopping from plantation to plantation during a given year, and were quite willing to forfeit a large fraction of their year's earnings simply for the joy of moving.

The breaking of contracts by laborers had a variety of causes. The one closest to the argument just critiqued had the worker leaving an employer because the debt incurred by drawing rations and provisions through the year was allowed to approach or even exceed the wages due at the end of the season. A worker who became aware of this had little to lose by breaking the contract and seeking employment elsewhere, where he or she would be unburdened by debt. It is not difficult to see that this behavior would be no less a problem for the planter hiring labor for a share of the crop.[3]

The common reason for the breaking of labor contracts was the ill-treatment of workers. Rough physical or verbal handling of labor by planters and their overseers, coupled with inadequate provisions of poor quality, were sure invitations to the free worker to run away. The labor management techniques of many planters seem to have undergone very little improvement in the first few years after Carl Schurz's observations in 1865. "The fact is," reported a bureau agent from Northern Florida in 1866, "the white man expects implicit obedience from the freedman: expects him to take the grossest insult with the calmness of a dumb brute." In 1867, General J. W. Sprague described the Arkansas planters' views concerning the management of free Negro labor: "The majority, however, seem to think that the only way to manage 'niggers' is to whip them and make them *know their places.*"[4] Much of the problem was caused by small landowners and managers who had been antebellum overseers on large plantations.

It should have been clear by the end of 1865 that successful planters were going to be the ones who invested in employee-employer relationships suitable to the management of free men and women. Many planters and their agents were either learning this the hard way or refusing to learn it at all. A planter confided to Trowbridge, "I should get along very well with my niggers if I could only get my superintendant to treat them directly. Instead of cheering and encouraging them, he bullies and scolds them, and

[3] See, for example, report of Ch. Raushenberg, Freedmen's Bureau agent in Cuthbert, Georgia, Nov. 14, 1867, printed in Lawanda Cox and John H. Cox, *Reconstruction, the Negro, and the New South* (Columbia, S.C., 1973), p. 340. *Tenth Census of the United States, 1880, Cotton Production I* (Washington, 1880), p. 356. *Montgomery Advertiser*, Sept. 19, 1866. Somers, pp. 60, 84. BRFAL, film M979, roll 44, p. 708, Arkansas.

[4] *BRFAL*, film M752, roll 37, p. 680; film M979, roll 23, "Report of J. W. Sprague to O. O. Howard," Jan. 1867, p. 137. Film M809, roll 18, pp. 460, 551, Alabama.

sometimes so far forgets himself as to kick and beat them. Now they are free they won't stand it. . . . The first I know, there's more bullying and beating, and there's more niggers bound to quit." The assessment of most northerners traveling in the South was that the great majority of southerners had not learned to treat Negro labor fairly.[5]

No doubt much of this is explained by the admissions of the southerners themselves. A candid young planter confided that "they've always been our own servants, and we've been used to having them mind us without a word of objection, and we can't bear anything else from them now." The problem ran deeper. Perhaps just as important was the existence of a crucial flaw in the recommended modes of punishment, which forebade corporal correction. If, under contract leasing for profit, the laborer and planter were bound to one another for a year, placing a disobedient hand in confinement with a bread-and-water diet might appeal to the employer during slack periods when all available labor was not essential to the plantation. During the busy season, time was money and an idle hand was an expense few planters willingly incurred. In January 1866 the bureau agent at DeValle Bluff, Arkansas, reported that "the planters as a general thing want force to be used in making them fulfill their contracts which I have told them repeatedly that I could and would not do." Charles Stearns's defense of the bureau-backed contract system, which he saw as a means of ensuring justice for both parties to a contract, reveals much: "If six men abruptly leave him [the employer], on a hot July day when his cotton is overrun with grass, it is poor consolation to him to be allowed to leave his cotton, and spend a week in pursuit of these men; even if he succeeds then in having them safely lodged in jail. That will not furnish him with other hands, or kill the grass, that is so rapidly destroying his cotton." For employers like Stearns and George Benham, the return of these men by military force seemed to be justified by the moral condition of the freedpeople. In the absence of prompt bureau action, and among men to whom public opinion gave approbation for the whipping of recalcitrant niggers, violence was to remain a significant disciplinary device. In the absence of a contract system, if laborers were free to quit an employer the only feasible punishment would have been the right to fire workers.[6]

Throughout the South, bureau agents were repeatedly forced to intervene between overseers and laborers. One agent in the sugar country west of New

[5] Trowbridge, pp. 367–68. *Joint Committee on Reconstruction*, testimony of Albert W. Kelsey, p. 3, Alabama. *BRFAL*, film M809, roll 19, pp. 771, 845, 930, Alabama; film M869, roll 34, pp. 274–76, South Carolina; film M826, roll 30, p. 323, Mississippi; film M1027, roll 28, p. 4, Louisiana; film M1048, roll 45, p. 147, Virginia.

[6] Trowbridge, p. 291. *BRFAL*, film M979, roll 23, Arkansas; also film M809, Alabama, pp. 990–91. Stearns, p. 108. Some bureau agents were ambivalent about the use of corporal punishment for precisely these reasons. For a revealing example of this ambivalence, see the report of Chaplain Samuel S. Gardner, agent at Selma, Alabama, in the *Final Report of T. W. Conway, Bureau of Free Labor Department of the Gulf*, July 1, 1865, p. 34. Some agents succumbed to the planters' requests: see *BRFAL*, film M798, roll 13, (GA), p. 169, film T142, roll 38, Tennessee. Andrews, pp. 23–24 (South Carolina). Dennett, pp. 291–92 (Alabama).

Orleans, "in several instances" had planters "fire overseers who dealt unfairly with laborers." Colonel Fred S. Palmer, Chief bureau inspector for Tennessee, reported that the effect of this failure to alter industrial relations was that many laborers were "willing to make a sacrifice of all their labor in order to be released from their contracts." In regard to the actual breaking of contracts, the vast majority of agents were consistent in their reports. Major General George H. Thomas perhaps exaggerated somewhat when he testified that "in every instance when the negro has become assured that he would be paid fair wages he has gone to work willingly, and has continued to work as long as the contract has been faithfully performed on the part of his employer." The chief inspector for the tobacco region of Virginia found that "almost all the cases that are referred to me" concerning troubles between labor and management were "attributable to something for which the *employer* is *responsible*." The list of such reports could be made as long as one wished. Much is no doubt explained by an exasperated Arkansas agent who complained that "all classes are ignorant of the practical free labor system, and very many do not trouble themselves to learn."[7]

The planters would no doubt have disagreed with the perspective of the bureau. In many cases the disagreement was not so much the result of conscious mendacity as a different interpretation of the facts. The bureau agent in Point Coupee Parish, Louisiana, explained that most difficulties were caused by the planters' complaints that the laborers would not work "as agreed to" in their contracts, while the freedmen claim "they are not treated with kindness or justice." The result was that when freedmen abandoned their contracts, planters viewed the act as one precipitated by some trifling cause. Major John W. Turner of Illinois offered a most balanced view:

> in those instances where there has been trouble, and the negro has gone off and left the place, I have found the employers to be impracticable men. As I said before, they could not forget that the negro had been a slave. They might not be disposed to punish them or beat them; but there was something in their conduct which excited in the mind of the negro that he had not accorded to him all the privileges of his new status, and the consequence was that the negro was discontented, was not so controllable, and was disposed to wander off and leave his work.[8]

The planting class and the "black power" historical school might have agreed with the analysis of the agent at Pine Bluff, Arkansas, who differed from the majority within the bureau:

> There are many trifling and indolent negroes who do not labor as required by the contract, but as a general thing this is submitted too by the employer on

---

[7] *BRFAL*, film M1027, roll 29, p. 3; film M752, roll 40, p. 10; film M1048, roll 46, p. 183 (emphasis in original). *Report of Joint Committee on Reconstruction*, p. 28, Mississippi, p. 13, John C. Underwood, U.S. District Judge, Virginia. Film M979, roll 23, p. 408; film M809, roll 19, p. 771, 931, Alabama; film M752, roll 37, pp. 766, 808, Florida; film M1027, roll 28, pp. 392–93, Louisiana; film M1048, roll 45, p. 91, Virginia.

[8] *Joint Committee on Reconstruction*, p. 5, Virginia.

account of the great scarcity of labor and the impossibility of supplying the places of hands discharged. The scarcity of labor in the whole of my district has placed the employer entirely in the hands of the laborer, and in almost every instance of violation of contract—the fault has been with the freedmen.

The evidential value of this statement becomes clouded when a few lines later the agent discloses that "most contracts are complied with by both parties." The quantitative qualifier is crucial. For it turns out that the agent's criticism was not about contract breaking as labor turnover but as labor discipline.[9]

An indication of the quantitative significance of the word "most" is provided by the data reported by other agents in the region and exhibited in table 8.1. The most striking aspect of these data is how few employment relationships were terminated during the planting season by employers or employees. Labor turnover of this minimal magnitude is seldom found anywhere. A major explanation is certainly the organization of the contract system and the active role taken by the Freedmen's Bureau as arbitrator and enforcer of contracts. If we make the assumption that contracts terminated by mutual consent reflect problems of industrial relations, the reason for premature termination of contract also fail to fit the requirements of the labor-market-instability view. These data, coming from an area where the excess demand for labor was exceedingly high, present a strong case against the labor instability argument, but how representative are they?

Although the data seem less reliable, South Carolina provides an excellent contrast for the purpose of examing this question. In the western portion of the state Major Walker, the sub-assistant commissioner, "estimated" violation rates of 20 percent for employers and 25 percent for employees. The discrepancies in the reports from the two districts that provided quantitative estimates, Abbeville and Edgefield, are of special interest. Captain Becker reported that of 1,031 contracts in Abbeville "there was not more than one percent violated by the employer," while "the freedmen have violated them at the rate of 10 percent." For Edgefield, Lieutenant Stone stated that "to the best of his belief," of 646 contracts "at least thirty per cent of the employers have violated them." Stone offered no estimate of the rate of violations by laborers, but he attributed "any breach of contract by the freedmen to their ignorance of the obligation in a written document, or a knowledge that he has been deceived in some particular point by his employer." Captain Becker's beliefs are worth quoting. According to him, employers' violations arose primarily "from the fact that the freedmen will sometimes fail to perform a fair amount of *labor*, disobedience of orders, or using insulting language." As for freedmen, their violations were due to "idleness and ignorance as to the use, purpose or meaning of a contract."

Whether we accept Becker's interpretation or Stone's is not important at this time. The crucial point is that in reporting on contract violations both were laying stress not on labor turnover but on labor performance within

---

[9] *BRFAL*, film M979, roll 44, pp. 692–94, Nov. 12, 1866.

Table 8.1 Termination of Employment Contracts in 1866, Arkansas

County	Contracts	Laborers	Contracts Broken	Broken by Employers	Broken by Employees	Mutual Terminations	% Contracts Broken
Phillips	650	4,149	23	6	17	—	00.5
Jacksonport (city)	483	—	18	—	—	—	—
Pulaski	—	1,140	30	6	9	15	02.6
Logan	111	—	5	3	—	2	—
Crawford	105	—	0	0	0	0	0
Hempstead, Pike	—	3,725	152	2	64	86	04.
DeVols Bluff (city)	350	1,720	120	—	—	—	06.9
Calhoun	160	—	10	10	—	—	—
Mississippi	471	—	"very few"	—	—	—	—
Clark	—	2,005	94	—	—	—	04.6

SOURCE: *BRFAL*, film M979, Arkansas, pp. 678–725, 1866. A contract usually contained the names of many employees. A broken contract represents one laborer.

the contract. A year later Stone would make this point perfectly clear. In 1867 conditions had improved so that by September "not more than 5 per cent" of freedmen had "violated contracts in material points." Stone then explained:

> It is proper to remark that by violation of contract I do not mean the temporary absence of a freedman for an evening or an afternoon to attend political meetings even though he might do so without the consent of the employer, but the absolute abandonment of work without just cause.

Stone continued, that, among employers, "not more than 3 percent have violated them as yet. When the crops are secured however, and the time for settling arrive there will be many complaints of unfairness." [10] It is this last sentence that leads to an explanation of the low turnover rates in Arkansas and the discrepancies in South Carolina.

The low rates reported by Captain Becker refer to the state of affairs up to and until the harvest and before time for payment. Major Walker's and Lieutenant Stone's higher rates include conditions after the harvest. The characterization of the southern labor market as chaotic and unstable throughout the year 1866 has been greatly exaggerated, if the word "instability" is intended to mean contract breaking as labor turnover. That temporary absenteeism and disputes over working conditions were common is incontrovertible, but these problems occurred irrespective of the form of payment. Indeed, to explain the destruction of a money wage gang system as the consequence of competition and labor turnover would be difficult in the face of the facts that money wages did not dominate the contract system and that intraseason labor turnover due to wage competition was minimal. On the basis of these arguments and evidence, there appears to be little theoretical or empirical support for the position that the gang labor system failed because of the instability of the labor force.

An important point has surfaced in this analysis. Previous historiography has been correct in giving attention to the problems of labor supply, but has laid too much stress on turnover, as opposed to day-to-day absenteeism, and the work effort supplied by a given force on a given plantation. The planters' initial demand for labor discipline similar to that of slavery failed immediately. This we have seen in the present chapter and the four preceding ones. Furthermore, recognition that share payments and postharvest money wages both function to immobilize labor during the year and economize on credit requirements reveals that planters recognized this failure quickly. The early choice of share payments over postharvest money wages was a concession that new work incentives were needed to replace the old compulsions. "Negroes do best when they have a share of the crop," explained a planter, "the idea of working for themselves stimulates them." [11]

---

[10] *BRFAL*, film M869, roll 34, p. 809, Major Walker's report; pp. 891–92, Becker and Stone quotations; pp. 374, Stone's 1867 report. There was of course much migration of labor to the Southwest, but migration and contract breaking should not be confused.

[11] Trowbridge, p. 392.

The employment contract in all its ramifications, whether explicitly stated or implicitly understood, is an exceedingly complicated arrangement. In addition to the form and expected amount of payment, both parties to the contract were concerned with the size of the work group, the amount of external supervision, the incidence of risk bearing, and the work incentives produced by the terms of the contract. The laborers could work in a small-to-large group, with little or much supervision, bear considerable or little risk, and have varying degrees of incentives for individual effort. Within any labor organization, the costs of negotiating these parameters and then enforcing the implied responsibilities of both parties may generally be termed transactions costs. In one sense, the search for efficient modes of minimizing such transactions costs played an important role in the evolution of agricultural organization. But in another sense, the very existence of these transactions costs depended upon social antagonisms between and among the laborers and employers.

The coming analysis will focus primarily upon cotton and to a lesser extent tobacco agriculture, but my remarks as to why sugar and rice production avoided the share system in favor of wage labor will provide insight into the significance of making a detailed examination of labor-management relations as a way of gaining an understanding of the nature of transactions costs.[12]

In Arkansas the low rates of contract termination reported for the planting season would also give way to an explosion of employer contract breaking just after and during the harvest. Bureau agents throughout the South would report that planters were deliberately attempting to induce their laborers to break contracts. The freedpeople corroborated these observations. A South Carolina freedman explained why contracts were broken: "De furst dif'culty about de matter be dis yer, we gits no meat; and de secon' is dat we gits de fum-tyin' too much." Another freedman ascribed diabolical intent upon the part of the planters. "There's a many *masters* as wants to git de colored peoples away, ye see; an dey's got de contrac's, an' dey can't do it, ye see, lawful so dey' buses dem, an jerks em up by de two fums, an don't give em de bacon an calls on em to do work in de night time an sun'ay, till de colored people dey gits on easy an goes off."[13] Under competitive conditions, this kind of treatment would cause the employer not only to lose his current work force, but to have difficulty recruiting replacements. In general, this argument is true, but only if the behavior is relatively idiosyncratic to a few employers. In fact, such behavior was all too common.

---

[12] This examination leads me to agree with Oliver E. Williamson that transactions costs are more important than technology in determining internal modes of organization. See Oliver E. Williamson, "The Organization of Work: A Comparative Institutional Assessment," *Journal of Economic Behavior and Organization*, vol. 1, 1980, p. 12.

[13] Andrews, *The South since the War*, pp. 203, 204 (my emphasis). For the testimony of white planters admitting to this behavior see pp. 204–5.

In order to understand why, without making an a priori ascription of evil intent on the part of the planters, the forces that were impinging upon the behavioral choices of both freedpeople and planters must be described. This will make it clear that the choice of labor contracts, by employers or employees, cannot be explained independently of the historical development of the postbellum economy.

Part III
Agrarian Organization

The experience that was had in this comone course and condition, tried sundrie years, and that amongst godly and sober men, may well evince the vanitie of that conceite of Platos & other ancients, applauded by some of later times;—that ye taking away of propertie, and bringing in comunitie into a comone wealth, would make them happy and flourishing; as if they were wiser then God. For this comunitie (so farr as it was) was found to breed much confusion & discontent, and retard much imploymet that would have been to their benefite and comforte. For ye youngmen that were most able and fitte for labour & service did repine that they should spend their time & streingth to worke for other mens wives and children, with out any recompence. The strong, or man of parts, had no more in devission of victails & cloaths, then he that was weake and not able to doe a quarter ye other could; this was thought injuestice. The aged and graver men to be ranked and equalised in labours, and victails, cloaths, & c., with ye meaner and yonger sorte, thought it some indignite and disrespect unto them. And for mens wives to be commanded to doe servise for other men, as dresing their meate, washing their cloaths, & c., they deemd it a kind of slaverie, neither could many husbands well brooke it. Upon ye poynte all being to have alike, and all to doe alike, they thought them selves in ye like condition, and one as good as another; and so, if it did not cut of those relations that God hath set amongest men, yet it did at least much diminish and take of ye mutuall respects that should be preserved amongst them. And would have bene worse if they had been men of another condition. Let none objecte this is men's corruption, and nothing to ye course it selfe. I answer, seeing all men have this corruption in them, God in his wisdom saw another course fiter for them.

Governor William Bradford
Plymouth Plantation

9

*Many of us cultivated farms last year withe the conservative people
and when the crop was ended they was so much indebted to others
they did not pay us for our labor, and many of their crops has been
taken by the Sheriff and sold from them before we got any pay*
                    Freedpeople of New Berne, North Carolina

Risk, Liability, and Possession

There is an important fallacy contained in the theoretical argument that the
planter bore all the risks of agricultural production by *guaranteeing* the laborer
a certain wage. In actuality the agricultural laborer was not guaranteed his
wage! Chapter 3 stressed the importance of the fact that a planter could fail
to realize a return on the crop large enough to pay all of the planter's
creditors. Since the wage hand expecting to receive a postharvest money
wage payment was one of the planter's creditors, it was possible that this
"loan" could also be defaulted. The laborer who contracted for a postharvest money wage was in effect buying, with labor services, a very risky
asset.

The years 1866 and 1867 proved to be two of the worst agricultural
seasons in the memory of southern farmers. Dr. N. B. Cloud of Montgomery, Alabama, writing for the United States Department of Agriculture,
described in detail the disastrous season of 1866:

> We had the most unprecedented amount of spring rain through the early sum-
> mer that has ever been known in the cotton states, culminating in the great
> flood of June 10, which almost entirely drowned out the cotton plant on the
> rich river and creek bottom lands. This long-continued, extremely wet weather
> not only injured the plant by a surcharge of water in the soil, thereby retarding
> its growth and rendering the plant more liable to succumb to the ravages of the

141

aphis (plant louse), but promoted the rank growth of grass and weeds that proved seriously injurious to the crop in the acreage "*turned out*" to grass.

After the rains came an unseasonable dry spell which at first had the promise of allowing what remained of the crop to be saved from grass and weeds, but the optimistically inclined were soon to learn that it was possible to have too much of a good thing. The welcome dry spell turned into a drought. "But this open dry weather, promising so propitiously in June," Dr. Cloud continued, "continued through July into August, and in many sections of the country to September, proving almost as destructive to the plant as did the rainy weather."[1] After the drought came more heavy rains and the boll worm and cotton caterpillar to destroy what was left. Conditions in 1867 were more of the same, but worse. These two disastrous seasons, occurring simultaneously with the inauguration of free labor, were to have a tremendous impact upon the southern economy and the black worker.

Early in 1866 there were ominous signs that the laborers and planters were destined to experience a particularly harsh introduction to the vicissitudes of free market agrarian capitalism. Many planters began the year with only enough working capital and credit to carry them through half the season. They gambled that by midsummer their crops would be promising enough to enable them to borrow more in order to continue to provide food and possibly to give a partial payment to their work forces. In the Mississippi Valley, Whitelaw Reid was prophetic:

> Nearly every man was overreaching his means. With capital to carry through a plantation of five hundred acres, he would be attempting a thousand. The result was that Negroes were ill paid and given cheap and scanty rations. If the Negro, dissatisfied with this specimen of the workings of free labor, broke his contract that was proof that "free niggers" would never make cotton without a system of peonage.

In South Carolina inadequate access to credit had forced the planters into what was probably the most extensive adoption of share payments in the South. The bureau agent in the Darlington district warned his superiors in May 1866 that "the planters have neither cash or credit to purchase" the provisions "required to supply labor till the harvesting of the crops." The agent advised that unless relief from the government came, "the plantations will be entirely abandoned by the freedmen, who will go in search of employment that will furnish them with present food."

The planters gamble had failed. In the same month bureau commissioner Robert Scott reported to Washington that in South Carolina "many planters who made a fair beginning, with the hope that some turn of fortune would carry them through, will be compelled to abandon the idea of raising a crop for want of provisions to feed the freedpeople working for them." From

[1] N. B. Cloud, "Cotton Culture in 1866," *Annual Report, Department of Agriculture, 1867* (Washington, D.C., 1868), pp. 191–92.

Halifax County, Virginia, an agent reported that on the tobacco farms where freedmen had been engaged for a share of the crops, "frequently neither party has any provisions, or credit," resulting in "much suffering" among the freedpeople.[2]

Six months after his prediction that trouble was imminent, General Scott, in his report for the year, explained the situation in South Carolina:

> The planters in many localities have made contracts which appeared fair and liberal; but the freedmen having neither provisions wherewith to feed their families, implements to cultivate the soil, nor seed to plant, were unable to fulfill their contract. Under these conditions it often happened that land-owners accused the freedpeople of laziness and neglect of the crops, while the latter held that the farmers were in duty bound to advance them the necessary provisions until the crops came in.

In September, a subcommissioner had reported why freedmen were breaking contracts: "With few exceptions, the planters who contracted with the freedmen, stipulated that the latter furnish seed and their own provisions. The little provisions from last years crop were consumed by May and the freedmen were compelled to seek work elsewhere. The crops being abandoned."

In 1867 the abandonment of crops was on a wider scale. Agricultural labor was actually thrown into involuntary unemployment. From Macon, Georgia, the bureau communicated its fear that "the inability of their employers to furnish supplies" would result in large numbers of workers being "thrown out of employment." In the Yazoo subdistrict of Mississippi it was reported that injury to the crops by flooding induced those freedmen "working for a share of the crop [to] exhibit great inclination to abandon the same and work for other parties who now offer high wages." In North and South Carolina, Virginia, Arkansas, Mississippi, and Louisiana, flooding of the crops and a lack of working capital caused the unemployment of laborers. Theodore Peters reported to the U.S. commissioner of agriculture that in Virginia, "the want of means both for paying hands and supporting the workforce of the plantation in food and forage has been the great drawback."[3]

Under these conditions, thousands of unemployed blacks were idling in the countryside and others were again crowding into the towns and cities.

2 Reid, pp. 290–91. *BRFAL*, film M869, roll 34, p. 507. *BRFAL, Synopsis of Reports*, May 1866, pp. 95–96. *BRFAL*, film M1048, roll 45, pp. 147–48; film M979, roll 23, Arkansas, pp. 161–62. Benham, *A Year of Wreck*, pp. 349–50, gives a particularly revealing description of this cause of broken labor contracts in Louisiana and Mississippi in 1866. Also see W. Miles Hazzard, "The Labor Question," *Southern Cultivator*, vol. 27, 1869, p. 206.

3 *BRFAL*, Synopsis of Reports, Nov. 21, 1866, p. 93; Sept. 8, 1866, p. 447; April 26, 1867, p. 113, June 1867, p. 164. Theodore Peters, *Report of an Agricultural Survey of the South: Monthly Report, Department of Agriculture*, June 1, 1867, pp. 195, 200–202. See *BRFAL, Synopsis*, 1867, p. 172, Louisiana. *American Freedman*, vol. 2, Jan. 1868, p. 364, North Carolina, Virginia; p. 251, South Carolina; p. 252, Louisiana. *BRFAL*, film M979, roll 23, Arkansas, pp. 184, 208–9.

The bureau in Georgia reported the reasons for the large amount of crime zealously documented by Reconstruction historians of the next generation: "Planters [are] unable to continue operations" and the "freedmen [are] now unemployed and are stealing."[4] By the end of the season, many of those laborers who had not been thrown into the ranks of the unemployed earlier were forced to join the more unfortunate. The short crops were forcing planters all over the South to default on their debts. From Arkansas the bureau reported, "Freedmen are being driven off to avoid payment of their wages."[5]

These actions served to make the establishment of harmonious relations between labor and employers all but impossible. George Benham, describing the conditions in the plantation areas along the Mississippi River at the end of 1866, looked to the future:

> The indirect loss was in the failure of the planters to meet their engagements with their laborers, and the demoralization resulting therefrom. The Negroes were badly enough demoralized when they were hired in the spring, but turned out, as they were now all over the South, to starve, in how much worse a condition would they be when the time came to gather them up for the purpose of making a new crop?

The freedpeople, whose past experience had taught them that their exmasters were wealthy men, could draw no conclusions other than that they had been intentionally defrauded of their wages: "our old masters have got heaps and heaps of money, and we knows it." "Never was the utter darkness of intellect of the recently enfranchised Negro more apparent than now. . . . If they were not fed and paid, it was because their employer did not want to do it—this was their conclusion; and so the break between them was widening."[6]

The failure to pay wages and shares, coupled with the inability of planters and freedpeople to establish a communicative relationship built upon trust, often led to violent methods of driving recalcitrant unpaid workers away from plantations. Complaints from freedpeople flowed into Freedmen's Bureau offices. Given the conditions, there was little the bureau could do. General John Sewall, after an inspection of the field of operations, reported that the driving of freedmen from plantations without their contracted pay was due to the "failure of the crops" which would "render it difficult for the employers to pay their hands, and discharge their other obligations." Sewall added that this circumstance "furnishes no excuse for violent dealings" with the freedpeople.[7]

---

[4] *BRFAL, Synopsis of Reports*, Oct. 1867.

[5] Ibid., Dec. 24, 1866, p. 48. P. 438, Alabama; p. 118, Florida; p. 68, North Carolina. In the Arkansas quote the term "wages" refers to share payments and money wages.

[6] Benham, pp. 402–6.

[7] *BRFAL, Synopsis*, Nov. 1866, p. 39; also p. 439, Alabama; p. 208, Georgia; p. 216, Arkansas; p. 429, Mississippi. *New York Times*, Oct. 31, 1866, p. 8. *BRFAL*, film M1048, roll 45, Virginia, p. 160.

Faced with the difficulties of feeding thousands of impoverished blacks and whites, some individuals found that what I have called the Tanzanian plan was not without defects. In Mississippi, where a year earlier whites had been complaining of the scarcity of labor and arguing that all the South needed to take care of its problems and provide for its people, black and white, was to be released of federal interference, one country newspaper demonstrated what states' rights meant for the blacks:

> It would be well, we think, for our citizens to encourage all freedmen, as they get out of employment, to go to Vicksburg. There is a free government boarding house there, called the "Freedmen's Bureau," and hence starvation may be avoided. . . . Hinds County would be infinitely better off could we get rid of ten thousand of our Negro population for the present winter.[8]

The field hands had learned that acceptance of *any* postharvest payment contract was in fact an agreement to extend credit and assume a great deal of risk. In light of this unwelcome discovery, new considerations had to be made concerning the advantages and disadvantages of various contracts. If the wage payment was not guaranteed, it was easy to see that a wage contract had a distinct disadvantage relative to a share payment. In a poor season, the laborer might receive only a fraction or, worse, no part of the promised wage. In a good season, the contracted wage was paid, but that was all. There would be no compensation to the worker during the landlord's profitable years for losses suffered when the wage was defaulted. A share payment did not have this fault. During poor seasons, the value of the laborer's fractional share might be low, but good seasons promised that this same fractional share would have a higher value. The *risky* money wage contract was an inferior asset relative to a share wage. The actions of state legislatures made this point even clearer.

In modern business practice it is taken for granted that the first claim on the assets of a bankrupt firm or employer is the property right of the firm's employees to collect their earnings. In the postbellum South, state legislatures recognized agricultural employees receiving a postharvest payment as creditors of their employer, de facto, by explicitly making the laborers' claims for their pay second to the landlord's rental claim, and in some cases second to the claims of merchants and factors holding liens on the employer's crop. In Alabama, recognition of the agricultural laborer as a creditor of his employer was made even clearer. General Sewall reported:

> The "stay laws" now in force in Alabama are oppressive to all classes of labor. It suspends the collection of debts for eighteen months, and leaves the poor man at the mercy of his employer for his wages. In many instances advantage is taken of this law to withhold the earnings of the whole season, and *no remedy exists.*[9]

---

[8] *Hinds County Gazette*, Nov. 29, 1867, quoted in Vernon Lane Wharton, *The Negro in Mississippi, 1865–1890* (1947, reprint New York, 1965), p. 123. For a similar expression in Tennessee see *American Freedman*, vol. 2, 1867, p. 251.

[9] *BRFAL*, Nov. 1866, p. 39 (my emphasis). This stay law was passed by the Alabama legislature on Feb. 20, 1866. Sewall complained that no remedy existed because under Andrew

Where legal redress was possible, the bureau often attempted to enforce the rule that the wages of laborers had a primary lien on the crop. For example, Arkansas had no lien laws covering the conflicting claims of agricultural laborers and employees in 1866. This absence of law or precedent allowed General Ord to take aggressive action in his efforts to protect laborers. Reporting for the period ending January 1, 1867, Ord stated that "to secure the dues of freedmen cotton has been seized and held by military power."[10] Federal confiscation of property on behalf of the laboring classes, black or white, was not something to be championed by the landlord class. Coincidentally or not, the precongressional Reconstruction legislature of Arkansas enacted a lien law giving a crop lien to employers for advances to laborers and to laborers for their pay just eight days after Ord's report. The general effect of this law would have been to put disputes about payment claims under the exclusive jurisdiction of the civil authorities if not for the fact that, in 1867, the government of Arkansas was put under military control.[11]

The congressional Reconstruction governments of the former confederate states were no less aware of the possibility of using political policy to effect economic outcomes than were the Democrats. The Republican-controlled state governments passed legislation in favor of their black constituencies. In Arkansas a few months after establishment of the Republican government in 1868 the legislature adopted the Freedmen's Bureau's precedent by creating for the laborers a first and "absolute lien upon the production of their labor" that "shall always be preferred to a landlord's" lien.[12] In January 1875 the reestablished Democratic government destroyed the priority of the laborer's lien by making the landlord's lien of highest priority unless the landlord explicitly endorsed the tenant's contract. In South Carolina much the same occurred. The Democrats, in 1866, established a prior lien for those making advances to farmers for agricultural purposes. The Republican government in March of 1869 reversed this by making the laborer's lien for a share or wages prior.[13] In Alabama General Swayne reported that the 1866 legislature had refused to make labor a lien upon the crop, and lamented the effect of the stay law on wages. By November 1867

Johnson's Reconstruction policy the acts of the Alabama legislature in 1866 were legal and the federal authorities could not interfere.

[10] *BRFAL*, Feb. 28, 1867, p. 80. For other examples of federal seizures of crops to protect wages see *New York Times*, Sept. 11, 1866 (Alabama); John Cornelius Engelsman, "The Freedmen's Bureau in Louisiana," *Louisiana Historical Quarterly*, vol. 32, 1949, p. 187; Raushenberg bureau report in Cox and Cox, pp. 349 (Georgia).

[11] *Acts of Arkansas*, 1867 (Little Rock, 1867), no. 122, approved March 3, 1867. Du Bois, *Black Reconstruction*, p. 546, and Kenneth M. Stampp, *The Era of Reconstruction* (New York, 1965), p. 144, discuss the implementation of military rule in the spring of 1867.

[12] *Acts of Arkansas*, 1868, p. 224–27, approved July 23, 1868. This law emphatically stated that "you cannot defeat a laborer's lien and no real estate shall be exempt from sale under execution of a laborer's lien."

[13] *Acts of Arkansas*, 1875. *Acts of South Carolina*, 1866 (Charleston, 1866), no. 4786, p. 380, approved Sept. 20, 1866. *Acts of South Carolina*, 1868–69, no. 147, p. 228.

General Swayne had taken matters into his own hands by issuing a general order creating a lien in favor of the wages of laborers and allowing attachment of crops about to be moved. In October of the next year Swayne forced the Alabama legislature to enact into law the conditions in his general order. Within two months the legislature amended this act by making the laborer's lien for wages or a share of the crop second only to the rent of the landlord.[14]

In state after state the Democrats and Republicans waged a continuous battle over this and related issues. The effect of landowner-oriented Democratic legislation was to reduce the security of postharvest money wages to an unacceptable level. In Georgia, where the laborer's lien for wages was inferior to the liens of landlords, merchants, and factors, if the planter's revenues "failed, or fell short, the factor took all; and the laborers, employed to a great extent on wages, sometimes lost all their pay, except what they had consumed during the year."[15]

In contradistinction to the problems of money wage workers, the laws and customs developing about the share system were making it relatively more attractive. The high cost and difficulty of obtaining credit have already been discussed. In view of these conditions, the commercial farmer sought to strengthen the institutional arrangements upon which a credit system could be based. One of the earliest steps was the formal recognition, through the legal system, of the practice of securing a loan by giving the lender a lien upon the borrower's prospective crop. This practice has been condemned by many and often indicted as the scourge of the postbellum period. It is now generally recognized that the practice was informally adopted upon a wide scale in the antebellum period.[16]

As early as December 1865, a farmers' convention in Alabama requested a law creating a lien for *landowners and merchants* advancing supplies to

[14] *BRFAL, Synopsis*, 1866, p. 438. *Acts of General Assembly, Alabama* (Montgomery, 1869), no. 31, p. 252–53, Oct. 10, 1868. Notice the timing of these events. Under Andrew Jackson's presidential Reconstruction policies in 1866 Swayne and Sewall were powerless to act on behalf of the freedpeople, because the civil governments were in power and had enacted lawful legislation dealing with the disposition of the cases. In 1867, with Alabama under military rule, Swayne acted by issuing his general order, and finally in 1868, with the state under Republican control, the bureau or military order became civil law. *Acts of the General Assembly of Alabama*, no. 122, p. 455–56, approved Dec. 28, 1868. In Texas, Assistant Commissioner Joseph B. Kiddo had anticipated Swayne by issuing a circular that awarded "laborers a lien upon the crop." see *BRFAL, Synopsis*, May 1866, p. 70, and also "Results of Emancipation in U.S.A.," *American Freedmen's Union Commission*, chairman Lyman Abbott (New York, n.d.).

[15] Nordhoff, *The Cotton States*, p. 109. The financial economist Eugene Fama terms the laws regulating priority claims to the assets of bankrupt firms "me first rules." For other examples of the effect of "me first rules" upon the laborer's payment see George Campbell, *White and Black: The Outcome of a Visit to the United States* (New York, 1879), p. 304, (North Carolina in the 1870s); Somers, p. 243 (Mississippi).

[16] Alfred H. Stone, "The Cotton Factorage System of the Southern States," *American Historical Review*, vol. 20 (April 1915). Woodman, *King Cotton and His Retainers*. pp. 36–40. See also Hammond.

farmers. The Alabama state legislature responded by passing a law "to give a lien on the crop and stock for advances to assist in making the crop." A study of this act is instructive for the purpose of understanding underlying economic and political phenomena of importance.[17]

The 1866 Alabama lien law allowed a "lien on the crop equal to the value of advances made by any person(s) to any person of the state."[18] There were two important implications of this law. First, any person with a legal claim of possession to a crop would be able to receive credit by hypothecating that crop. Second, the wording of the law allowed merchants and agents other than the employer to extend credit on the basis of a crop lien to anyone with a claim to a prospective crop. The access to credit made possible by the lien law enabled an employer paying for labor with a share of the crop to reduce the need for cash to pay wages throughout the year and to reduce financial risk by dispensing with a fixed wage obligation altogether and placing the credit lien for the laborer's subsistence on the laborer's share of the crop.

The point is of fundamental importance. An employer using the standing wage system could avoid the need for cash to pay wages during the employment period, but would have to obtain cash or credit to furnish the laborers with food and clothing. Workers whose only source of income was the promise of a money wage at the end of the year would have little chance of getting credit of their own. A creditor would require security, and since the security for the wage was the crop, which belonged to the employer, the employer would have to sign a lien on the entire crop and borrow for the workers, with the intention of deducting the loan from future wages. With share payments, not only would the employer not have to make wage payments throughout the year, but the need for credit to feed and clothe laborers could also be avoided, for they could borrow on the security of their own share of the crop. In addition, the employer's liability and therefore risk was reduced, since the employer's share of the crop would not be encumbered should the value of the tenant's share prove insufficient to cover advances for subsistence.

The process was explained by Mississippi planter Samuel Postlewait. He adopted share payments, he said, because he could "force his hands to feed

[17] A. C. Bear, writing in the *Montgomery Advertiser*, Dec. 15, 1865, quoted in William Warren Rogers, "Agrarianism in Alabama, 1865–1896" (Ph.D. diss., University of North Carolina, 1959), p. 21. On Oct. 20, 1865, South Carolinian Henry William Ravenel noted in his private journal that agricultural operations required some means of credit based upon the coming crop: *The Private Journal of Henry William Ravenal*, ed. Arney Robinson Childs (Columbia, S.C., 1947), p. 256. Recent studies of the postbellum economy by some economists have tended to deemphasize the impact of the lien laws and the credit system. This tendency is a mistake because these institutions had productive and distributive effects upon the economy. See Higgs, *Competition and Coercion*, DeCanio, *Agriculture in the Postbellum South*, and Reid. These economists deemphasize the credit system by default, ignoring it in their formal models of the economy; this approach implies that the credit market was not only competitive, but functioned under conditions of complete efficiency in the economists' meaning of that term.

[18] *Acts of Alabama*, 1866, no. 12, Jan. 15, 1866.

and clothe themselves" through the crop lien by securing "supplies from local merchants" and reducing his "costs for rations." The system had been widely adopted on the Sea Islands, where it was reported that planters, arriving upon the islands too late for an early planting, had not "the means where with to plant" and could not obtain credit from "capitalists [who] were afraid to advance money" because of the "unsettled condition of the Islands." The result was that only a few of the planters with "the means agreed to advance provisions to the freedmen," while the rest "simply contracted with the freedmen for a share of the crop leaving them to work and manage it as they pleased."[19]

The plan was not without drawbacks. The employer sometimes lost control over the laborer, since it was the merchant and not the landlord who controlled the laborer's food supply. This problem, however, could be avoided if the laborer's independent loan contract with a merchant was coupled with a codicil that made the landlord an intermediary between merchant and laborer. For example, freedman William Boyd of Carteret County, North Carolina, working upon the farm of Lockhart Gibbs, entered a loan contract with merchants Isaac Ramsay and Sons and agreed to "deliver to said Lockhart Gibbs as *agent* of said Ramsay and Sons out of his half of the crop of cotton when gathered, ginned, and baled a sufficient quantity to pay and satisfy the amount of his indebtedness."[20]

For this reason planters and merchants, when paying shares, preferred to place the lien upon the entire crop in the planter's name. Through this method the planter controlled the allocation of supplies to laborers but could still assess the value of the loan and therefore the liability of laborers' supplies upon the field hands through the latter's share of the crop. For the laborer this practice was no improvement over standing wages. As long as the entire crop was covered by a lien signed by a planter who controlled that crop, the employees' earnings could be taken by the employer's creditors. The problem could be circumvented only if the share laborer were, in a legal sense, not a mere wage laborer but part owner of the crop and vested with the right to dispose of his own part of the crop.

In their quest for this property right, black field laborers in the early Reconstruction period were to achieve a remarkable success whose historical importance should not be neglected because of the reversals enacted by Democratic legislatures in the late Reconstruction period. We might begin with the state of Alabama, whose local Freedmen's Bureau, under the direction of General Wager Swayne, has recently come under sharp attack by historians for alleged preferential treatment of employers.[21] By 1867 the

[19] Quoted in Ronald L. F. Davis, "Good and Faithful Labor: A Study in the Origins, Development, and Economics of Southern Sharecropping, 1860–1880" (Ph.D. diss., University of Missouri, Columbia, 1974), pp. 113–14. *BRFAL*, film M869, roll 34, p. 812.

[20] *BRFAL*, film M843, roll 43, "Lien Contract," March 1867, p. 1001 (my emphasis). See ibid., film T142, roll 69, "Lien Contract" between Ernest Wiedermann, merchant, and freedmen Thomas J. Cox and S. Jones, Jan. 4, 1867, Hempstead County, Arkansas; film M826, roll 50, "Lien Contract," Panola County, Mississippi, p. 771.

[21] Wiener.

Alabama bureau was enforcing the share laborer's ownership rights to a share of the crop. Bureau agent W. E. Connolly described the state of affairs in the southeastern section of the Alabama plantation belt in 1867:

> Applications are constantly made at this office by freedmen or by whitemen in behalf of freedmen, to have "that part of the crop belonging to the laborers according to contract" released from mortgage, attachment, or levied on the *entire crop*, by parties who have claims against the employer. . . . and even when the employers agreed to "*find*" the laborers, giving them a *less* portion of the crop, the *entire crop* is sometimes seized under mortgage, or attachment, to pay the debts of the employer. In all such cases, orders have been issued from this office, forbidding the operation of any mortgage attachment, or other legal process, upon that portion of the crop which belongs to the laborers according to the contract, leaving it subject to their control.

In Demopolis, Marengo County, Major Pierce's November 1867 report corroborates the view that in Alabama the bureau was enforcing the position that share laborers owned their crop. "We have several cases of civil officers attaching crops including freedmen's share for debts of the planter for rent of land, supplies, etc." Major Pierce related, "and have given notice in such cases that the crops of the freedmen will be protected by military authority if necessary."[22]

It is possible that in some cases the planters attempted to use the bureau position to their own advantage. For example, when five bales of cotton produced upon his land were confiscated by the local sheriff for debts incurred prior to the war, a Morgan County, Georgia, planter appealed to General Davis Tillson upon the grounds that the cotton "did not belong to me, but to my freedmen" who had recognized it "as their property before [it was] removed from my premises." The six freedmen produced a contract stating that they had worked for the year 1866 for a share of the crop and upon that basis, made "claim" to the cotton "as our distributive share of the crop." The problem in the case did not involve the freedmen's right to the cotton, but the local bureau agent's suspicion that the planter had forged the share contract in an attempt to retain the cotton for himself.[23]

Since the civil authorities were often uncooperative when conflicts over crop shares developed between laborers and planters, the bureau was often forced to resort to military action. The extent of these difficulties is indicated by the burning down of the courthouse in Green County, Alabama, where 1,800 suits for past wages due were kept on file. In the same county the sheriff resigned from his office because "he would not attach white mens cotton at the suit of 'niggers.'"[24] One straightforward way to circumvent these problems was to simply seize the crop and thereby ensure that the laborers received their share.

[22] *BRFAL*, film M809, roll 18, p. 160–61, Eufaula, Barbour County. Ibid., p. 49. Major Pierce was in charge of Marengo, Sumter, Green, and Pickens counties, p. 72, 1868; p. 427 Huntsville, Madison, Co.; p. 611–12, Montgomery.

[23] *BRFAL*, film M798, roll 15, 1866, p. 454–56.

[24] Ibid., film M809, roll 18, p. 85; film M752, roll 40, p. 12–13; Tennessee.

With respect to crop seizures, share laborers possessed a distinct advantage over wage hands. Laborers who had contracted for a postharvest money wage had to wait until the crop was sold or bring legal suit to hold the crop for the wages before it was sold. To take the latter action wage laborers would have to possess prior knowledge that the planter intended to default their wages. Since wage workers had no legal claim on the crop itself, the obligation of bringing suit to prevent removal of the crop would have to rest on the workers. In the civil courts litigation was both expensive and balanced unfairly against labor. In September 1868 the Florida bureau complained to Washington of more crop failures and the refusal of planters to pay their hands "unless compelled by expensive litigation." In Louisiana the state legislature granted labor a first lien upon the crop for wages, but the bureau report to General Howard considered it worthless, because exercising the right of lien required "the posting of a bond that freedmen could not afford to pay."[25] Even when the crop was carried off without the laborers being paid, the share hand had an advantage over wage workers, because it was possible for a hand with a contractual claim to part of the crop to secure his legal fees and court costs "by giving a lien on his share of the crop."[26]

Because share contracts specifically gave the worker a claim to the crop, bureau agents frequently resorted to seizures for the benefit of the laborers. Agent Charles Raushenberg explained in 1868 that in the "frequent" cases of "arbitrary disposals of the freedmens' portion of the crop" in and around Albany, Georgia, he "generally succeeded by taking steps towards levying attachments on the cotton in preventing it from being sold or shipped," forcing the employer to make settlements at the bureau office. Major C. W. Pierce, in describing similar methods of protecting the crops of Alabama share hands, added his "fear that some laborers who are working for wages will not receive their pay unless some extraordinary means are used to secure them." The agent for the district about Panola County, Mississippi, explained the greater possessory right of the share laborer about as clearly as anyone:

> I have invariably stopped the removal of the crop untill the time when it should be ready for market if the freedmen were at work for a portion then taking their portion, and seeing that they got all due them if so desired, or allowing them to take all due them and dispose of as they choose; if they were at work for stated wages, have obliged the attaching party to give good, and sufficient security for the payment of the labor on the place.[27]

[25] *BRFAL, Synopsis of Reports*, Sept. 1868, p. 454; *Report of the Commissioner* (Freedmen's Bureau, 1868), p. 37, Civil War Pamphlets, Yale. For bureau complaints of discrimination in civil courts see *House Executive Document No. 329*, 40th Cong., 2nd sess. and *The Ku-Klux Conspiracy: Report and Testimony taken by The Joint Select Committee to Inquire into the Condition of Affairs in the Late Insurrectionary States*, 42 Cong., 2nd sess. 14 vols. (Washington, 1872), pp. 30, 32, Tennessee; p. 1549, Alabama.

[26] "Report of C. Raushenberg" in Cox and Cox, p. 348.

[27] Ibid., p. 349. "Report of C. W. Pierce" *BRFAL*, film M809, roll 18, p. 49. Also ibid., p. 959, Tuscaloosa, Alabama; p. 836, Selma, Dallas Co., Alabama; pp. 611, 620, Mont-

The geographical diversity of this policy, and perhaps the infectious toxicant provided by close proximity to the Sea Islands, where the principle of justice to free labor was attaining feverish heights, is illustrated by an example from Chatham County, Georgia (Savannah). In January 1866 Clotaire Gay, lessee of M. H. Williams's Rice Hope plantation, entered a share contract with twenty-five freedmen to raise rice and cotton and to cut and sell timber. The crops were to be divided equally, after deduction of costs, under supervision of Gay and James M. Simms, a Negro who was acting as the laborers' business agent. Low yields led to an unprofitable year for all concerned. After an attempt by Gay's creditors to attach the entire crop for debts, Gay was arrested for fraud at the instigation of the freedmen. Under bureau adjudication it was decided that the freedmen's share of the crop should be divided prior to attachment by Gay's creditors.[28]

We may well imagine the thoughts of Louisiana planter Thomas Ogden when he received the following letter from the local bureau agent in February of 1868:

> Complaint [has] been made at this office by Freedman Benjamin Olfred working on your plantation called Mill Creek in this Parish [West Feliciania] that you have shipped or sold eight bales cotton their share of the crop made on said plantation during the year 1867. This cotton pointed out to me the shipment mark is F and there is a cross mark on each bale put on by the Freedmen as I am informed by Ben, the cotton is now seized by me in the possession of Mr. J. W. Brady subject to my order for settlement with your employees for the year 1868. You will come forward and file bonds in this office for the amount of your indebtedness to said Freedmen.

Perhaps J. W. Brady was no less surprised, if he received a letter similar to the one sent to merchant J. F. Irvine in November 1867.

> You will take the (1338)lbs of cotton being (4) bales delivered to Whitman Brothers by the Sherriff being a part of the (1951)lbs of the freedmens share of the crop as you have been selected as their shipping agent, and hold the proceeds of the same subject to a settlement by me with the parties interested and ship for sale the (1951)lbs for the freedmen.

Agent Joseph S. Rabb enforced a similar order upon an unsuspecting Mississippi planter:

> *Isaac Hughs*, colored, complains that [he] cannot obtain settlement of his interest in [the] crops of 1867, on the "*China Hill Place*," in Claiborne County, Mississippi, or his just rights and interests in said crop. Therefore, no more

gomery, Alabama. Film M826, roll 30, "Report of Operations, Panola District," Nov. 11, 1867, p. 988.

[28] John W. Blassingame, "A Social and Economic Profile of the Negro in Chatham County, Georgia, 1865–1880" (Manuscript and notes, 1971), pp. 14–16. The case can be found in *BRFAL*, "Rice Hope Plantation Records, 1866," box 192, entry 287. See *BRFAL*, film M798, roll 13, p. 265, Georgia.

cotton or other crop or property in the crop of 1867, will be removed from said place, or from the possession of any party or parties holding the same, or in any manner disposed of.[29]

As a general rule, reliance upon the legal process and bureau crop seizures placed a very poor second to the laborers' demand that their share of the crop be divided in kind either in the field or at the employer's gin. In the fall of 1866 bureau offices were flooded with laborers like Virginia tobacco hand George C. Welch who "wanted to know if he could not have all produce divided on the ground, after being gathered." A similar complaint came from a South Carolina field hand who charged that his employer was "violating his contract by trying to compel him to gin all of the cotton before he gives him his share."

The triumphant but problematic arbitration of a similar case by a Florida agent is both representative of agent decisions and illustrative of the weaknesses inherent in the attempt by individual agents to gain for the laborers this right. It seems that a rather daring Floridian named Thomas, working for a share of the crop on the premises of the Jackson County sheriff, complained because Sheriff Ellis "wished to take the undivided cotton to gin at a place which Thomas objected to and they quarrelled." After a rather turbulent dispute the agent was able to report that "the Sheriff" approached him "and signified his willingness to abide by my decision which was that the entire crop should be gathered and divided at that time."[30]

All planters were not as amenable to this plan of division as the Florida sheriff. Lieutenant Louis Stevenson probably described the situation correctly when he charged that Virginia tobacco planters "who intend to do rightly divide the crop as fast as gathered," for only "to the man who will try to escape a fair division" [is] "delay advantageous." A general policy was required, and it may well have come through the initiatives of the South Carolina freedpeople.

In the fall of 1866, responding to "daily complaints" of unfairness in the Williamsburg district, General Robert K. Scott sent Lieutenant Colonel A. E. Niles to investigate the problem. Colonel Niles voided the share contracts in the area, and on the "statement of Negroes" he laid down two general rules which were to give a tremendous impetus to the development of the share system. The second of these decrees, which General Scott informed the governor of South Carolina had the authority of O. O. Howard himself, was that share hands could "demand a division as soon as the crop is housed—and neither party has any right to sell or dispose of any part of the crop without the consent of the other."[31]

   [29] *BRFAL*, entry 1895, vol. 436, p. 50. Ibid., p. 34, Nov. 16, 1867; ibid., Dec. 21, 1867, p. 126, also pp. 42, 46. See *BRFAL*, film M826, roll 30, p. 162, Lauderdale District, p. 396, Vicksburg, Warren County, Mississippi 1867; ibid., entry 2114, "Register of Complaints," pp. 12, 13, Holly Springs, Marshall County, Mississippi, 1867.

   [30] *BRFAL*, entry 4259, "Complaints 1866 and 1868," Rocky Mount, Virginia, 1866, p. 32. Ibid., entry 3150, Chester, South Carolina, 1866, p. 32. Ibid., film M757, roll 37, p. 1017–18.

   [31] *BRFAL*, film M1048, roll 46, p. 124, Nov. 30, 1866. *BRFAL*, *Synopsis of Reports*, Dec. 19, 1866, p. 148.

This action stemmed from a written complaint to O. O. Howard from the freedmen of Edisto Island, demanding the dismissal of Assistant Commissioner Scott. The freedmen had argued that Scott, in ordering that the "colored people of Edisto" store all crops "in buildings furnished by the landowners, who [were] authorized to put it under *lock* and *key* until all the crops be gathered in," was "not faithfully discharging the high functions of his office." One month later the freedmen wrote Howard, withdrawing their request and stating that Scott "has in compliance with your request, visited Edisto, and made a satisfactory explanation." Scott, in an October 21 speech on Edisto, reaffirmed his friendship for and support of the Negroes and sent an officer to investigate wrongs and arbitrate differences to the satisfaction of the freedpeople.[32]

General Howard's decision provided local bureau agents with much of the federal authority they required. Colonel Fred S. Palmer was merely stating general bureau policy when in 1867 he reported that in Tennessee it was the "duty" of the bureau to "divide the crop as per agreement. Sell the portion belonging to the freedmen, make settlement" and finally, "pay the remainder to the freedmen."[33]

One consequence of this policy was the great frequency in 1867 labor contracts of clauses which stipulated that "the share belonging to the" laborer "shall not be sold without the knowledge and consent of the" laborer, or more frequently that division occur for "corn when cribbed" and "cotton when ginned and baled," or for all "crops as they are gathered."[34] How successful the policy was is difficult to ascertain. The scores of reports from agents describing their personal attention to the division of crops supports the conclusion that the practice was widespread. On the other hand, it must be remembered that the number of agents available to do this exacting and time-consuming work was undoubtedly too small to meet the needs of all contractors. Then again, it is likely that agent W. J. Parmon of Jackson County, Florida, who explained to his superiors that he could not venture far into the countryside to divide crops "as directed" because it "would be decidedly dangerous and unhealthy," was not alone in this decision.

Nevertheless, for the laborer the possibility of some protection based upon ownership rights in the crop was surely better than the prospects of a suit for wages in the civil courts. Where the bureau was active, freedmen like Nathan Garrett of Bolivar County, Mississippi, could feel secure with the information that he could make a complaint that his employer was

[32] *BRFAL*, film M757, roll 37, pp. 408–12, Oct. 2, 1866; ibid., p. 462, Nov. 3, 1866. Both Scott and the freedmen stated that similar problems and complaints over the division of crops were being made throughout the state.

[33] *BRFAL*, film T142, roll 40, Sept. 1867; ibid., Louisiana, entry 1499, vol. 223 1/2, pp. 1–100. M869, roll 34, p. 962, Dec. 31, 1866, South Carolina.

[34] *BRFAL*, film M843, roll 34, "Labor Contract" between J. L. Praither [employer] and Aleck Hoover, Mecklenburg County, North Carolina, Dec. 29, 1866, p. 843. Ibid., film M979, roll 41, "Labor Contract," Mrs. E. R. Wright [employer] and Freedpeople, Lafayette County, Arkansas, Jan. 2, 1867, p. 176; film M826, roll 50, "Labor Contract," William Walker [employer] and Freedwoman, Oktibbeha County, Mississippi, Jan. 1, 1867, p. 693.

"about to ship [his] share of the cotton crop against his will, when it is provided in the contract that it must be divided at the gin."[35]

Close observers of early labor market conditions in the South were acutely aware of the reasons for the popularity of share contracts among laborers and planters. In Alabama, summing up the difficulties that had occurred during 1867, Wager Swayne cut through to the essential issues. The share system "in lieu of money wages" was "an arrangement preferred by the laborers as more secure, and by the planter from his inability to pay until the crop was prepared for sale." The point was supported by the agent at Greenville, who in June 1867 surmised that the prevalence of "the system of working upon shares" was "probably induced by the fact that last year a great proportion of the laborers were defrauded of their wages." In the same year a Mississippi planter, writing in the *Hinds County Gazette*, declared that the joint crop contract was necessary and beneficial for both freedmen and planters. The planter who paid shares would have the advantages of reducing his own risk and "could approximate an estimate of his annual portion of the crop, as a certain basis of credit, free from any incumberance of lien for labor." The main advantage to the freedmen was

> The certainty of receiving their wages at the end of each year. As the crop is gathered, they can separate their portion and have it marked, stored, and sold under their direction, and the proceeds applied only to the payment of their debts. This will be appreciated by many who have not yet received their wages for the past two years' labor.[36]

Testifying before the congressional committee investigating Ku Klux Klan activities, C. M. Hamilton, who had been assistant subcommissioner of the bureau for the Western District of Florida, provided a summary of the entire issue. Upon assuming his post in 1866, Hamilton found that the standing wage contract far too often resulted in planters' contriving "fraudulently [to] drive workers away without pay." Hamilton's interrogator understood the contradictions involved between the principle of free labor and the standing wage system. "If they were paid wages by the month, it would make no difference to the laborers, would it, when they went off, so their wages were paid them? You changed to an interest in [the] crop?" Hamilton answered that he had voided standing wage contracts in favor of share payments for two reasons: he felt that an interest in the crop would provide incentives to work, and it would "insure a compensation for labor performed."[37]

[35] *BRFAL*, film M757, roll 37, pp. 839–40, Sept. 11, 1866; film M826, roll 30, p. 365, Sept. 1867. The insistence of both Negroes and the bureau upon this contractual stipulation is probably one of the many reasons for the reduction in bureau-approved contracts after 1866.

[36] *BRFAL*, film M809, roll 2, "Swayne Annual Report, 1866," p. 129; roll 18, p. 279. *Hinds County Gazette*, Dec. 6, 1867, reprinted in Cox and Cox, p. 353–54. Also see *American Farmer*, vol. 11, no. 7, Jan. 1868, pp. 218–19, reprinted from *Texas Almanac*.

[37] *Ku Klux Klan Investigations*, 42nd Cong., 2nd sess., no. 41, pt. 13, pp. 282–86. In Arkansas, General Ord had advocated shares in 1867 for these reasons: *BRFAL*, film M979, roll 23, pp. 202–3.

These forces, when coupled with the crippling financial disasters of 1867 and 1868, created a situation that drove many of those planters who had not already done so to adopt share payments. The problem of obtaining financial credit, difficult enough in 1866, was severely exacerbated in the next few years. From the Alabama plantation belt, Major Pierce provided a typical summary of the year 1867: "the crops of this year are a failure, planters are bankrupt, and the Commission Merchants, from whom they obtained their supply of provisions on a deed of trust on the crop, being in most all cases loosers to a great extent, the farmer cannot procure credit for another year." The decision to resort to payment by shares was no longer merely attractive but imperative. As one Alabama planter had put it earlier, "the losses of the year [1866] absorbed all the ready money of the country— actually crippled and disabled many—rendered the payment of wages the next year an impossibility; in consequence the Negroes in 1867 worked for shares of the crops."

By 1868 these problems had even penetrated to the large planters in the Natchez, Mississippi, district. The local agent reported: "I find that the planter who is unable to feed any labor has in many cases, leased to the freedman his land, which the latter agrees to cultivate for a share of the crop." The planters' motivation for this arrangement is made even clearer by the agent's explanation that some of the laborers depend "upon their own resources for food; others upon the daily expectation of receiving their bounty money, while a few, have what they believe means enough to last until gardening can come to their relief."[38]

Unable to obtain credit from private sources, thousands of planters turned to the government for financial assistance. The bureau was authorized to make loans on the security of the growing crops. The bureau, in taking liens on the crops for loans of supplies, performed the same function as the merchant-creditor. In its desire to strengthen the economic position of the laborer, it provided what was likely the final push which made payment by a share of the crop a permanent institution.

In Florida, the bureau in 1868 was authorized to issue "rations each month to Negroes occupying and cultivating at least ten acres of land." This program had precisely the same effect as the lien system, which enabled planters to avoid the cost of feeding share-laborers who could pledge their crops to an obliging merchant. Financially destitute planters hastened to put their workers under share contracts so that they could qualify for the rations. "One local agent who lacked enough provisions to meet all requests said that the landowners seemed more disappointed than the freedmen who would have eaten the food." In Louisiana a similar bureau program authorized direct loans to planters. Brevet Major General Edward Hatch provided a impetus to the demand for and supply of share contracts by issuing an order which stipulated that planters receiving loans from the bureau

---

[38] *BRFAL*, film M809, roll 18, pp. 53–54. Loring and Atkinson, *Cotton Culture in the South*, p. 106. *BRFAL*, film M826, roll 32, p. 189, Feb. 12, 1868.

must turn over to freedmen their share of the crop so that freedmen, with the assistance of the bureau, might dispose of it themselves. Planters failing to do this would be required to process their entire crops through the bureau.[39]

From 1866 until the middle 1870s, when Democratic-controlled state legislatures would begin to redefine the legal position of the share laborer, the impact of the lien laws and the presence of third-party lenders meant that in practice the choice of a share rather than a postharvest wage involved not an acceptance of more risk, as is usually claimed, but a shifting of risk from the employee and employer to the merchant. The bureau's 1866 ruling that the shareholder was the legal owner of his share of the crop provided a strong stimulus to the use of share payments. Since the possibility of bankruptcy always existed, no payment system was devoid of risk. However, the wage worker who received no pay was forced to prosecute an expensive and risky lawsuit against the employer for the wages owed. The share hand had a right to a portion of the crop, and the responsibility of instituting the lawsuit for unpaid debts was placed upon the merchant and planter. While the overall risk inherent in agricultural production played an important role in the ultimate choice of labor payment form, the practical effects of the sharecropper's possession of a share was more fundamental. This possession made the share worker liable for consumption debts, but that liability was limited to the sharehand's own debts and independent of the planter's debts. And the limited liability made possible by ownership reduced the risk incurred by both planter and laborer. These circumstances explain the overwhelming adoption of shares in the first three years after the war.

This point and the process by which it unfolded demonstrate that a labor market based on money wage contracts failed to evolve not because the field hands provided an inherently unstable labor supply. This failure occurred because the financial position of too many planters was too weak for them to make a reasonably periodic payroll, and because free labor, after a disastrous experience, wisely declined to extend credit to planters on such risky terms as a pseudo-guaranteed wage to be paid with a lump sum at the end of the season. The entire evolution provides a superb lesson in the force of the law.

[39] George R. Bentley, *A History of the Freedmen's Bureau* (New York, 1974), p. 144. *New Orleans Daily Picayune*, Sept. 6, 1868. For the same policy in South Carolina see *BRFAL*, film M869, roll 36, pp. 3–5, 125, 1868; *BRFAL*, Louisiana, entry 1826, vol. 390, pp. 160–65, 180.

10

Nother thing I members when I was a little boy; dat dey was vidin de corn atter de s'render. Dr. DeGraffenreid measured de corn out to all of em whut was share han's. He'd take a bushel an give em a bushel. When he mos through he'd throw a ear of corn to dis one, and give himse'f a ear; den he break a ear in two, an he take part an give dem part. Dat was close measurin, I tell you.

Oliver Bell

Of Justice and Collectivity

Black workers demanding a share of the crop during the last half of the 1860s were avoiding the standing wage, but they were not escaping the obligation to labor in coordinated groups. There are two compelling reasons for taking a skeptical view of the established belief that by 1867 or 1868 sharecropping by tenant families was the dominant form of labor organization on cotton plantations. First we might pause to consider that in Great Britain, as elsewhere, the inverse movement from decentralized to centralized production was often characterized by intermediate industrial formations, such as the subcontracting by a central capitalist to groups of worker-collectives often found in such industries as mining, iron, and ship-building.[1] The point is wholly theoretical, but if one believes that southern agriculture bypassed an intermediate stage, it would be of some interest, if not importance, to ask what differences in social and technological structure were responsible. When the second reason for skepticism is noted the extent of this doubt becomes very great. Virtually every student of the period agrees that the vast majority of planters initially attempted to organize free labor upon a model as close as possible to the antebellum

---

[1] Clapham, *Economic History of Modern Britain*, vol. 1, Ashton, *Coal Industry of the Eighteenth Century*. J. U. Nef, *The Rise of the British Coal Industry* (London, 1932). G. C. Allen, *The Industrial Development of Birmingham and the Black Country* (London, 1929).

organization. When this attempt was made with share payments the result was the collective contract.

Once it is understood that planters, early after the war, began working their laborers in gangs and paying them a share of the crop, there is an easy first step which provides a partial picture of the evolution of sharecropping on the family plan. It is a well-known proposition of economic theory that when the payment received by each member of a group depends upon the performance of every member of the group, there is an incentive for each member to shirk his or her performance. In general, if the members of the work group are not of uniform ability the more able will tend to do less work than otherwise, because the additional effort they might exert results in increased output which must be divided among all workers. Thus workers abler than the average do not receive what they consider their fair share. Workers of lower ability than the average have an incentive to reduce their effort, because the reduction in the crop will be shared by the entire group and not just the individual responsible. The end result is a great amount of labor stinting, dissension within the work gang, and an overall smaller crop for both the labor gang and the employer.[2]

The best-known firsthand description of the evolution of family share farming from the work gang system is entirely consistent with the predictions of the theory of the "free rider." Here is how David C. Barrow, Jr., described the transformation of his father's plantation in Oglethorpe County, Georgia:

> For several years after the war, the force on the plantation was divided into two squads, the arrangement and method of working of each being about the same as they had always been used to. . . . The laborers were paid a portion of the crop as their wages, which did much towards making them feel interested in it. . . . After a while, however, even the liberal control of the *foremen grew irksome*, each man feeling the very natural desire to be his own "boss," and to farm to himself. As a consequence of this feeling, the two squads split up into *smaller and then still smaller* squads, still working for part of the crop, . . . until this method of farming came to involve great *trouble* and *loss*. . . . The crop was frequently *badly worked*, and in many cases was *divided in a way that did not accord with the contract*.

The key phrases are italicized. They identify four central problems. In order of appearance these are: the laborers' desire for autonomy in their work habits, the gradual decentralization of labor organization, poor work performances by groups, and discord over the distribution of earnings.

Barrow's account explains that the troubles described led to a complete subdivision of the plantation into small family units "who are responsible only for damage to the farm they work and for prompt payment of their rent."[3] Barrow's interpretation that the reason for the quarreling was the

---

[2] In modern economics this phenomenon is generally known as the "free rider" problem. See Oliver E. Williamson, *Markets and Hierarchies*: *Analysis and Antitrust Implications* (New York, 1975), chaps. 2 and 3.

[3] David C. Barrow, "A Georgia Plantation," *Scribner's Monthly*, vol. 21 (April 1881), pp. 832-33 (my emphasis).

"natural desire" of each man to farm for himself, be "his own boss", and escape the "irksome control" of the foreman has no doubt provided strong evidence for the traditional explanation of the rise of family sharecropping. As a general or "macro" overview this interpretation, relying entirely upon the freedpeople's known and overwhelming desire for family farms, is adequate to an extent. But a complete understanding must also encompass important details of a largely microeconomic nature. These details help illuminate questions of some importance.

If, for the moment, we restrict our attention to the Barrow plantation, the question which must be answered is, How long did this process of scaling down work group size take? To accept family sharecropping as generic before the end of the 1860s is to ignore all the economic and social forces tending to support a more continuous attachment to the past. We would have to believe that the freedpeople immediately understood that the collective share contract could become family sharecropping and possessed the power to force this upon the planters in a rather brief encounter, or that the planters themselves had minimal objections to the total decentralization of production. In the actual course of events, both learned their lessons through an involved process of trial and error.

In an illuminating essay on postemancipation conditions on the Barrow plantation, J. William Harris, while stressing the transfer of power between landlord and laborers, provides valuable information relevant to our question. Harris informs us that in 1868 the Barrow's Sylls Fork plantation "was still worked with gang labor," the workers being divided into two large squads under the supervision of a "white former overseer." The total disintegration of this squad organization to family tenancy on the Barrow plantation is a process that would have to be traced from 1868 through 1876, when parts of the "Home Place" were still farmed under smaller squads called "Companies."[4] This span of time suggests that either the freedpeople were concerned with other aspects of labor organization, in addition to total independence, or that the resistance of Barrow to absolute decentralization was much stronger than has been supposed, or both. What were the problems and concerns of the two parties, and how representative was the Barrow experience?

Much dissension was caused by distributional concerns. Captain W. Miles Hazzard considered these problems so detrimental to productive efficiency that he recommended complete abandonment of all postharvest payment schemes. One defect was that a division of the crop in the presence of the laborers could give them occasion to contemplate the disparities in the returns to capital and labor. Captain Hazzard's description of the division of the rice crop on his South Carolina plantation provides fundamental insight into the problems of management-labor relations with collective share contracts:

[4] J. William Harris, "Plantations and Power," in *Toward a New South? Studies in Post-Civil War Southern Communities*, ed. Orville Vernon Burton and Robert C. McMath, Jr. (Westport, Conn., 1982).

But when this process was continued . . . and one bin was piled up nearly to the ceiling, and the other lowered almost to the floor; and the great pile was seen to be the portion of one man, and the small one that of fifty, no logic known to the human mind was capable of persuading the fifty men that they had not been cheated.[5]

Perhaps even more problematic were the considerable problems inherent in the collective distribution of the laborer's share. Captain Hazzard elaborated:

No contract that embraces a settlement at the end of the year; will ever prove satisfactory to the laborers. It is an element of discord. If simple in its terms it leads almost inevitably to *inequality* of compensation as between the *laborers themselves*. If equality in this respect is attempted, the system becomes complex, and the laborers do not understand it, and suppose themselves wronged by the very justice that is dealt out to them.[6]

The simple terms alluded to were those stipulating that each laborer in the work group was to draw an equal share of the final product. In 1866 this plan was adopted by many planters. We find, for example, tobacco planter William D. Royston of North Carolina instructing five male hands that their share of the crops was one-half, "to be equally divided among" them. Equal in effect, but more imaginative in the use of language, were the 1867 terms agreed to by fourteen males and Thomas W. Steele of Pulaski County, Arkansas. All agreed that "when said crop is ginned and baled, then it shall be divided at the gin house on this plantation share and share alike." An indication of the problems experienced by Captain Hazzard in his attempt to maintain this simplicity is provided by the complaint of field hand James Brooks that Mr. Duval Prochain of Point Coupee Parish, Louisiana, had "violated the agreement" made with Brooks "and others, by placing in the field a woman and a boy, who are to share equally in the crop with the others." In adjudicating the matter, bureau agent Thomas H. Hopwood decided that the "woman and boy shall have at [the] end of the year, a fair allowance for their labor only."[7] This recourse to distributive justice upon the basis of individual productive efficiency created new problems. Thomas Crenshaw of Butler County, Alabama, probably merely postponed the difficulties until the actual division when he made the ambiguous stipulation in his contract with ten men and eight women that "one half of the remainder of the corn and cotton that is made, is to be divided among said laborers according to merit."

A widespread attempt at clarity was the practice of classing the hands on an efficiency basis prior to the signing of the contract. This, in the words of John Summer, Jr., of Ashley County, Arkansas, meant that each would "draw pay in proportion to the labor performed," ensuring "that all that

---

5 *Southern Cultivator*, vol. 27., Jan. 1869, p. 54.

6 Ibid. (my emphasis).

7 *BRFAL*, film M843, no. 34, p. 581, "Labor Contract," 1866. Ibid., film M979, roll 41, p. 490. *BRFAL*, entry 1826, vol. 390, "Letters, 1866–1868," Louisiana, April 18, 1866, p. 16.

work may get what work may earn." In most cases hands were placed in one of four or five efficiency classes, ranging in the first case from a full or first-class hand to a quarter-hand. In early years the planter most often made the determination of a worker's classification. This procedure promised fairness only if the planter (or his or her agent) had complete knowledge of the laborers' individual capabilities, as when all were the planter's former slaves.[8]

Even if we presume that the classification of hands was agreed to by the laborers, there existed a final problem that may well have been insoluble for many planters unversed in the principles of elementary algebra. Captain Hazzard's dismissal of the classificatory system on the basis of complexity was probably seconded by one candid planter, who appealed to the editor of an agricultural journal to solve his problem of dividing $546.83 among four hands who had been classed and worked as follows:

1. full hand, worked 12 mos.
2. 3/4 hand, worked 11 mos., 6 days
3. 1/2 hand, worked 10 mos., 9 days
4. 1/4 hand, worked 9 mos., 3 days.[9]

The disparities in work time illustrate the difficulties that arose over the problem of providing individual, and therefore group, incentives for adequate levels of work effort. The experience of missionary Charles Stearns provides singular confirmation of the transmuting power of the "free rider." In 1868 Stearns and his partner began to pay their laborers share wages, while continuing to work them in overseered gangs. The result was

> that a large portion of those engaged in this enterprise, would inevitably shirk their portion of the work, leaving it to be performed by those more honest; and yet would claim an equal share in the profits with others. This of course, created dissatisfaction among the industrious ones, and jangling and disputes followed; until it was apparent that those most needed in such an enterprise, preferred working for themselves.

In giving advice to others who were contemplating employing freedpeople on the share plan, Stearns advised that "at present, the very best that can be done, is to furnish land to small squads of five or six." Stearns attributed these conditions to his belief that the freedmen were not at the time sufficiently advanced intellectually or morally to have the "amount of business talent, and self-abnegation" required for working upon the cooperative plan.

Charles Stearns was correct in his belief that the freedmen possessed too much "natural selfishness" to work under a cooperative plan without internal friction. But in this respect the freedpeople were simply demonstrating that

---

[8] *BRFAL*, entry 104, vols. 1 and 2, nos. 128 and 129, Alabama. Ibid., film M979, roll 43, p. 486.

[9] *Southern Cultivator*, Oct 1868, quoted in Thompson, p. 293. See also *New York Times*, Jan. 6, 1869, p. 7.

their social development was, at least in one category, well prepared for market capitalism.[10]

Much earlier, on the Sea Islands, experience with the collective contract had been similar. In his second report to Treasury Secretary Chase, Superintendent Edward Pierce noted that under the "gang system" "nothing is found to discourage faithful laborers so much as to see the indolent fare as well as themselves." The more astute plantation supervisors had abandoned the practice of raising the subsistence crops by "working a common field" with "the people in a gang," because, as Edward Philbrick observed, such a system "would be only a premium on deceit and laziness, and would fail to call out the individual exertions of the people." The new arrangement resulted in "each man" getting "twice as much as he used to when they worked and shared in common."

In Alabama, agent W. E. Connelly, late in 1867, was chronicling the failure of the large collective share contract. Connelly found that "the plan of working fifty to one-hundred freedpeople on one plantation" was "proving unsuccessful" because the people "now being free to think and act for themselves felt their individual responsibility for their conduct" giving cause for "domestic quarrels [to] arise." Charles Raushenberg found that laborers employed under collective share agreements "frequently not only become discontented with the employer but with each other, accusing each other of loosing time unnecessarily and of not working well enough to be entitled to an equal share in the crop."[11]

These socioeconomic forces were no doubt responsible for a large portion of the problems centering on the lack of labor discipline during the early years of freedom. One way of avoiding the difficulties is to closely measure the work performance or output of each individual, basing the share of the crop on individual productivity. This the planters and freedpeople attempted to do by scaling down the size of work gangs into smaller units commonly

---

[10] Stearns, pp. 516–17. Stearns had desired to try the "cooperative plan" in 1867, but had been dissuaded by his partner, who preferred money wages (p. 170). It is probably no coincidence that share wages were adopted in 1868, after two successive crop failures put the partnership in extreme financial distress: see pp. 150, 258–59, 195, 200–202. In 1867 Stearns obtained a loan from two colored men in Augusta who approved of his work with the freedpeople (p. 200).

Stearns also doubted that southern whites were sufficiently advanced to work upon the cooperative plan. He should have included northern whites. Actually, the experiences reported about work gangs of freedpeople under the share system are similar to attempts to organize cooperative agriculture in the late nineteenth- and twentieth-century forerunners of the Israeli kibbutz. For example, in a cooperative established in Merhavia in 1911, "there was conflict not only between workers and management, but also among the workers themselves, who quarrelled over differential wages": H. Darin Drabkin, *The Other Society* (New York, 1961), p. 64.

[11] "Second Report of Edward Pierce," Port Royal, June 2, 1862, in Frank Moore, p. 319. Pearson, pp. 94, 108–9, 111. BRFAL, film M809, roll 18, "Report of W. E. Connelly," August 8, 1867, pp. 134–5. Raushenberg in Cox and Cox, p. 342.

These wartime developments on the Sea Islands probably explains the fact that they were exceptions to the general rule. Family sharecropping was general there by 1866. See BRFAL, film M869, roll 34, p. 814.

called squads. The smaller the size of the work group the closer an individual's own work effort will be related to his or her share of the crop. W. H. Evans, president of the Pomological and Farmers' Club of Society Hill, South Carolina, advised in 1869 that "the essence of the share system consists in dividing the hands into families or small bodies. Unless this is done so far from stimulating industry it discourages it." [12]

Reducing the size of the work group confronted the planter with an obvious question: what is the most efficient work group size under the share system? Charles Stearns had singular evidence that the optimal group size might be the family. One year Stearns worked a group containing several hands under the share system on a piece of land adjoining and of the "same quality" as that of one Levi Reed. Reed and his wife "obtained nearly three times" as much cotton as did the group, which was working double the amount of land. "His cotton was half as high again as mine, and entirely free from weeds, presenting a beautiful appearance, while mine was choked with weeds." Charles Stearns's philanthropic solution was to put the matter of labor organization to a vote, and not surprisingly the freedmen were "unanimous in favor of each man having twenty-five acres of his own." [13]

Few planters voluntarily adopted the democratic method of making business decisions. Most were convinced that the most efficient work group size for blacks, under any payment system, had been achieved with the antebellum gang. This appears to be related to the controversy concerning the existence of greater productive efficiency on large plantations as opposed to smaller ones in the cotton South. Roger Ransom and Richard Sutch have correctly pointed out that the issue for postbellum agriculture was not whether there existed economies of scale on large plantations, but whether the work gang system would have been more efficient than family tenancy. Viewed in this way the question concerns not economies of scale but, as was indicated above, the productively optimal organization of a given size labor force and land area.

How the planters viewed this question can be conveyed by the observations of Union general Lorenzo Thomas, who described the incentive problem as it occurred on a plantation on Lake St. John in Louisiana. The plantation was worked by Negroes who fed themselves and received one-half the crop. "But that system is found not to work well, for the Negroes want someone over them to direct them. They complain of each other, saying that this one will not do as much work as the other." It is interesting that General Thomas's statement that the "Negroes want" supervision admits of two interpretations. The more obvious is to interpret the use of the word "want" as Thomas's assessment that because of their own shortcomings the

---

[12] *Southern Cultivator*, 1869, p. 55. For more on the "free rider" problem and the response to it, see *Dept. of Agriculture Report*, 1867, p. 424; *Southern Farmer*, vol. 2, 1868, p. 110; *BRFAL*, film M798, roll 13, Columbia, Georgia, p. 135; film M1027, roll 28, Louisiana, p. 277; *Monthly Report, Department of Agriculture*, Jan. 1870 (Washington, 1870), pp. 9–10.

[13] Stearns, pp. 517, 516. Edward Pierce and Edward Philbrick agreed. See Frank Moore, p. 319; Pearson, p. 94.

workers required outside direction. The second interpretation would have the workers themselves desiring a leader with authority to maintain work discipline. This is consistent with an argument given by Joseph E. Stiglitz, that in work situations where individual members of an organization have an incentive to shirk their labor effort, each may voluntarily submit to some group supervision which increases the payment received by all.[14] Both interpretations are relevant to the discussion at hand. If, as is most likely, General Thomas intended the first interpretation, he would have found nearly unanimous agreement from planters.

In their study of slavery, Robert Fogel and Stanley Engerman argue that the well-organized gang system, compelling a "steady and intense rhythm of work," was fundamental to the "superior efficiency" of large plantations. In a more general context, Stephen Marglin has reminded us that in the precise sense of more output for equal inputs, a factorylike organization need not be more "efficient" than a decentralized organization to be more profitable for the capitalist. When comparing free and slave labor the point is obviously crucial. The planters, very cognizant of the coercive basis of their profits, would only have disagreed with those who argued that slavery was not necessary to induce profitable amounts of labor from blackpeople.[15]

It is here that the real argument over the productivity of free and slave labor must be joined. Those that argue that the productivity of the slave plantation could not have been "replicated" with free wage labor are of course correct. But the issue of the comparative productivity of a centralized versus a decentralized agrarian organization is not the all-or-nothing juxtaposition of slave gangs versus family farms, but something between. "Free" wage labor with more supervision than family sharecropping might well have been more profitable for both parties. In asking why such a system never developed it is not correct to use the fact that free whites were loath to work in antebellum slave gangs, nor to assert that freed blacks would not accept a wage-based system in a postbellum South that never really attempted such a system. It is highly likely that *if* a wage labor system had been attempted on an extensive and prolonged scale, the crucial difficulty would have been the inability of a great percentage of ex-masters to adopt management techniques suitable for free labor.

Professors Ransom and Sutch's conclusion that the work gang offered no productive gains and that this "facilitated the rise of tenancy by removing

[14] Lorenzo Thomas, "Testimony," *Joint Committee on Reconstruction*, Louisiana, 1866, p. 141. Joseph E. Stiglitz, "Incentives, Risk, and Information: Notes Towards a Theory of Hierarchy," *Bell Journal of Economics*, vol. 6, no. 2 (Autumn 1975), pp. 571–72.

[15] Fogel and Engerman, p. 204. Stephen Marglin, "What Do Bosses Do?", *Review of Radical Political Economy*, Spring 1974. Also Gavin Wright, *The Political Economy of the Cotton South* (New York., 1978), p. 7 and chap. 3 for a good discussion of the economies- of-scale debate. Professor Marglin's somewhat overstated thesis, that coercive maintenance of an intense labor discipline in early factories was a sufficient profit incentive to create those factories, provides needed redress to the overstated modern position that the factories' existence *required* machine production. Early historians such as Marx and Paul Mantoux still seem to offer a balanced position.

what would otherwise have been a serious economic objection to the fragmentation of production" does not agree with the actions of the planters.
Given the incentive structure provided by share wages, their resort to the
squad system must be interpreted as an attempt to maintain centralized
control of their workforces in a manner as close to the gang system as possible. In addition, the prolonged period of transition in the face of these
adverse incentives is direct evidence that planters did have a serious
economic objection to the fragmentation of production. In 1869, a writer
in *Hunt's Merchant Magazine* explained that

> when emancipation occurred the planters made great efforts to associate the
> laborers together on their large plantations, but the system has completely
> broken down and given place to the squad system, where from two to eight
> hands only work together, in many instances a single family. The squad system
> on large plantations is much less productive than the old system of associated
> labor, as there is no concert of action and fair division of labor.[16]

The major result of the decentralization of production, as seen by the
planter, was the concomitant loss of factory organization and, with it, factory discipline. These were the two elements that made the work gang
attractive. Under the gang system, specialization and division of labor were
achieved by forming different gangs which could be set to work at specific
and distinct tasks. The most discussed aspect of the gang system was its
ability to extract a coordinated and steady work pace from the individual
members of the gang under the direction of the driver, whose job was to set
and maintain that work pace. Frederick Law Olmsted described how work
discipline was maintained with a gang of women ploughhands on an antebellum plantation:

> Twenty of them were plowing together, with double teams and heavy plows.
> They were superintended by a male Negro driver, who carried a whip, which
> he frequently cracked at them, permitting no dawdling or delay at the turning;
> . . . they are constantly and steadily driven up to their work, and the stupid,
> plodding, machine-like manner in which they labor, is painful to witness. This
> was especially the case with the hoe-gangs. . . . moving across the field in
> parallel lines, with a considerable degree of precision. . . . A very tall and
> powerful Negro walked to and fro in the rear of the line, frequently cracking
> his whip, and calling out . . . "Shove your hoe, there! Shove your hoe!" But
> I never saw him strike anyone with the whip.[17]

The antebellum driver was able to coerce a degree of labor discipline that
would not be possible after emancipation, but the organization of workers
into centrally controlled squads promised more coordination and labor
discipline than was possible with single family production units. In this
regard the attempt to organize laborers into squads of multiple families was
a decision that a system composed of a few minifactories, although worse

[16] Ransom and Sutch, p. 76. *Hunt's Merchant Magazine*, vol. 61, 1869, p. 199.
[17] Frederick Law Olmsted, *A Journey in the Back Country: In the Winter of 1853-4* (1860,
reprint New York, 1907), pp. 83-84.

than one large factory, was more productive than one composed of many domestic units, where neither factory organization nor labor discipline could be maintained. In order to examine this hypothesis it is necessary that we determine in some detail both the internal structure of the squad system and the relationship of squads to one another and the central employer. These questions are significant both for American history in general and Afro-American history specifically. With a few exceptions our knowledge of labor organization and of the structure of the work life of the freed-people during the Reconstruction period is woefully scarce.[18]

David Barrow's response to the labor-management problems caused by free labor was typical of the large planter. In their 1868 survey of the cotton states, Boston cotton brokers Loring and Atkinson provided a summary of the relation between employer and employed from their 1,000 pages of survey data:

> There are two systems of employing laborers at the South, by share and by wages. In the share system the usual arrangement is that the laborer receives half the corn and cotton when he "finds" himself. If rations are given they reduce the share of the laborer to one-third or one-quarter. At times on the share system, the labor goes against the land, the laborer and land-owner dividing all other expenses between them. In the above cases, the usual practice is for the employer to supply cabin and fuel for his laborer (with a garden patch free of rent, for his vegetables, etc.,) and tools, seed, teams and feed for the same. But frequently the laborer bears half the cost of seed and fodder—in some instances, even half the cost of repairs of tools.

This description of employing laborers by a share conforms perfectly to the family sharecropping system. However, it is not family sharecropping, but share wages by work groups, as is made perfectly clear in a footnote describing an alternative to the above system:

> Yet another system has been pursued, this and last season, by some planters—that of fencing off the plantation into lots of say fifty acres each, with cabins a quarter of a mile or so apart, placed in each lot near the woods, so that the laborer can keep hogs, and fronting on a main avenue. Each lot is cultivated by a squad of eight to ten laborers, who own the teams, and feed themselves—the landowner feeding the mules—and receiving one-half the crops, in the cultivation of which *he has supervision*. It is presumed that the

---

[18] The few notable exceptions are Davis, "Good and Faithful Labor"; Williamson, *After Slavery*; C. Mildred Thompson's superb study, *Reconstruction in Georgia*; Paul S. Taylor, "Slave to Freedman", *Southern Economic History Project*, Working Paper no. 7, Jan. 1970; Michael Stuart Wayne, "Antebellum Planters in a Postbellum World: The Natchez District, 1860-1880" (Ph.D. diss., Yale University, 1979). Ralph Shlomowitz, "The Transition from Slave to Freedman Labor Arrangements in Southern Agriculture, 1865-70" (Ph.D. diss., University of Chicago, 1978), and idem, "The Squad System on Postbellum Cotton Plant-ations" in Burton and McMath, op. cit., is most useful for its treatment of the economics of the squad system. My own thinking on this problem has been crystallized through discussing both points of agreement and disagreement with Professor Shlomowitz.

fact of the laborer owning the teams insures better treatment, as the negro is frequently accused of cruelty to animals when he has no pecuniary interest.

Some such arrangements as these have been found to work very successfully indeed, and we find the "tenant system for freedmen," as it is called, in operation to a greater or less extent in all parts of the South, and growing in favor. It is a sort of connecting link between the old gang plantation system and the small farm or peasant system.[19]

This paragraph clearly indicates that in 1867 and 1868 some planters paying shares began to pare the work gang into smaller work units denoted squads. There is no mention of family sharecropping, and indeed, if family sharecropping were the predominant mode of organization about which the share system was built it would hardly make sense to stress that the planters were "fencing off the plantations into lots." The movement toward squads of freedmen working assigned lots of their own was a movement from the work gang, not a movement from family plots.

The reports of agricultural operations coming from bureau agents in the crucial years 1867 and 1868 corroborate the point that in these years freedpeople working for a share of the crop on plantations were generally doing so under the squad organization and not family tenancy. The agent in the district about Tuscaloosa, Alabama, where most freedmen were working on shares, reported in December 1867 that "there is a few F.M. [freedmen] renting ground for half the crop, with mules and provisions furnished." In the same year, agent W. E. Connelly reported that in the area of Marengo County, "a few farmers have determined to divide their plantations into small farms of 100 acres" next year which "they will rent or lease to industrious freedmen." From central and northeastern Mississippi an agent reported that "some of the largest planters, have informed me that they intend next year [1868] to rent out their plantations in small lots to freedmen." An agent in the tobacco region of Virginia provided what seems to be his estimate of the average size of squads in 1867. He reported that the usual payment was for one-fourth of the crop, "to be divided among five laborers, and at that proportions when more than five hands have been employed." That the five or more hands were not members of a nuclear family unit is apparent from the fact that their share was to be divided among them proportionally.[20]

---

[19] Loring and Atkinson, pp. 25–26. Note the statement that this squad system was known as the "tenant system for freedmen." This imprecise terminology must be responsible for many incorrect suppositions by historians that contemporary references to the tenant system meant sharecropping by individual nuclear families.

[20] *BRFAL*, film M809, roll 18, p. 958, Dec. 27, 1867; ibid., p. 134–35, Eufala, and p. 149, Aug. 1, 1867. Ibid., film M826, roll 30, p. 230, Aug. 31, 1867 and p. 451, Sept. 1867, Natchez. Film M1048, roll 46, p. 445, Jan. 31, 1867, Burkeyville, Virginia, and p. 398. New Kent and Charles City Counties, Virginia. See Raushenberger in Cox and Cox, p. 342, Nov. 14, 1867, and pp. 351–52, Nov. 10, 1868, Albany, Georgia. For an example of a southerner recommending abandonment of share gangs in favor of family tenants see *Anderson Intelligencer*, Jan. 23, 1867, South Carolina, in Chartock, p. 95. *Hunt's Merchant Magazine*, 1869 (see n. 16 above). The reports of the Alabama agents are confirmed by the *Montgomery Daily Advertiser*, which

The payment of shares to squads composed of fewer workers than comprised the work gang was the representative labor system of cotton plantations into the 1870s. Robert Somers, during his trip through the southern states in 1870–71, remarked, "The Negroes toiled in gangs or squads when slaves, and they toil necessarily, though under much less control of the planter, in the same form still." This statement was made while Somers was touring plantations in Alabama, but the context of his writing makes it very clear that the squad system is to be taken as the generic organizational form of labor on cotton plantations throughout the South. This reading of the text is substantiated by numerous references to the squad system throughout Somers's book, and in sources dealing with other states. In a very informative article published in 1871 an anonymous southerner verifies the report of Robert Somers:

> The following is generally the actual method of carrying out contracts based on a part of the crop: the freedmen are divided into squads of four, six, eight or ten, as the case may be, and a certain portion of land is assigned to each squad and planted in cotton, corn, potatoes, peas, etc., and the produce divided according to the special terms of the contract, when the crop is gathered.[21]

The statistical data shown in table 10.1, calculated from a sample of 425 share contracts embracing 570 distinct work groups for the years 1867 and 1868, verify these contemporary observations. The average work group sizes of 9.13 and 6.74 on plantations for the two respective years is very consistent with the previously cited testimony that share squads in the late 1860s and early 1870s were usually from 2 to 10 laborers in size. In the more relevant year, 1868, 83 percent of the observed groups fall within the cited range. Both this statistic and the existence of work groups of much larger size are consistent with Mississippi planter Sam Postlewaite's 1869 remark that the three squads of 16, 17, and 21 hands which composed his labor force were "large squads."[22]

---

on May 14, 1867, reported "you do not see as large gangs together as of old times, but more frequently squads of five or ten in a place": quoted in Peter Kolchin, *First Freedom: The Response of Alabama's Blacks to Emancipation and Reconstruction* (Westport, Conn., 1972), p. 46. On the Alabama squad system see Fleming, pp. 722–3. *New York Times*, Dec. 11, 1866, p. 2, Georgia.

[21] Somers, pp. 120, 121, Alabama, and pp. 146–47, Mississippi. Southerner [pseud.], "Agricultural Labor at the South," *Galaxy*, vol. 12 (Sept. 1871), p. 331. *Monthly Report, Department of Agriculture*, Jan. 1870 (Washington, 1871), clearly implies that the squad system was the most prevalent form throughout the cotton region (pp. 9–10). For more evidence of the squad system in the period 1866–71 see *Report of the Department of Agriculture 1867*, p. 419, Georgia, p. 422, Mississippi, p. 424. Loring and Atkinson, pp. 10, 132; *Hunt's Merchant Magazine*, vol. 61, 1869, p. 199; *Southern Cultivator*, vol. 27, 1869, p. 54–55, 58–59, *Nashville Daily Press and Times*, Jan. 7, 1869, in Georgia; p. 181, Mississippi. Taylor, *Negro in Tennessee*, p. 148. *DeBow's Review*, 1869, p. 609.

[22] Cited in Shlomowitz, "The Squad System." The sample used for these statistics is much smaller than the total number used in this study. The diminution is due to factors in many contracts which make them unreliable for a determination of work group size. For the year 1868 the range 2–10 is quite close to the range measuring one standard deviation from the mean, 1.81–11.73.

Table 10.1 The Size of the Share Work Group on Cotton Farms

	Sample Size	Minimum	Maximum	Average	Median	Mode	% Family	% Male with Non = repeating Surnames	Standard Deviation
Plantations, 1868	134	2	35	6.74	5	4	19	23	4.91
Plantations, 1867	195	2	43	9.13	7	4	9	16	7.84
Small farms, 1868	127	2	9	3.52	3.3	2	36	29	1.45
Small farms, 1867	114	2	9	4	4	3	33	18	1.53

SOURCE: *BRFAL*. States represented in sample are Alabama, Arkansas, Georgia, Louisiana, Mississippi, North Carolina, South Carolina, Tennessee. For further description of data and estimation procedures see Appendix A.

As informative as these averages are, they fail to provide the kind of representative information that would give precise knowledge of the sizes of work groups. The pitfalls involved in using statistical abstractions like the "average" are well known, not the least being the fact .13 and .74 workpeople do not exist in any real sense. Also, while 6.74 would seem to be too high for the "average" number of working hands in a nuclear family, it is not impossible that our "average" size work group represents not a squad in any meaningful sense but a typical share tenant nuclear family. Fortunately, the sampled contracts include the surnames of contracting workers, often the workers' ages, and other information which allows a determination of whether or not the entire group forms a nuclear family. For 1868 we find that 19 percent of all work groups are composed of a single nuclear family. Therefore, if we define a squad as a group of laborers drawing its members from two or more nuclear families, and receiving their earnings from the same specified crops, we find that 81 per cent of the 1868 work groups upon large farms or plantations are squads. A good relative measure of the numerical unimportance of the family work unit in 1868 is given by the fact that 23 percent of the squads and 20 percent of the entire sample are work groups composed entirely of males, all of whom possess different surnames.

In both years the modal squad, the work group size which occurs most frequently, is four members. This squad size might be a good candidate for that elusive concept, the "typical squad", but the modal squad only accounts for 16 percent of the sample. Perhaps a more reasonable approach is to give up the search for exactness and simply say that the "typical squad" contained something between three and eight members. These work groups run the gamut from those composed of a single family, to combinations of families, to those made up of workers none of whom are related in any obvious way.

At the risk of being tedious, a few examples of squads within this "representative" range might be in order. In 1868 Nathan Bass of Floyd County, Georgia, trustee for Valeria Bass, worked at least three squads. Two of these were composed of

>   Harrison Little
>   Henry Little
>   Jink Millen
>   Liddy Thompson

and

>   Bluford Cochrain
>   Barney Glover
>   Efraim Leak

each working for one-third of the crop raised by their group. In the same year and state, James Neal of Pike County hired

>   Jesse Mims
>   Allen Lovely
>   James Nelson

> Phillis Phillips
> Isam Caldwell
> Bob Neal

who were to share one-half of the crop they produced. In 1868 J. R. Williams of Henderson County, Kentucky, contracted with

> Henry Legon, his wife and nine-year-old son
> D. Norris, his wife and fourteen-year-old son

for one-half the corn and tobacco, which was "to be divided equally between Legon and Norris" when harvested. An 1867 example: Mrs. E. P. Alexander of Baton Rouge, Louisiana, hired three squads, of four, five and six. The largest squad on her plantation was composed of

> Thomas Green
> Patrick Brown
> Duke Billins
> Wilson Jordan
> Catherine Green
> Lucy Thomas

who shared equally the one-half of their joint production. For a final example, Virginia tobacco planter Edmund W. Hubard in 1870 hired several work groups, among which was the eight-man squad of

> Jim Trent
> Jacob Trent
> Dempster Jones
> John Deane
> Madison Hembrick
> John Booker
> Going Braxton
> Henry Brown

A portent of things to come was the allotment of a separate parcel of land on the same plantation to Bowman Brown and "his two sons and a daughter as his hands."[23]

That the collective contract and squad organization of the labor force was mainly a plantation phenomenon is consistent with common sense and the data presented in table 10.1. Where the employer's land holdings would only support two to eight laborers, limiting the work group to a single family often became almost a necessity. In the counties composed largely of small farms, where better than one of every three work groups is a nuclear family in the 1868 sample, we begin to come across contracts of a different variety. The contract between John Watson of Greenville, South Carolina, and

---

[23] *BRFAL*, entry 998, vol. 1, no. 345, Georgia files. *BRFAL*, "Kentucky Assistant Commissioner Labor Contracts," 1868, box no. 42. *BRFAL*, entry 1499, vol. 223½, Louisiana files, pp. 2–6. Thomas, "Plantations in Transition," p. 534.

Mrs. Stanley, a freedwoman, and her three sons and one daughter for one-third of all crops on the land cultivated by them is not that atypical of the contracts found on small farms in 1866.[24]

These data tell a great deal about agrarian organization in the early Reconstruction South. The gang system on cotton plantations was an institution of the past by 1868. The evidence, both statistical and testimonial, supports the conclusion that the gang system's replacement, the small-scale collective contract, was far from being dominated by family-based share-cropping. What the data fail to tell is how the laborers' work lives were organized. What did the squad organization mean to and for the field laborers and the planters?

II

If the hypothesis of planter resistance to complete decentralization is correct, there are four characteristics which employers would have insisted upon as being essential aspects of postbellum plantation management. Foremost of these would be their insistence upon retaining complete managerial control over allocational decisions, such as the crop mix, methods and patterns of work, and the financing and marketing of the crops. In addition, if the move from the work gang was a reluctant concession, employers would attempt to keep the productive efficiency of different work groups as equal as possible.[25] The third requirement would be a desire to maintain a purely capitalistic relationship between the squad and employer as well as within the squad. This could best be done by the *employer's* selection of a foreman or squad leader who, vested with managerial control over the squad, would act as the agent of the planter. Finally, insofar as there existed work activities which the employer believed to have considerable economies of scale related to large group coordination, such activities should be organized under the gang system.

Opposed to the planter's wishes were the ambitions of the laborers for self-autonomy. In a very real sense, any relaxation of the employer's ability to insist upon these key elements represented a transfer of social power to the freedpeople. The point and the dynamics of the ensuing struggle are relevant to a range of issues. Our examination will span the rest of this and the whole of the next chapter.

In Chapter 3 it was pointed out that there exists a formal isomorphism between the final form of the southern agricultural tenancy system and early industrial organizations in various English industries. Here it shall be found useful to note that in many cases the intermediate organizational forms connecting totally decentralized and centralized organizations in

[24] *BRFAL*, film M869, roll 34, p. 586. For small farms, see for example, *BRFAL*, film T142, roll 66, "Labor Contracts" 1866, Gibson County, Tennessee; *BRFAL*, entry 220, vol. 1, nos. 160, 161, "Labor Contracts," 1866 Franklin County, Alabama, Alabama Files.

[25] This condition would manifest itself empirically by the allotment of equal-size acreage to each squad.

various industries had their counterparts in the Reconstruction South. As a tool for understanding the working of basic economic forces, the historical development of the British coal industry is an example of structural isomorphism that can aid understanding of the present problem.

By the seventeenth and eighteenth centuries, besides some large mining companies operated by capitalists (in every sense of that term), the organization of part of the British coal industry had taken a familiar form. Often an independent collective of workmen (colliers) would contract with the proprietor of a mine for their mutual benefit. In other cases a small entrepreneur with limited capital could contract with the proprietor and secure labor by hiring a gang or company of workers. In the latter case the middleman capitalist would negotiate a work contract with each gang's leader, who was called a charter master. Such contracts might give the lessee(s) and workmen a share of the total output—all the output remaining after payment of a fixed rental quantity to the proprietor—or a fixed piece rate based on a standard quantity unit.[26] Historian Thomas S. Ashton, noting the important point that the workmen "had a substantial measure of control over the conditions of their working life," described the function of the proprietor as being limited to provision of capital, delineation of general techniques of work, and finding markets for selling the output. "The system," remarked Professor Ashton, "was not unlike the 'merchant capitalism' of the domestic textile and metal-working trades."[27]

The process of industrialization eventually ended these democratic workers' collectives in favor of a more capitalistic relationship within which the cleavage that developed between the workers and the gang leader provided both incentive and opportunity for the leader to enforce a more intense work routine upon the men. Whether this transference to more capitalistic forms of industrial organization was entirely due to some natural process of Darwinian competition, whereby "the more powerful personalities among the workers," through accumulation, "became masters," or occurred because the mine owners, interested in the benefits of increased work discipline, intervened, is open to question.[28] In the postbellum South,

[26] For example, a John Collyer and eight others contracted in 1593 with the proprietor of the mineral rights in the manor of Kippax, near Leeds, to work an existing pit. The proprietor was to provide working capital in the form of ropes and picks and receive one-third of all the coal obtained. In 1670 Sir Francis Blake, a proprietor in Northumberland, contracted with seven "hughers" who were to deliver to his manager every fourth bowl of coal they dug. In 1637 four "colers" contracted with the Bishop of Chester to dig coal and cannel for a year. The work group was to supply candles and bellows, sharpen their own tools, and hire additional men to wind the coal up the shaft. In turn the bishop was to provide baskets and ropes and "build them up a hovell." The men were to receive 8d for every quarter (about one-third of a ton) of coal mined. A farmer, James Osborne, in 1609 subcontracted two pits, one for an annual money rent and the other for a specified quantity of coal. For these examples see J. U. Nef, *The Rise of the British Coal Industry*, vol. 1 (London, 1932), pp. 414–15, 424. For further examples from the eighteenth century consult Ashton, *The Coal Industry of the Eighteenth Century*, chap. 7, pp. 101–10.   [27] Ashton, *Coal Industry*, p. 100.

[28] Ibid., pp. 113–14. Professors Ashton, Clapham, and Nef seem too ready to accept the Darwinian explanation of an independent entrepreneurial emergence without an examination of alternatives.

many landowners, perhaps too impatient to wait for the realization of the evolutionary process, sought to perpetuate the cleavage which had previously existed between field hands and drivers.

For the planter, the disintegration of completely centralized control involved a loss of the kind of labor discipline that could be maintained with close and continuous supervision. It would be surprising if the prerogative of supervision was easily given up, and indeed the evidence implies that initial moves to decrease the size of work gangs perpetuated close supervision through the practice of assigning a foreman over each squad.[29]

Peter A. Parker of Holmes County, Mississippi, whose hands agreed to place themselves "under the old plantation rules and regulations for our control and management," was most explicit upon this point. The laborers agreed that "Parker shall have the right to divided us off into squads placing one of our number over us as foreman or squadman." The wording suggests that Parker took the initiative in scaling down the size of work groups. A more typical illustration of the origins of the squad system is the 1868 contract between cotton planter David Hinton of Wake County, North Carolina, and sixteen male freedpeople who "each for himself agree[d] that they will work under the control of a foreman to be selected at any time or times by said Hinton." Similarly, tobacco planter Willis Lewis, M. D., of the same state, in 1867 agreed with twelve males that "the hands are to work in two parties, six in each party." Furthermore, it was "mutually understood that Henry Hall is my headman or foreman who will manage my farm in my absence and any hand refusing to obey him will be discharged forthwith."[30]

That the relationship between the foreman and the other workers in these early squads was capitalistic should be clear from the hierarchical cleavage. Any doubt in the Peter Parker case should be dispelled by the fact that Parker's elaborate system of supervision and labor discipline required that fines assessed upon laborers for absenteeism and various offenses be paid to himself and the foreman in equal shares. The arrangement was indicative of the contractual specification of the employer's control over the methods and hours of work. John N. Roberts of Beaufort, South Carolina, prescribed that on his Mount Hope plantation "it is understood that the hands are to work in gangs of eight full hands the foreman of each gang to do three quarter hands work and J. N. Roberts is to make up the quarter hands

---

[29] Ralph Shlomowitz's view that squads were "semiautonomous and self-regulating" democratic collectives must be qualified to some extent. The evidence indicates that in the early stages of its existence as a numerically significant institution, the squad was primarily a nondemocratic organization. Since the squad system was a transitory development, lasting about ten years as a system, we will find that it was a purely democratic organization for perhaps one-half of its life. Professor Shlomowitz finesses this problem by *defining* the squad as a democratically organized work group: Shlomowitz, "The Squad System," p. 272.

[30] *BRFAL*, film M826, roll 50, "Labor Contract—Peter Parker and Freedpeople, 1866," p. 198, and, "Labor Contract—Peter Parker, 1867," p. 730. *BRFAL*, entry 2819, box 53, North Carolina, "Labor Contract—David Hinton, 1868." *BRFAL*, entry 2797, vol. 1. no. 209, North Carolina, "Labor Contract, Dr. Willis Lewis and Freedpeople," pp. 273–77.

share to foreman." What the foremen were to do during the other quarter of time is sufficiently explained by the fact that someone had to keep tallies and enforce the rule that lost time was fined at the rate of fifty cents a day if excused, two dollars if "AWOL"; hands were punished with dismissal and forfeiture of crop share if "AWOL" two days or more.

An early example of decentralization to the squad system is documented in the 1866 contracts of a Madison County, Tennessee, planter who contracted independently with various groups of workers. The agreement with one of these groups, composed of three different families, and its foreman Anderson, was typical. Anderson was "to make a hand amongst the rest," and "all the workers agree[d] that Anderson" should "come in and have 1 hands part say 1/6 part of all that is made." That special compensation went to Anderson is evident from the fact that Anderson's one-sixth was apportioned from the entire crop under employer Fuller's guidance, and then the remaining workers were allowed jointly "to collect their 1/3 from what is left." Furthermore, the hands were enjoined to obey Anderson without "murmuring or contention," this last point being enforced by Anderson's power to fire any workers who "became impudent."[31]

The squad organization was often initially adopted without the planter abandoning gang coordination in those work activities where it was deemed most productive. On Alabama plantations, "squads were united to hoe and plough and to pick the cotton, because" it was supposed "they work better in gangs." Dr. Willis Lewis, after agreeing that the tobacco was to be worked by two equal parties, added "that the twelve hands are not to be divided until the crop is pitched." James Rawlings, "principal-superintendant" of a Desha County, Arkansas, plantation, contracted with a number of squads in 1867, stipulating "that until the land is ready to lay of they will work as one gang." A. R. Chisolm worked two plantations in the Beaufort District of South Carolina with several squads composed of six hands each. Chisolm's detailed management instructions to squads required that "all plantation work which shall accrue to the benefit of all concerned in this plantation shall be performed by us the undersigned jointly with the rest of the contractors." The tools and work stock were "to be worked by the ploughman of the squad of six and by him kept in good order."[32]

The motivations of the best of the employers who made an early adoption of the squad organization was explained by the president of the Agricultural Club of South Carolina. This planter's experience recommended "that it is best to work several farms on the same plantation—a small parcel of hands

---

[31] Peter Parker contract (see note 33 above). *BRFAL*, South Carolina files, "Labor Contract John N. Roberts," 1867. *BRFAL*, film T142, roll 66, "Labor Contract Between T. Fuller—Anderson and Freedpeople,"; also "Labor Contract Between T. Fuller—Barker and Freedpeople."

[32] Testimony of contemporary Alabama planters, cited in Fleming, *Civil War and Reconstruction in Alabama*, pp. 722–23. "Labor Contract of Dr. Willis Lewis" of North Carolina (see note 31 above). *BRFAL*, film M979, roll 41, "Labor Contract, James Rawlings," Jan. 7, 1867, p. 176, and film T142, roll 72, "Labor Contract of James Rawlings," Jan. 3, 1867. *BRFAL*, South Carolina files, "Labor Contracts, A. R. Chisolm," 1867.

to each farm—and I have adopted the plan of appointing over each farm a freedman of experience, ability and fidelity, as chief or superintendent. This position is given as a reward for promptitude and good judgement." The explanation would probably have satisfied the owner of several thousand acres in the lower Mississippi Delta who, in 1867, had divided his work people into "gangs of twenty," each supervised by a "negro foreman" who made one of the twenty and was responsible for reporting to the "general overseer" all cases of "absence or idleness."[33]

This organization of labor gave the planter as much control as he could expect within the confines of the collective share contract. The employer chose the size of the work group, its foreman, and in most cases the classification of each worker's efficiency and therefore distributive share. As already stated, such power in the hands of a benevolent master who had extensive knowledge of the capabilities, friendships, and enmities of all of his laborers may have been acceptable. In general, the employer could not be expected to have this much information. Since the adoption of the squad system was an attempt to avoid the extreme incentive problems incurred with large share wage gangs, it would be important that the internal organization of squads be constructed on the basis most likely to avoid the more obvious difficulties of the gang system. Groups of workers with widely disparate abilities who were relatively unknown to one another could hardly be expected to avoid serious incentive problems. The difficulty in forming desirable groups results from the phenomenon known in the economics literature as "asymmetric information." While the planter generally could not know which laborers were likely to work together peaceably and efficiently, the laborers themselves would possess very good information of this kind. Workers allowed to select their own co-workers would naturally tend to choose those known to themselves to be good laborers whom they could trust in a cooperative endeavor. With every worker attempting to associate with the best laborers possible the process should, under competition, lead to groups based upon fairly homogeneous abilities and mutual trust.

The fact that planters could no longer enforce stability in the composition of work groups exacerbated these problems. With labor free to move, the employer who attempted to assemble work squads in a piecemeal fashion by assembling groups of strangers invited all the incentive problems inherent to the collective contract. In March 1868 Mrs. Sam S. McMillan and John Lipscomb practiced just such a policy. Their labor contract required that contractors work "in company with such other freedmen as may have been or may hereafter be employed." We may speculate with confidence that this recruitment policy provides explanation for the cancellation of the names which appear in a column under the heading "Sam's squad." We might also wonder whether the agreement "that all lost time from sickness or otherwise

[33] Peters, *Report, 1867*, p. 20. *DeBows' Review*, vol. 4, 1867, p. 272, quoting from *New York Evening Post*.

shall be deducted in order that justice may be done in the division" of crops was sufficient to prevent dissatisfaction among the remaining five men and one woman listed under the heading "Abram's Squad."[34] A widely used method of eliminating this difficulty was to subcontract part of the plantation to an individual freedman who was charged with the responsibility of hiring his own hands. The practice was reported to be a common arrangement in the cotton fields of the Virginia lowlands during the early 1870s: "The farmer generally rents his land to some colored man of enterprise and character for half the crop. The tenant hires his own force and oversees them. The proprietor furnishes implements and teams. The tenant feeds the stock and 'hands.'" The concise but highly descriptive statement of a Georgian explains this system about as well as could be desired: "Landowners are beginning to contract with '*boss*' freedmen, and let them get hands, and work such and such areas, and look only to the boss."[35]

The size of the subcontractor's work force could vary from a small squad working a modest farm to a gang composed of numerous squads. An example of the former, with an egalitarian distribution of labor income, is provided by the 1867 contract between freedman Mark and employer Thomas Thompson of Hinds County, Mississippi. Mark was to have one-third of the crop, "to be equally divided between" himself and the three men and two women he had hired. Mark had "bound himself to furnish five hands besides himself." That there was a limit to the extent of the egalitarianism is suggested by the requirement that the workers "being willing to make those five hands do now agree to bind themselves to labor for the said Mark" and "by the direction and management of" Thomas Thompson. The subcontract could be of a classic capitalistic form, as when James Collins of Sumpter County, Georgia, leased a large tract of land to freedman William Styles. Styles, who was to receive one-third the crop, hired eleven men and contracted to pay each one a wage ranging from $40 to $250 a year. It could take a not so pure but still capitalistic form, as in freedman Griffin Obediah's contract with Paul A. Kleinpeter of Baton Rouge, Louisiana. Obediah agreed to cultivate Kleinpeter's plantation for one-half the crop. Boss Obediah was to pay all the expenses for his eight hands, who were to receive as a group one-third of his half share.

Such arrangements were quite common in the rice-planting districts of Georgia and South Carolina. General George P. Harrison, lessee of Louis Manigault's Gowrie plantation, divided the land into five equal sections of seventy-eight acres. Each section was put under the supervision of a Negro "foreman" who was charged with the responsibility of hiring his own squad of about ten laborers. At the end of the year the members of each squad

[34] *BRFAL*, film M826, roll 50, p. 1093. "Abram's Squad" consisted of George Sanders, Abram Rolley, Phillip King, George Suten, Cora Listin, and John Rolley.

[35] *Richmond Enquirer*, Sept. 17, 1872, p. 2. *Rural Carolinian*, vol. 1, 1869–70, p. 511 (emphasis in original). The Georgian resided in Elbert County, a neighbor of David Barrow's Oglethorpe County. See Williamson, p. 148. *American Farmer*, 1877, vol. 6, no. 2, p. 46.

were allowed to divide the net proceeds of one-half the crop they had produced among themselves.[36]

Under these kinds of arrangements the use of words like "overseer" and "boss" was more than the choice of a familiar phrase. The black entrepreneur contracting to operate a small farm or large plantation for a profit above the earnings of his hired hands was in much the same position as the antebellum overseer. In a period when most planters claimed to be losing money when paying half or a third of the crop to labor, it is difficult to believe that a system where contractors had to pay labor from their half or third share could do anything but degenerate to the harsh practices of labor driving often associated with the subcontractors' attempt to eke out a meager profit margin.

Both the wide adoption of the subcontract and the primary function of the subcontractor are indicated by the large number of contracts made with one party who was to hire, feed, and control the hands. James Rawlings, the plantation superintendent mentioned earlier, instructed squad leader Charles Henderson that he was to "see that his hands are out in good time and that they conform with plantation regulations and are subject to [the] control of [the] agent or principal on the place." George Spinlock of Copiah County, Mississippi, agreed "to superintend the laborers and conduct the business upon the farm as foreman," being "accountable for the conduct of each and everyone" of his hands.

Typically, the primary employer was charged with furnishing the fixed capital, such as land, work stock, tools, and "quarters to Gilbert and his men," as was agreed in the contract between a Kentucky tobacco planter and subcontractor Pleasant Gilbert. In addition to the requirement that he "preserve good order amongst the hands under him" or simply "control his squad," the squad leader, like Alabamian Jackson Hambrick, was required "to give his attention, honestly and faithfully to the cultivation" of his 250 acres. In addition, Hambrick was "to feed and clothe himself and pay such hands as he may employ to work."[37]

The use of the subcontract, here as elsewhere, was a natural strategy for employers who had no experience or desire to indulge in the distasteful labor of hiring and managing, on a day-to-day basis, laborers whom they

[36] *BRFAL*, film M826, roll 50, "Labor Contract, Thomas Thompson," Jan. 3, 1867, p. 846. *BRFAL*, entry 711, vol. 2, no. 138, Georgia, 1867, p. 24. *BRFAL*, entry 1492, vol. 223½, Louisiana, 1867, pp. 20–21. *Manigault Plantation Records*, "Visit to Gowrie and East Hermitage Plantations." *BRFAL*, entry 287, "Labor Contracts, Georgia," boxes 192–93. Williamson, pp. 132–38.

[37] *BRFAL*, film T142, roll 72, "Contract between James Rawlings and Charles Henderson," Desha County, Arkansas. Film M826, roll 50, "Contract between Joseph G. Sessions and George Spinlock," Jan. 1867, p. 835. *BRFAL*, Kentucky Assistant Commissioner Labor Contracts, box 42, "Contract between W. M. Griffin and Pleasant Gilbert," 1868. Film M826, roll 50 "Contract between P. J. Briscoe and Henry Briscoe," Jan. 11, 1868, p. 1049. Film M979, roll 41, "Contract," employer unknown, squad leader "Louisiana," Arkansas, Jan. 1867, p. 399. *BRFAL*, entry 116, vol. 1, no. 82, Alabama, "Contract between Joseph Hambrick and Jackson Hambrick," 1867.

viewed as unruly and insolent. "Little or no intercourse is thus held between General Harrison and the mass of the Negroes, and provided the work is performed, it is immaterial what hands are employed," Louis Manigault noted with no little approval.[38]

The failure of the scheme to survive in any significant sense was due to the people's adamant insistence that they would not "be driven by nobody." "I don't want no driving, either by black man or white man" was a universal assertion.[39] This familiar point has often been used to explain the adoption of family sharecropping. There is no doubt that the emphasis upon the desire of blacks to escape the driving system is well placed. The question is to what extent this desire affected work organization. A clue that the question is more complicated than is usually understood is given by Harriet Beecher Stowe's testimony that the freedpeople were habitually locked into the old way of doing things, and actually refused to change from the gang system: "They clung to the old ways of working,—to the gang, the driver, and the old field arrangements,—even where one would have thought another course easier and wiser."[40]

What the freedpeople disliked was not working in gangs per se but the master-slave relationship associated with it during the slave regime. It was the intensely close supervision of the overseer and drivers armed with threatening whips, exacting compulsory labor, that was objectionable. Successful planters had to invest in employee-employer relationships more suitable to the management of free labor. Available evidence suggests two things: many planters did alter their management techniques, and the freedpeople were initially not so concerned with working family plots as share tenants as they were with payment by a share.

That the freedpeople had an intense dislike for working in gangs under drivers and overseers, like so many other common beliefs about the freedpeople, needs to be subjected to critical study. A problem with the thesis is that the evidence brought forth to demonstrate its validity seldom distinguishes among an inherent dislike for gang labor, overseers, and drivers; dislike for a particular overseer and driver; or a reaction to inept management of freed labor on the part of particular foremen. In fact, it would seem that the best conjecture is that most of the conflicts between fieldhands and foremen fell into the latter two categories.[41]

When freedpeople complained of the presence of overseers it was often explained to them "that overseers, or leaders of gangs, were necessary in free labor, and are employed in all parts of the world." A compromise agreed to was that the planter would not "insist upon the same ones they

[38] Manigault, "Plantation Journal." For some examples of the "elsewheres" see A. H. John, *The Industrial Development of South Wales, 1750–1850* (Cardiff, 1950), pp. 76–80; G. C. Allen, *The Industrial Development of Birmingham* (London, 1929), pp. 160–67.

[39] Pearson, p. 113. See also Chapters 4 and 5.

[40] Stowe, *Palmetto-Leaves*, p. 290. Also see Stearns, p. 517. Leigh, pp. 30, 55–56.

[41] See Chapter 8.

had under the old system."[42] One problem was that in any given region the experienced drivers and overseers would have been far too familiar and unloved by most laborers in the region. To those planters truly committed to the need for adaptation of laborers and employers to the new order, not only was this apparent, but the alteration of the title "driver" to "captain" or "foreman" was more than mere window dressing. South Carolina rice planter Barnwell Heyward based his methods of managing free Negroes upon these principles. Heyward had the practical acumen to realize "that the freedmen would naturally resent working under the guidance of the same drivers who had managed them in the days of slavery." Barnwell Heyward also believed that the planter must take on more of the responsibilities of the overseer. "The old plan of management," he felt, "must be changed; the day of the driver had passed."

"The only way to make planting, or as you would say, 'farming,' a profitable business now," counseled another planter, "is to divide your force into squads of eight or ten hands each, and have a white man to every squad, not to drive, but to lead."[43] In the same spirit of practicality, J. J. McCarty of Lauderdale County, Mississippi, guaranteed his share laborers that "no overseer shall be about on said plantation other than" McCarty "and one of the said laborers that may be selected for that purpose."[44] Whether or not McCarty retained the right to select the foreman, as did Barnwell Heyward, is not known. If he surrendered this right to the laborers, he would have been joining many planters who, in taking this step, became unwitting participants in a movement toward agrarian democracy.

A significant variation on the previously described method of decentralizing production is furnished by the 1867 contract between Arkansan John B. Burton and eighty-five women and men. For our present purposes the clause of most interest informs us that it was "agreed on all sides that the plantation [was] to be divided into four equal parts and that the laborers [were] to select their leaders and divide themselves into four equal companies." Within the same week John Sumner, Jr. was entering a contract with eighteen women and twenty-three men; he agreed that "the hands may be divided into one or more squads just as they may deem best." These squads were allowed to "choose one or more of the best informed men as their foreman." B. F. Ward of Butts County, Georgia, struck a middle ground when he "divided them into squads and families, or let them make selections of their own co-workers," requiring only that the "heads of squads" be "good practical farmers."[45]

[42] *Joint Committee on Reconstruction*, p. 41, South Carolina; *BRFAL*, film M1027, roll 28, p. 385, Louisiana.

[43] Duncan Clinch Heyward, *Seed from Madagascar* (Chapel Hill, 1937), p. 157; *Southern Farmer*, vol. 2, 1868, p. 110, cited in Shlomowitz, "Transition," p. 114.

[44] *BRFAL*, film T142, roll 71.

[45] *BRFAL*, film M979, roll 41, "Labor Contract," Lafayette County, Arkansas, Jan. 1867, p. 176. Ibid., "Labor Contract," Ashley County, Arkansas, Jan. 1867, p. 486. *Report of Department of Agriculture, 1867*, p. 419.

Given their decisions to offer and accept collective share payments, both decentralization and the use of worker self-selection in the composition of work groups had benefits for both parties. Even the adoption of the procedure of subcontracting to an individual squad leader was a recourse to this principle of worker's self-selection, lending some credence to the conjecture of one student of the history of industrial organization that "to allow a certain amount of independence and self-determination to the subordinate workers in a great concern may be merely an intelligent bit of organisation from above."[46]

How often the planter recognized these advantages on his own is difficult to determine. In any case, when managerial foresight failed, the laborers exercised their new rights by demanding a voice in the determination of work group structure. It may be that complaints about gang labor and overseers on places like the William Mercer plantation, near Natchez, where the people demanded the "right to work in family squads under their own supervision," were solely attributable to a recognized demand for family tenancy. This supposition must be tempered by an understanding of the incentives problem under the collective contract and the fact that the family squad was generally composed of several families working jointly. It seems more reasonable to accept Ronald Davis's conclusion that during these early years the freedpeople had little recognition of what the share squad might become, but were attempting to find those working arrangements which would prove most remunerative. If so, such demands conform well to the obvious presence of the "free rider" in the testimony of a South Carolina plantation superintendent that "universally" the freedpeople "swear that they will not work in gang[s], i.e., all working the whole and all sharing alike."[47]

Even the existence of self-selecting work groups must not lead us to presuppose an immediate loss of supervision which, at least indirectly, worked in the employer's interest. An early description of the work routine of self-selecting share groups on south central Louisiana plantations provides the evidence:

> The modes of working them are varied. Some oversee and manage them, other[s] have them divided in small gangs with a trusty leading man at the head, and the rest are expected to keep up to him. On some places where they work for a share of the crop they are divided in this way and allowed to select some leader from among themselves.

It is a reasonable hypothesis that a similar organization was the outcome of an objection by a group of Port Royal freedpeople not to the presence of drivers, but to the continuance of the old drivers, whom they argued should become field hands while old field hands took their places.[48]

[46] Clapham, p. 178.

[47] Davis, p. 105. Pearson, p. 112. See Letter of John Murray Forbes, *Boston Advertiser*, June 10, 1862, cited in Rose, p. 141. *New York Tribune*, Jan. 14, 1867, p. 5 (North Carolina).

[48] BRFAL, film M1027, roll 28, Feb. 1866, St. Martin Parish, p. 172. Ibid., March 31, 1866, St. Mary Parish, p. 371. Also *Montgomery Daily Advertiser*, May 14, 1867, cited in Kolchin, p. 46. Pierce, "Report to Secretary Chase," in Frank Moore, p. 305.

The degree to which the planter could establish a disciplined and regular work pace rested upon the amount of authority vested in the squad leader. Examining the perspective of the planter leads to the conclusion that the early disappearance of the squad system was due to its failure, as an institution, to provide that discipline. If squads became autonomous, democratically organized units, there would probably have been little loss to planters in a complete reduction to a household production system. The reasons are not very difficult to ascertain. Once the workers gained the power to select their own foremen and co-workers, the authority vested in the leader came from the workers themselves. Under these conditions the planters' desire to maintain a well-marked line of cleavage between the work squads and their leaders was doomed to fall victim to a natural desire for a more egalitarian distribution of authority within the work squads. Although a detailed examination of this point must await the next chapter, it should be noted that the workers' desires were not without precedent. In early nineteenth-century England, large construction contractors often subcontracted much of their work to independent gangs, who, in the first half of that century, were still composed of two types. John Sharp, a small contractor, explained: "there are in some works what are called butty-gangs; there they are all alike, and one receives the money and shares it among the others. In other places one man takes the work and employs men and receives the benefit of it himself." The men, it was explained, "preferred the butty-gang, where all were equal"; as one witness articulated disputes over earnings could be arbitrated "with an odd blow on the head and a quart of beer or two extra."

Whether the industry was mining, shipbuilding, or construction, we find that the presence of these democratically organized work groups invariably led to the principle of self-selection of co-workers. David F. Schloss, near the end of the last century, listed the fundamental ideas of cooperative work:

(A) that each group of workers is to be associated by their own free choice, (B) that these associates shall work under a leader elected and removable by themselves, and (C) that the collective remuneration of the labor performed by the group shall be divided among all its members (including this leader) in such a manner as shall be arranged upon principles recognized as equitable by the associates themselves.[49]

[49] Clapham, vol. 2, p. 408. David F. Schloss, *Methods of Industrial Remuneration*, 2nd ed. (London, 1894). These collectives were an early form of the butty-gang. Why the men preferred them is explained by a description of the subcontracting system which the butty-gang was destined to become:

> Such a gang boss would then hire 15 to 20 men on day wages; and if he could keep them up to a rapid rate of work, he could make a very profitable business of working the coal by contract. His main object being to keep the men going continuously, working the coal as fast as possible, he would also work in hewing coal himself, and often with prodigious energy whilst at it, so as to set the pace. His success depended very largely on his capacity for "driving" his men. In every coalfield the system was objected to by the miners; and by a series of strikes it was abolished in one part of the country after another.

H. Stanley Jevons, *The British Coal Trade* (London, 1915, reprint 1969), p. 334.

From the seminal treatise of the Webbs to David Montgomery's analysis of the attempts for "worker's control in America," these ideals have permeated the thought of serious students of the laboring classes. Their early presence among the freedpeople casts doubt upon Sir John Clapham's inference that they are attributable to "claims rooted deep in the traditions of an old and proud skilled trade."[50] Rather, the Freedmen's preferences imply strongly that the leading ideas of cooperative work arise very naturally when workers are free to voice their preferences.

After their initial introduction, the picture that emerges is that share squads were primarily composed of self-selecting workers who, pooling their individual resources, elected their own leaders through some democratic process. Under these circumstances, the leader's primary responsibility seems to have been as spokesman for the group. Journalist Edward King's description of squad organization on cotton plantations in Louisiana and Mississippi during the middle 1870s conforms completely with the arrangement made by two squads of white workers on a Louisiana sugar plantation in 1874, as reported by the *Daily Picayune:* "One of their number was placed at the head, with power of attorney from the others, authorizing him the right to make all purchases, negotiations, etc., and transact all business pertaining to the crop interest." Under this cooperative system the leader ceased to be an agent of the planter. The consequence was that the squad, as a group, was able to "work about as they please" with "greater independence than the day-laborer usually enjoys elsewhere."[51]

Planters lost power when squad leaders ceased to be their agents; one attempt at circumventing this was to generate competition between the various squads on a plantation by offering a bonus to the most productive group. One historian, who described the small collective as a transitional stage between the gang system and family tenancy, went so far as to characterize it as an organization of "competitive groups." Given the extent of the practice, this characterization may be quite descriptive. A Texan who divided his share laborers into "sections" and promoted competition through the use of a "premium system" tried this strategy in 1866. The premium was probably similar to that in the system of an Alabamian "who hired sixty freedmen at moderate wages, divided them into six gangs of ten each, and offered a premium of three hundred dollars to the gang which should produce the greatest number of bales." Some planters followed the practice of an Edgefield, South Carolina, planter who, in 1871, divided his hands "into squads of three" men. "The General," it was reported, "stimulates them to work by getting up a rivalry between different squads."[52]

[50] Sidney and Beatrice Webb, *Industrial Democracy* (London, 1902). David Montgomery, *Workers' Control in America* (Cambridge, Mass., 1979). Clapham, vol. 2, p. 178.

[51] Edward King, *The Southern States of North America* (London, 1875), p. 273. *New Orleans Daily Picayune*, March 22, 1874. *Southern Cultivator*, 1869, p. 59. Nordhoff, pp. 64, 71. Somers, p. 147. M. S. Wayne, pp. 196–203. R. L. F. Davis, pp. 114–16.

[52] Sheldon Van Auken, "A Century of the Southern Plantation," *Virginia Magazine of History and Biography*, vol. 59, July 1950, pp. 345–65. *New Orleans Daily Crescent*, Feb. 24,

We may speculate that for the planter or merchant bestowing the advances, the strategy of placing special financial responsibility upon the squad leader would be attractive if the leader had some special status among the squad members. For instance, the squad often possessed working implements and stock of its own, and since the assets of individual squad members would seldom be identical, it would be natural for the landlord or merchant creditor to prefer to deal directly with the squad member who had the most to lose and could therefore be expected to exert considerable pressure upon his colleagues to give a work performance at least as great as his own.[53] Though the planter desired that the squad leader have as much authority as possible over his co-workers, and the squad member with the most to lose, and perhaps gain, would have an incentive to see that the group worked hard, it must not be supposed that no friction over the matter of authority within the squad existed. No matter how unequal the inputs to production provided by various members might be, the squad system was still a cooperative arrangement and distributional and incentive problems would invariably arise. The universal method of disposing of this problem seems to be the practice of creating work squads under conditions where natural relationships provide a squad leader with the most authoritarian status possible. Building squads around a core group related through kinship bonds provides such relations. In squads composed solely of members of an extended family the chosen leader would often be a paternal figure: the father over his children, an uncle or aunt over nephews and nieces, or even an eldest brother over his siblings.

Charles Stearns, who counseled that squads worked best when composed of "all members of the same family or related to each other," was extremely aware of the virtues of an authoritarian paternal figure as a motivating factor in the interests of labor discipline: "The industrious ones have no notion of working hard, while others are listlessly performing their task; and I cannot possibly blame them. One man, this year, felt obliged to give his own son a tremendous beating, for not performing his share of the labor." The obligation was often contractually specified. Disciplinary instructions given to parents contracting for a single family unit sometimes went well beyond those required of ordinary foremen. Thomas Ferguson, who sharecropped with his stepson, son, and two daughters, agreed to "control [his] family and make them work and make them behave themselves." Jack and Leonora Seals obligated "themselves to govern their children strictly, and to enforce due obedience from them to said employer." The perceived advantages of a paternal figure are made clear in a report from the manager of a Madison Parish, Louisiana, plantation in

---

1866, p. 5, quoting the *Texas Intelligence*. Trowbridge, p. 448. *New York Tribune*, May 31, 1871, cited in Shlomowitz, "Transition," p. 117. Barrow, "A Georgia Plantation," p. 831. *Report of the Department of Agriculture, 1867*, p. 419.

53 *BRFAL*, film M826, roll 30, p. 679. Ibid., entry 2121, "Complaints—Brookhaven, Mississippi," 1868, p. 187.

1878 explaining to his employer why a squad of two young men had been dissolved:

> Jim and York had to be sepeated [*sic*], they have gathered very little over their rent. Youngsters like them have to have a leader. Jim is located between Abe Thoms and Ashwood. York is with Bob Miles who raised him. Zed I don't know what will become of him, but some settled man, or tenant will take him.[54]

In addition to its inherent advantages for promoting internal discipline, the squad based on bonds of kinship offered other securities based on mutual trust and the possession of intimate information about one's colleagues. These reasons are sufficient to explain Reily and Rebecca Dixon's parental concern for the welfare of their three teenage sons, who were "to be allowed to work in the same squad with Daniel Webster their uncle." The ability to recruit needed labor during peak periods, such as harvest time, when the demand for labor would exceed the size of the squad, was an advantage impressed upon the mind of Robert Somers. "A strong family group, who can attach other labor, and bring odd hands to work at proper seasons, makes a choice, if not always attainable, nucleus of 'a squad'," he observed. Planters like James A. Gillespie, who "contracted with squads usually centering around a family unit, often including members distantly related, or simply people who got along together fairly well,"

---

[54] Stearns, p. 517. *BRFAL*, film M826, roll 50, "Labor Contract," Oktibbeha County, Mississippi, p. 932. Ibid., "Labor Contract," Monroe County, Mississippi, p. 647. M. S. Wayne, p. 204. Drawing on his fieldwork studying the harvesting of sugarcane with migrant labor by sugar factories in India during the 1970s Jan C. Breman describes the organization of labor:

> The seasonal migrants are taken on exclusively in *koytas* [three-person work teams]. . . . workers who make up this team usually belong to the same household. . . . the agreement made is only with the head, the actual cutter. In principle the migrant workers can offer their services to different *mukadams* and take up the most attractive offer. However, in practice the broker operates as far as possible in a familiar area and prefers to take on workers who are already known to him. Some of these will be relatives, neighbors. . . . the workers also prefer to go with a *mukadam* who is known to them; in strange surroundings they are wholly dependent upon him, both for their sustenance during the harvest and for the payment made at the end of it. . . . Their choice is determined by the amount of trust in the *mukadam*; the reputation of the factory . . ., and their acquaintance with other *koytas*. . . . Apart from the risk the mukadam runs with the recruitment, he must also be able to rely on those he contacts to perform adequately in the harvesting. . . . The factory speaks of a bond of trust they suppose to exist between the *mukadam* and his *koytas* but this is an altogether too idyllic picture of the way in which the former exerts his authority. The pressure which he brings to bear is chiefly of a noncontractual nature and is mostly or wholly ineffective if a personal relationship is lacking. . . . Another excellent way to safeguard against nonfulfillment is to make *koytas* liable on their side. The migrant workers are therefore made to stand as guarantee, for each other or for their close relatives for the cash advanced to them.

(Jan C. Breman, "Seasonal Migration and Cooperative Capitalism: The Crushing of Cane and Labor by the Sugar Factories of Bardoli, South Gujarat," *ADC-ICRISAT Conference on Adjustment Mechanisms of Rural Labor Markets in Developing Areas*, Hyderabad, India, Aug. 1979). Also compare Nef, p. 427, with the financial trust problems indicated below, note 58.

were submitting to the demands of the Negroes as well as their own profit motives.[55]

The premium placed upon kinship units by both the suppliers and demanders of labor provided a stimulus for the dissolution of the squad system. Among the odd hands called to work at proper seasons, most conspicuous was the freedwoman. The plethora of testimony complaining about the absence of black women from *regular* fieldwork strongly indicates that laboring squads in the cotton fields must have been overwhelmingly composed of males.[56] There was perhaps one important exception to the disappearance of the black woman in the field with hoe in hand or behind the plough. The anonymous southerner writing in *Galaxy Magazine* in 1871 noted sardonically:

> it has been stated that many planters possess more open land than they have force to cultivate. We have also seen that many women do not engage to work regularly in the field, and it so happens that freedmen sometimes own a horse. When these circumstances are concurrent it occurs, maybe, that the proprietor rents land to his own hired freedmen, receiving from him one-third of the corn and one-fourth of the cotton produced, the labor being performed by the freedman's wife and horse.

Charles Nordhoff, referring to the North Carolina freedpeople, observed, "where the Negro works for wages, he tries to keep his wife at home. If he rents land, or plants on shares, the wife and children help him in the field."[57]

Where the laborer worked a plot of land on his own accord, the profit incentive and the sociology of work relations called for a greater supply of labor from the entire family. This response could not have escaped the notice of all planters indefinitely. If sharecropping on the family plan induced women to work in the fields more frequently, it was a method of increasing the scarce labor supply. The possibility was noted by an observer in 1869:

> in the end we think it [sharecropping] will actually increase the amount of labor, as the man who is cultivating a number of acres for himself, in part, will command the services of his wife and children in case of need. In this way a

[55] *BRFAL*, film T142, roll 72, "Labor Contract between John Brand and Reily and Rebecca Dixon for their three children," Phillips County, Arkansas, Jan. 20, 1866. Somers, p. 120. R. L. F. Davis, p. 114. *The Farmer*, Nov. 1866, pp. 421–22, cited in Shlomowitz, "Transition," p. 111. Rose, *Rehearsal*, p. 359.

[56] For a fair sampling of this testimony see Loring and Atkinson, pp. 4, 13, 14, 15, 16, 20, 75, 92, 106, 115, 137; Nordhoff, pp. 38, 72. Only 25 percent of the laborers in the 1868 sample of share squads were female. For a systematic discussion of the contraction of the black labor supply see Chapter 12.

[57] "Agricultural Labor at the South," *Galaxy*, p. 333. Nordhoff, p. 99; also *New Orleans Daily Times Picayune*, Jan. 15, 1874; Stearns, pp. 517–18. *New York Times*, Jan. 17, 1870, p. 3, quoting *Charleston News* of Jan. 11. *BRFAL*, film M826, roll 50, "Labor Contract between J. Colbert and Joe Blythe—Harrison Atkinson," Noxubee County, Mississippi, Jan. 1, 1867. Ibid., film M869, roll 43, "Record of Labor Contract," Monck's Corner, South Carolina, Feb. 2, 1868. Fleming, p. 722.

large force of laborers, now withdrawn from this department of industry, will be returned.[58]

This had to be weighed against the possible losses due to less organization and a dispersal of the management talent of the best squad leaders. But in the end this objection may not have caused much of a problem. The very competitiveness of the agricultural economy meant that not much capital was required to set oneself up as a nominally independent renter or small landholder. In reasonably prosperous years the lien law would enable just those squad leaders with the most talent and industry to break from the cooperative squad system, where the fruit of their work was shared not only with the planter but possibly with less able workpeople who refused to be driven in democratic squads.[59]

If the David Barrow plantations are again taken as a benchmark, it can confidently be stated that by the year 1876 the dominance of nuclear family tenancy had been established. The assertion is consistent with the sketchy evidence that has been assembled for the presence of squads after 1872. The strongest evidence comes from the pens of two traveling journalists touring the Mississippi Valley. Charles Nordoff reported in 1875 that Louisiana share laborers "usually work in Squads." A year earlier Edward King offered the following description of agrarian organization in the Natchez area: "There are many plans of working large plantations now in vogue, and sometimes the various systems are all in operation on the same tract. Sometimes the plantation is leased to 'squads,' as they are called."[60]

Unfortunately, these remarks give no definite information as to the numerical significance of the squad organization. Edward King's words, and his later discussion, imply that by 1874 squads were neither a curiosity nor the obviously dominant method of organizing labor. Nordhoff's remarks imply dominance, but the quotation must be viewed with care if taken as evidence. The problem is that some planters used the term "squad" quite loosely. In the antebellum period it was apparently quite common to refer to any work unit smaller than the planter's conceptualization of the proper size for a gang as a squad. That such terminological practice continued into the postbellum period is evidenced by the existence of labor contracts where a wife and husband, working alone for a share of the crop, were explicitly denoted a "squad." That the term was used interchangeably for the "squad system" and the family unit is shown by the remarks of a Mississippi planter who explained that "under the share plan, each man wants his land and team to work his family or squad to himself; thus on a

[58] *Hunt's Merchant Magazine*, 1869, p. 273. Brooks, p. 47.

[59] The phenomenon of the most desirable members of a group contract entered into by heterogeneous individuals self-selecting themselves out of the contract and leaving the least desirable is called "adverse selection." It occurs in diverse markets, such as the selling of insurance, used cars, and the buying of labor. For a pathbreaking discussion of this problem see George Akerlof, "The Market for Lemons: Qualitative Uncertainty and the Market Mechanism," *Quarterly Journal of Economics*, vol. 84 (1970).

[60] See Harris, pp. 253–54. Nordhoff, p. 72. King, p. 273.

plantation of any size there might be a half dozen squads." At any rate, at this stage of development the squad had become something very different from what most employers had intended, so different that the use of one name to denote all of the various work organizations discussed here is at best confusing. In fact, contemporaries referred to these work groups by a number of alternative names, each of which is much more descriptively appropriate for the democratic workers' collective. Various groups were referred to as companies, clubs, and associations. It is probably no accident that the smaller work groups operating on the Barrow plantation, without foremen, during the 1870s were denoted "companies" and not "squads," as had been the practice before the decentralizing process began. Probably, through force of habit, the most widely used term remained *squad*.[61]

The squad proper continued to be used through the Reconstruction period, but references to it after 1875 tend to be framed in a manner which suggests they were becoming unusual. Such is the case when freedman Henry Adams, testifying to the unjust treatment of laborers in 1878, emphasized with clear disapprobation that the landlords on one plantation "were working [the] families in conjunction" upon shares.[62]

That some such cases should exist is not at all surprising. The speed of adjustment varied across plantations. Samuel Postlewaite, in 1869, ruefully wrote to his uncle that "this was the last year" the freedpeople would be willing to "work in gangs." His uncle, James Gillespie, resisted the laborers' requests for separate houses outside the old slave quarters until 1874. Such persistence testifies to the planter class's basic agreement with South Carolinian Henry Ravenel that the antebellum "organization of labor" was best for "preserving discipline." Ravenel, like many others, resisted as long as he could the Negroes' insistence that under the new regime "they would not want a driver or overseer," and desired a "piece of land to work for their own use."[63]

This chapter has examined a central theme pertaining to the fragmentation of the postbellum plantation. But the socioeconomics of the "free rider," inasmuch as it exhibits a strong tendency towards individualism among the freedpeople, obscures concurrent developments within the squad system which led to the freedpeople's attempt to redefine agrarian work organization within the collective consciousness of the work squad. Here I have only considered these issues of democratic worker control peripherally. In the next chapters it will be argued that this attempt to reformulate the basis on which the plantation was organized played an important role in

[61] See the switch in terminology in the quotation from J. William Harris cited in note 4 above.

[62] Phillips, *American Negro Slavery*, pp. 55, 60, 237, 268. *Southern Cultivator*, 1867, pp. 461–62, in Shlomowitz, *Senate Report No. 693*, 46th Cong., 2nd sess., vol. 7, p. 117.

[63] M. S. Wayne, pp. 195, 169–70. Childs, *Private Journal of Ravenel*, pp. 212, 216. See Thomas, "Plantations in Transition," for Thomas S. Watson's 1867 refusal to give the Negroes a form of sharecropping, p. 227. *New York Times*, Jan. 7, 1870, p. 2, on the resistance of Georgia planters to *family* sharecropping.

providing the landlords a reason to dissolve the squad system. The so-called compromise that led to the diffusion of family sharecropping was the product of a long and involved process. To ignore the economic factors in this struggle would be a serious mistake. Yet the motivating presence of ideology and of the sociology of the family may, in the final analysis, have been the crucial determinant. If it is true that one of the freedpeople's primary sources of market power was the postwar reduction in the supply of female labor, the then Governor William Bradford's description of the effect of abandoning the gang labor system at Plymouth plantation may contain more than one lesson for American historians:

> they begane to thinke how they might raise as much corne as they could, and obtaine a beter crope then they had done, that they might not still thus languish in miserie. At length, after much debate of things, the Govr (with ye advise of ye cheefest amongst them) gave way that they should set corne every man for his owne perticuler, and in that regard trust to them selves; in all other things to goe on in ye generall way as before. And so assigned to every family a parcell of land, according to the proportion of their number for that end, only for present use (but made no devission for inheritance), and ranged all boys and youth under some familie. This had very good success; for it made all hands very industrious, so as much more corne was planted then other waise would have bene by any means ye Govr or any other could use, and saved him a great deall of trouble, and gave farr better contente. The women now wente willingly into ye feild, and tooke their little-ons with them to set corne, which before would aledg weaknes, and inabilitie; whom to have compelled would have bene thought great tiranie and oppression.

The Reverend James Lynch, when making his stump-speaking tour of the southern states during the first few months of freedom, was probably unaware of how deeply his words penetrated the national fabric when he declared, "we have met here to impress upon the white men of Tennessee, of the U.S. and of the world, that we are part and parcel of the American Republic." [64]

[64] Governor William Bradford, *The History of Plymouth Plantation*, in Walter Blair, Theodore Hornberger, Randall Stewart, James E. Miller, Jr., eds., *The Literature of the United States*, 3rd ed. (Glenview, Ill., 1953, 1971), p. 64. *Colored Tennessean*, Aug. 11, 1865. Ransom and Sutch base their entire analysis of the development of the postbellum southern economy upon the large reduction in labor supplied by women and children. In Chapter 12, I make a quantitative and qualitative assessment of this phenomenon.

11

*We have been attempting to manage free labor upon the old plan-
tation system, and when we have failed we have been satisfied to
ascribe the failure to defective labor or defective systems.*

W. H. Evans

*These people seemed to grasp intuitively the most complicated prob-
lems, and the most advanced doctrines in the great questions as to the
remuneration of labor. Only just emancipated, they at once take
ground, to which the laborers of the old world seem to have been
struggling up through all the centuries since the abolition of serfdom.*

Harry Hammond

The Labor-Managed Economy

At the close of the year 1871 noted South Carolina planter and agricultural
journalist D. Wyatt Aiken appraised the state of southern agriculture by
writing that "no question has so perplexed the southern planter for the past
six years as that of labor." The practical difficulty as seen by Aiken was

> that neither capital nor any other power has been able to control labor in the
> South: They [Negroes] have demonstrated to us the power of concert and the
> effect of union. Are we too blunt to learn? If not we should at once set about
> profiting by their example. This can be most effectually done by adopting
> throughout the cotton states the same general system of labor contracts.[1]

It would be easy to interpret Aiken's plea for agricultural employers to
combine by adopting "the same general system of labor contracts" as
another example of a planter seeking a means to extract labor at subsistence
pay. To do so would be to miss much of importance. Aiken was in fact call-
ing not for a cartel-enforced minimum rate of earnings to labor, but for
uniformity in the *form* of payment and organization of labor. An unwritten
aspect of the postbellum period is the attempt by many planters, after a few
years of unsuccessful adjustment to free labor, to come to grips with the

---

[1] D. Wyatt Aiken, "Labor Contracts," *Rural Carolinian*, vol. 3, Dec. 1871, no. 3, p. 113.

new social and industrial relations made imperative by emancipation. An antebellum slaveholder whose values were rooted in the southern code of paternalism and the moral and intellectual superiority of white men, Aiken was in many ways typical of the set of planters who in making this adjustment sought also to elevate the economic welfare of the laborers.

A respected antebellum planter, Aiken had served as a colonel in the Confederate Army. One of the publishers of the *Rural Carolinian*, he was a well-known authority on agriculture and a leader in the attempt to promote the adoption of scientific methods in the practice of southern agriculture. Aiken objected strongly to the presence of Negroes in South Carolina politics. In a letter to the *Abbeville Banner* he referred to the 1868 South Carolina Constitutional Convention as "that mongrel convention" and called upon whites to actively resist federal reconstruction and Negro equality: "Does policy demand that we should still further crouch before the roaring lion? Does not self-preservation goad us to beard him in his den?" The Freedmen's Bureau agent in Aiken's district, whom Aiken had publicly challenged to a duel, suspected the colonel of being involved in Ku Klux Klan activities. Nevertheless, in 1870 Aiken became one of the first South Carolinians to successfully operate his plantation under the system of renting individual tenancies to families of freedmen for a fixed quantity of cotton per farm.[2]

Earlier in 1871 a gentleman of Charleston had expressed opinions similar to Aiken's. The remedy for the labor problem, as this writer saw it, was "to be found in a well regulated system of labor, and in a judicious organization among planters and landowners" that would purchase a "just and equitable system of free labor." The writer wistfully admired the advantages of the slave system but, along with Aiken, recognized that free labor required new management techniques. Heeding the advice of the *New Orleans Tribune*, he enquired of his fellow planters, "Cannot freedmen be organized and disciplined as well as slaves? Is not the dollar as potent as the lash? The belly as tender as the back?"[3]

Believing that a free labor system must recognize the dollar, D. Wyatt Aiken advised that the most important element to be recognized by employers was that "all contracts are or should be reciprocal interests, and the one most advantageous to the employer should be most remunerative to the employee, and *vice versa*."[4] Capitalists, the profit-maximizing time-thrift kind, were seldom heard stating such heresies. But at the time that Aiken wrote there were traces of a movement taking place among sagacious northern industrialists who had imported from Europe what were then

[2] *BRFAL*, film M869, roll 34, pp. 0148–51. *Rural Carolinian*, vol. 3, Dec. 1871, no. 3, pp. 114–15. One of Aiken's activities as a Red Shirt in the South Carolina elections of 1876 is described by Richard T. Greener, a black professor at the University of South Carolina and the first black known to graduate from Harvard College, in Dorothy Sterling, *The Trouble They Seen* (New York, 1976), pp. 466–67.

[3] *Rural Carolinian*, vol. 2, July 1871, no. 10, p. 572.

[4] Aiken, "Labor Contracts," pp. 113–14.

rather radical ideas about labor management. Aiken, with the enunciation of his principle, had set forth the primary basis of this new philosophy.

In 1886 a relatively small group of men gathered in New York City to discuss, among other problems, the labor question. Few of these men are now remembered, and none are as famous as the commercial and financial industrialists who amassed fortunes in the late nineteenth century. However, these men were as much a contributing factor to nineteenth-century economic growth in America as the industrialists were. The American Society of Mechanical Engineers contained many of the men responsible for the design and implementation of the labor management and production schedules of the manufacturing concerns throughout the country. Henry R. Towne, one of the society's most esteemed members, spoke to the issue of labor conflict, arguing that "the final and most perfect solution" of the labor relations problem must be found in "a system in which the interests of the employer and employee, not only in the results of labor, but in the entire economy of production, are harmonized and united."[5] There are several aspects of this labor question which will interest us, but to these employers productive efficiency and a proper system of incentives for labor to provide that efficiency were paramount.

Whether he worked in agriculture or manufacturing, in America or Europe, the inefficient, low-quality service provided by the wage laborer was heavily indicted by the employers. A few examples reveal the tenor of thought. To D. Wyatt Aiken the shortage of labor was a result of the relaxed discipline of free labor: "Our labor though abundant, has practically been scarce because inefficient." "The innate laziness of the negro, as a general rule, induces him to expend just so much of his muscle as in *his* judgement repays his employer. If paid five or twenty dollars per month, he is satisfied if his work returns just that much for the investment, never dreaming that increased wages should improve the quality as well as the quantity of his labor." John T. Hawkins, president of the Campbell Printing Press and Manufacturing Company in Taunton, Massachusetts, was in perfect agreement with Aiken. "The one undeniable fact that the average workman or laborer today considers it to his gain that he does the least instead of the most he can do in a given time, is one of the fundamental causes of the present labor troubles." The reputation of the nineteenth-century agricultural worker in Europe fared no better. A German writer expressed this universal complaint: "Every practical farmer knows how imperfectly agricultural work is done by hirelings of all sorts, and how little what goes by the name of superintendance is able actually to effect in securing good execution of work."[6] We might produce further examples, but the point has been made elsewhere as well as here.

[5] *Transactions of the American Society of Mechanical Engineers*, vol. 8, 1886, p. 293.

[6] Aiken, "Labor Contracts," p. 114. *Transactions*, p. 277. German writer quoted in Sedley Taylor, *Profit-Sharing* (London, 1884), p. 94. The engineers were as a group extremely vocal in their concern over this problem. See *Transactions*, pp. 269–94 and 630–62. E. L. Godkin, "The Labor Crisis," *North American Review*, vol. 216, 1867, pp. 210–12.

E. P. Thompson has reminded us that this preoccuption of employers with the work intensity of laborers is a product of the development of "a greater sense of time-thrift among the improving capitalist employers."[7] In the early stages of market-oriented enterprise the employers' time-thrift manifested itself by largely coercive practices which were necessary to accommodate the low wages they insisted were necessary for profitable business. Slavery, of course, is the paradigm form of this manifestation. But the use of physical coercion necessitates sanction from the state. This dictum was being impressed upon the minds of those southern planters who were restrained by the Freedmen's Bureau, much as it had been taught to English textile mill owners who, earlier in the century, appealed to their government for the power to indenture orphan children because "free laborers [could not] be obtained to perform the night-work, but upon very disadvantageous 'terms to the manufacturers."[8] When the employers no longer have recourse to blatent forms of coercion it becomes imperative that the work discipline they deem necessary to profitable enterprise come from the psychological needs of the workers themselves. So much has been written upon this subject that it might seem we should be able to pass over it quickly, but it turns out that with the problem as we now pose it there exist some important subtleties which warrent close attention.

For the rising bourgeoisie it is religioin and in particular the Protestant ethic which is given a large measure of credit for effecting this inner compulsion for work and accumulation. But E. P. Thompson, writing of the English working class, has rightly asked how a religious appeal for inner discipline could be attractive to a "forming proletariat." What in Protestantism, in its relationship to capitalist accumulation, could appeal to a group "whose experience at work and in their communities favored collectivist rather than individualist values, and whose frugality, discipline or aquisitive virtues brought profit to their masters rather than success to themselves?"[9] In 1865, the former masters of American slaves were well aware of the relevance of Thompson's question. The antebellum slaveholder had been no stranger to the theoretical possibilities of "the transforming power of the cross." Generations of masters had been inspired with the idea of eliciting voluntary obedience and faithful work from slaves through the instrument of religion. "They had a white man," recalled an ex-slave, "that would come over every fourth Sunday and preach to us. He would say, 'Be honest, don't steal, and obey your marster and mistress.' That was all the preaching we had down in Mississippi."

The frequency of services, and the religious sincerity of the masters, varied throughout the South, although the gospel delivered to the slaves seldom changed. But the elasticity inherent in the interpretation of Scrip-

[7] E. P. Thompson, "Time, Work-Discipline, and Industrial Capitalism," *Past and Present*, no. 38, Dec. 1967, p. 78.

[8] *Reports of the Society for Bettering the Condition and Increasing the Comforts of the Poor*, vol. 4 (London, 1805), Appendix 1, p. 2.

[9] Thompson, *Making of the English Working Class*, p. 356.

ture, combined with important social and personal needs of the individuals in a slave community that remained antagonistic to pursuit of individual gain, led, as has been convincingly argued by recent historical scholarship, to a religion that diverged widely from the intentions of the masters.[10] What this suggests is that spiritual forces alone may be insufficient without the added inducement of material reward. But even further than this, material reward too will be insufficient without the relevant community's sanction of behavior oriented toward this goal. Thus a slave who directed his best efforts toward material reward offered by the master had to align himself with that master's interests and would invariably run the risk of being labeled a "white folk's nigger." Although such deviants certainly existed, this behavior was not the norm in slave society. However, freedpeople are not slaves, and the shackles broken by that fact may stem from within a slave community as well as without.[11] The freedman who stood to gain by his own efforts without costs to the rest of the community might become much more receptive to the ideology of individualism.

To recognize that the effect of religion upon the work ethic of the laboring classes is dulled by the inability of the laborer to profit by increased work effort is to take the first step toward understanding what D. Wyatt Aiken and Henry R. Towne meant by "reciprocal interests" between employees and employers whose relationship would be "harmonized and united." If the virtues embraced by Protestantism were most beneficial and effective when applied to the individual entrepreneur, then perhaps the laborers could be made into entrepreneurs in their own limited right.

Payment by piece rates is the employer's first step in this direction. A mid-nineteenth-century English pamphleteer, whom Marx trenchantly labeled an apostate, provides a picture of payment by the piece that would have earned the approbation of many an employer:

> The system of piece-work illustrates an epoch in the history of the working-man; it is *halfway between* the position of the mere day-laborer depending upon the will of the capitalist and the co-operative artisan, who in the not distant future promises to combine the artisan and the capitalist in his own person. *Piece-workers are in fact their own masters*, even whilst working upon the capital of the employer.

For Henry Towne, "the theoretical correctness" of payment by a piece-rate system was founded in the principle that "every man is working for himself"; "that," he counseled, "is the ideal condition in any industrial establishment."[12] Workers who were their own masters, it was not doubted,

---

[10] Rawick, *American Slave*, vol. 19, p. 198. On this point see Blassingame, *The Slave Community*; Genovese, *Roll Jordan Roll*; and Lawrence W. Levine, *Black Culture and Black Conciousness* (New York, 1977).

[11] According to Weber, one of Protestantism's important contributions to the development of capitalism was its part in the development of new social mores which gave communal sanction to profit making.

[12] John Watts, *Trade Societies and Strikes, Machinery, and Co-operative Societies* (Manchester, 1865), quoted in Karl Marx, *Capital*, vol. 1 (New York, 1967), p. 551 (my emphasis). *Transactions*, p. 644.

would be not merely susceptible to but in perfect consonance with the principles and ideas of the industrialists themselves.

Payment by the piece, despite its clear advantages over mere time wages, also had some disadvantages. Let us consult Mr. Towne again: "In piece work, the workman is interested in making his own efficiency as high as possible, but has no interest whatever in the economy with which he uses the plant. On the contrary, his interests and those of the manufacturer are opposed to one another. If a workman, by the wasteful use of tools, can increase his product and wages, he naturally does it. There is obviously a point where the loss from this wastefulness will more than offset the increased product. Piece-work, pure and simple, has, therefore, this element of antagonism between workman and employer." Compare Charles Babbage, speaking of European labor in 1832:

> A most erroneous and unfortunate opinion prevails amongst workmen in many manufacturing countries, that their own interest and that of their employers are at variance. The consequences are, that valuable machinery is sometimes neglected, and even privately injured, that new improvements introduced by the masters do not receive a fair trial, and that the talents and observations of the workmen are not directed to the improvement of the processes in which they are employed.[13]

The southern analogue to Mr. Towne's argument was manifested in the frequent complaint that share-laborers were wasteful of tools and abusive to work animals. A bureau agent who witnessed the problem gave a more favorable explanation than did most planters when he reported that the freedmen in their "earnestness" to work were "working down their [planters'] stock."

The similarity of results was a consequence of the share system's being a generalized version of payment by piece rates. The laborers received a fraction of the price of each pound of cotton, bushel of rice, or hogshead of sugar produced. For the engineers and the most efficient and progressive planters the solution of these problems was in a system of profit sharing between employers and workers. "In no other way," cautioned Towne, "can you so completely identify the workman with the proprietor, and interest him, not merely in the product of his own work, but in the product of his fellows around him and in the entire economy of the place, the economy of material and everything which tends to reduce cost." Charles Babbage advised that "it would be of great importance, if, in every large establishment the mode of payment could be so arranged, that every person employed should derive advantage from the success of the whole; and that

---

[13] *Transactions*, pp. 645–46. Charles Babbage, *Economy of Machinery and Manufactures* (London, 1835), p. 250. Babbage, a Cambridge professor of mathematics, is recognized by many as the father of ideas which led to the modern computer. Work on his own computer, which he called a "calculating-engine," induced him to visit, over a ten-year period, many factories in England and on the Continent in order to become familiar with contemporary mechanical arts. That experience led to the writing of this book, which in many ways is remarkable.

the profits of each individual should advance, as the factory itself produced profit."[14]

The admission that the interests of workmen and employers were not only not in harmony but were opposing was a break from an American ideological tradition, which for years had managed to ignore a passage in Adam Smith's *Wealth of Nations* which indicated that the opposing interests of labor and capital might hinder the "invisible hand's" performance in maintaining free markets. Republican social philosophers such as E. L. Godkin were especially concerned that this conflict would lead the workmen "to consider themselves a class apart, with rights and interests opposed to or different from the rest of the community." Godkin saw the inevitable triumph of the employers as a dangerous threat to democratic institutions.[15]

A generic conceptualization of this problem had been anticipated by the English utilitarians. In recognizing the possibility that the egoistic actions of the individuals composing a state might never harmonize through any natural process, the utilitarians "argued that in the interest of individuals the interest of the individual must be identified with the general interest." This precept was called "the principle of the artificial identification of interests." It is from this perspective that Republicans like Godkin viewed the problem and offered profit sharing as its perfect solution:

> We need besides this, to deliver us from the dangers to which the traditions of feudalism and the forcing system [wage labor] have exposed us, the elevation of the working classes from the condition of hired laborers, toiling without other aim than to do as little as possible, and without other reward than fixed weekly wages, into that of partners dependent for the amount of their compensation on the amount of their immediate production, and stimulated by self-interest into diligence and carefulness, and into the study and comprehension of the whole industrial process of the laws which regulate the relations of labor and capital, production and distribution,—or in other words, into playing in society the part of men, and not machines.

To argue that profit sharing would identify the interests of employer and employee for the common good did not supply a method by which artificial identification could be implemented. In general this would be a problem for the legislature, but Godkin deemed legislative action inappropriate because in this instance it violated the principle of free contract. The only alternative was moral suasion. Employees and employers would have to be convinced that the proposed common interest was also in their individual interest.[16] Theoretically speaking, profit sharing should have been successful. In practice the identification of interest it inspired worked too well.

Profit sharing failed to become the panacea that its advocates envisioned. One reason for this may be the motives of the employers themselves. While it is essential to see that those employers who were willing to try this system

14 *BRFAL*, film M1027, roll 28, p. 395. *Transactions*, p. 290. Babbage, p. 251.

15 Halévy, *The Growth of Philosophical Radicalism*, p. 16. Godkin, p. 210.

16 Halévy, p. 17. Godkin, pp. 212–14. See Chapter 6 for further discussion of Godkin's views.

were among the most benevolent, we must not lose sight of the fact that profit sharing almost invariably carried with it an underlying paternalism that in practice severely limited the impact of this benevolence. It would be an error to ascribe to this search for the optimal labor payment system any intent upon the part of employers to make independent economic agents of the laborers. To the employer the optimal payment system was the one which maximized employer profits. Consider the statements of the employers themselves. In July 1869, A. S. Cameron & Company, manufacturers of steam pumping machinery in New York City, adopted a plan of profit sharing to supplement fixed wages. In explaining the plan to his employees Mr. Cameron was quick to point out that the "system of co-operation" was adopted as "a practical business movement, and not as a charitable one." After an explanation of the method of sharing the motivation for profit sharing was revealed:

> What we expect in return is that each person will work as though the establishment belonged to himself, and see that each of his fellows does the same—saving every minute of time, and every speck of property. We most earnestly desire that all lazy persons shall be pointed out to us for prompt dismissal, so that those who work hard and honestly to advance their interests will not be called upon at the end of the year to divide the fruits of their industry with the sluggards.

Another American example was the Peace Dale Manufacturing Company of Peace Dale, Rhode Island. A real innovator, Peace Dale was the first American woolen manufacturer to successfully operate with power looms. In 1878 the company initiated a plan to divide "surplus profits" among its employees as an addition to wages earned. The laborers were enjoined to eliminate waste and develop thrift and industry. After a few years of mediocre results the men were reindoctrinated with the conditions of the compact: "Every weaver who makes a mispick, every burler who slights her work, every spinner who makes a needless knot, in short, every person who makes waste of any kind, of course, makes the amount to be divided smaller, by making a loss to the concern; and we think a manifest improvement is evident."

The message was similar throughout the South. Laborers who were blessed with an interest in the crop were enjoined to be watchful of shirking co-laborers. And above all, as G. W. Harvey of Carroll County, Mississippi, required in his 1868 contracts; "they shall be obedient and respectful in their deportment, active and industrious in their habits and careful and saving with all that belongs to or appertains to the interest of the place." A common general clause in early North Carolina share contracts required that laborers "having an interest in the crops to be raised, shall labor for their protection or preservation, if it becomes necessary, either from fire, flood, or tempest, or any other unforeseen casualty, in the night time or on Sunday, just as a prudent proprietor ought to do."[17]

[17] Carroll D. Wright, "Profit-Sharing," in *The Seventeenth Annual Report: Massachusetts*

To metamorphose the workers into individualists, entrepreneurs in their own right but only nominally so, was no mean trick. The primary problem that profit sharing failed to overcome was its tendency to inculcate in the workmen a feeling of proprietory claim that transgressed the boundaries set by the employers. To the manufacturer the epitome of this predatory transgression was the cooperative firm. For the planter it was the laborer turned independent renter. For both it betokened loss of industrial control, an increase in the number of competitors, and decreasing profits.

In 1865 the *Spectator*, a London periodical, praised the benefits of profit sharing in the "Wirework Company of Manchester": "the first result was a sudden decrease in waste, the men not seeing why they should waste their own property any more than any other master's, and waste is, perhaps, next to bad debts, the greatest source of manufacturing loss." This same article deemed it necessary to warn unsuspecting employers of the dangers of excessive experimentation in cooperation. "The Rochdale experiments" with industrial cooperatives, "important and successful as they were, were on one or two points incomplete. They showed that associations of workmen could manage shops, mills, and all forms of industry with success, and they immensely improved the condition of the men, but then they did not leave a clear place for the masters."[18]

Even if cooperation did not lead to laborers' sharing in the ownership of the means of production, the employers were fearful lest it lead to a labor-managed economy. One of the objections to profit sharing singled out as meriting special rebuttal by Sedley Taylor, England's most avid supporter of the system, was: "If workmen are once permitted to share in the profits of a concern, they will presently insist on overhauling its books, and even on thrusting themselves into its business management." Taylor, much as had Charles Babbage fifty years earlier, gave his own game away by expressing a "personal conviction that in proportion as participating workmen feel themselves qualified by improved general education and more thorough general knowledge to exert some share of influence on the management of the concern in which their own interests are bound up, they will gradually acquire the power of exerting that influence." The employers were in no way ready to accommodate worker interference with managerial prerogative.[19] Those firms adopting profit sharing were careful to inform workers that the hierarchical cleavage between management and labor could in no way be violated. In 1879 a Parisian firm initiated profit sharing and in the general regulations governing the system advised the workers that management was to preserve "its authority over the employees of the company." Another Parisian firm in 1872 warned the workers that "the participants do not possess the right of intermeddling in any respect with the

---

Bureau of Statistics of Labor (Boston, 1886), pp. 172–73, 182. *BRFAL*, film M826, no. 50, p. 1042. *BRFAL*, film M843, roll 43, "Labor Contract between John H. Caldwell and Peter Caldwell," 1867, Mecklenburg County, North Carolina, p. 841; also pp. 843, 793.

[18] Marx, p. 331. Marglin, "What Do Bosses Do?"

[19] Sedley Taylor, *Profit-Sharing* (London, 1884), p. 67. Godkin, p. 197.

bookkeeping."[20] In spite of the employers' insistence on retaining complete authority in the firm, the evidence indicates that the workers found it difficult not to interfere with managerial decisions that impinged directly upon those who, as it were, worked for themselves.

These difficulties are endemic to firms where the incomes of the workers depend directly upon the profits of the establishment. Within the genre, the paradigm is the labor-managed firm. Both economic theory and practical experience in the modern worker-managed firms of Yugoslavia or the Israeli kibbutz suggest that labor's participation in profits leads to a number of difficult problems. In contrast to the capitalist employer's desire to maximize profit, the workers in a labor-managed firm will desire to maximize not profits per se but net earnings for each laborer. Except in some special market environments these different objectives will lead to different managerial strategies. These differences invariably lead to conflict between labor and management in firms where profits are shared. Generally, the labor-managed firm will maintain a smaller labor force than will the profit maximizer because fewer laborers allow a larger share for each worker. In addition, the worker's interest in firm profits makes him or her more reluctant to leave the firm. For the employer this is a strong point when economic conditions require large output and employment, but not when maximum profit requires a decrease in the labor force. In practice the laborers tend to take a more pessimistic view than management when considering the investment of current funds in projects with either an uncertain payoff or a payoff far in the future.[21] In addition, if the workers possess heterogeneous skills, jealousies and discontent may arise over the distribution of the profits. Finally, the labor-managed firm will often be much more willing to trade profits for a more amenable work environment than will the capitalist.

These frictions were not unknown to nineteenth-century industrialists who experimented with profit sharing. The Peace Dale Company had rejected an original plan of employee shareholding as impracticable because of "a strong objection" to giving "any employee any other claim to retention in employment than his record of efficient service." After six years the company became concerned about employee dissatisfaction with investment policy: "Some will no doubt argue, that the reason why there are no profits is because of the heavy expenses incurred for improvements. The remark has been made, that the 'new finishing room will eat up all the bonus this year.'"[22]

---

[20] Sedley Taylor, pp. 163, 170. Godkin, pp. 191-92.

[21] Technically these differences between labor and management or simply between laborers are caused by the absence of a complete system of markets for spreading risks. If individuals are unable to purchase insurance against the contingency of personal income losses, those with the least wealth will generally wish to take fewer risks. Labor-managed firms are discussed in J. Vanek, *The General Theory of Labor-Managed Market Economies* (Ithaca, 1970); Benjamin Ward, "The Firm in Illyria: Market Syndicalism," *American Economic Review*, vol. 48, 1958.

[22] Wright, pp. 178, 182; also see pp. 211–13, 215–30. Also William E. Barns, ed., *The Labor Problem* (New York, 1886), a compendium of a symposium published in the periodical

These problems were most acute in agriculture and, because of the recent relationship between master and slave, particularly exasperating to both parties in the postbellum South. The *Spectator*, writing about England in 1865, gave as its final reason favoring managerial authority an argument that fit the situation in the postbellum South far better: "And lastly, cooperation among workmen is not so consonant to the national genius as cooperation between masters and men—limited monarchy having got into our bones."[23]

Few landlords found fault with this position. Planters resorting to shares and various profit-sharing systems took special care to specify in their contracts that the status of the laborer was equivalent to that of a wage hand, with the employer retaining all control and decision-making power. Typical language is found in an 1868 Arkansas contract, wherein a six-man share squad was told, "this contract gives the management and control of this place exclusively to the party of the first part [landlord]." More explicit language was used by B. W. Lauderdale of Shelby County, Tennessee, in 1866: "This contract is not to be considered a partnership said freedmen, are but hired laborers, as much so as if they were payed so much money by the month. They are to labor exclusively under the direction and control of the said B. W. Lauderdale." David Hinton of Wake County, North Carolina, informed his 1868 share hands that "general management" of the John Lewes plantation, of which Hinton was agent for Lewes's heirs, was "under discretion of John E. Kirks," overseer. The sugar planters of Louisiana had similar views about the status of share workers. In 1872 R. R. Barrow worked two plantations comprising 220 acres of sugar cane and 240 of corn by the share system. Supervision was the province of Mr. Barrow. The *New Iberia Sugar Bowl* reported that "the manager has full control of the place and crops, the same as if the hands were hired." There were, of course, exceptions to this general rule. For example, B. J. Harris and Lee Shelton, freedman, both of Lynchburg, Virginia, in 1866 "entered into a co-partnership to work a part of the said Harris farm." Lee was paid one-half the tobacco and corn and was obligated to "find himself, family, and hirelings." The latter arrangement was, however, very infrequent.[24]

Self-declaration of the planter's autonomy over the process of production proved to be an insufficient force. As early as the summer of 1866 a bureau agent in Jefferson County, Florida, reported that "in some cases where they have part of the crop they think they ought to have the management of them, and get into conflict with their employers." The conflict spread and became a permanent source of trouble for the operation of the share

The Age of Steel, in which the views of manufacturers, employees, and others upon all of these issues are given.

23 Quoted in Marglin, p. 73.

24 *BRFAL*, film T142, no. 72; ibid., no. 71. *New Iberia Sugar Bowl* quoted in *New Orleans Daily Times Picayune*, Sept. 11, 1872. *BRFAL*, North Carolina entry no. 2819, box 53. Also *BRFAL*, film M826, nos. 49, 50, Mississippi; film T142, nos. 66–72, Tennessee.

system. The protests of the planters themselves are adequate to establish this. Consider the observations, and perhaps experience, of a well-informed southerner: "Another objection to this plan of sharing the crop is, that the freedman enjoying an interest in the crop not only conceives himself priveleged to neglect it, but considers that he has a right to a voice in its management, and sometimes takes it upon himself to disregard instructions. With him 'my crop' and 'I'm gwine to do so and so' are common expressions even in regard to firm interests." Colonel Aiken prognosticated the ruin of the South as long as planters continued to "admit into agricultural co-partnerships those who, because they can hold the plough, or use the maul, claim to be *representatives in the firm.*" In 1878 a correspondent to the *Southern Cultivator*, arguing for a system of disciplined laborers under money wages, complained that "negroes left to their own judgement, and their own volition must fail, for with a very few exceptions they have neither; and where you work" on shares, "they are beyond cavil *co-partners*, and they have a right, and in the fullness of their conceit, exercise that right, to have a say so in everything." In Alabama the *Montgomery Advertiser* reported in 1871 that a statewide convention of planters declared that it was time "to discard the ruiness policy of doing a partnership business in the way of planting."

Bureau agent Charles Raushenberg provides us with a detailed itemization of the points of disagreement between landowners and share laborers:

> The majority of complaints that have been made at this office by both races have found their origin in contracts, where freedmen received as compensation for their labor a certain share in the crop. The majority of the plantations in my division were worked under such contracts. The freedman claims under such contracts frequently that he has no other work to do but to cultivate and gather the crop, that being a partner in the concern he ought to be allowed to exercise his own judgement in the management of the plantation, that he ought to be permitted to loose time, when it suits his convenience to do so and when according to his judgement his labor is not needed in the field, that he ought to have a voice in the manner of gathering and dividing the corn and cotton and in the ginning, packing and selling of the latter product—while the employer claims that the labor of the employee belongs to him for the whole year, that he must labor for him six days during the week and do all kinds of work required of him whether directly connected with the crop or not, that he must have the sole and exclusive management of the plantation and that the freedman must obey his orders and do all work required as if he was receiving money wages, the part of the crop standing in the place of money, that the laborer must suffer deduction for lost time, that if he does not work all the time for him, he is not bound to furnish him provisions all the time, that the crop must be gathered, divided and housed to suit the convenience and judgement of the employer and that the share of the employee must be held responsible for what he has received in goods and provisions during the year.

"The greater part of them have worked for a share of the crop," explained the agent in East Baton Rouge Parish, Louisiana, "and considering themselves as partners of the planters would not pay proper attention to his orders, but insisted upon working as they pleased and when they pleased." James DeGrey, in charge of a neighboring parish, East Feliciania, was even more explicit: "Those who worked on shares, did just as they pleased, informing the planters that they were working for themselves and knew best what they were doing." How far this encroachment upon managerial domain could extend is revealed by the contract of freedman Richard Baudrey of Desoto County, Mississippi, who underscored his managerial independence by having written into his 1867 share contract that it was "understood that the party of the 1st part [landlord] shall not interfere with the party of the 2nd part [Baudrey] in raising this crop."[25]

The emancipated laborers, under a system they considered cooperative farming, expressed the same desire for a reign of labor management as other workers did in similar circumstances. True to the predictions of economic theory, a major complaint of employers was "the difficulty of discharging hands when they become inefficient or refractory." This was correctly perceived to be a problem that was "inherent" in the method of payment by shares. The chief reason was inadvertently explained by S. D. Mangum of Adams County, Mississippi, when he insisted that he had "been compelled to discharge" five disobedient men "without paying them off" because "they were working on shares and I cannot tell what their portion will be." The quandary often led to bureau arbitration or, in later years, the civil courts. The decisions meted out in Mangum's home state chronicle the erosion of the squad leader's authority as well as the inconsistencies in the policies of local agents that give some legitimacy to planter complaints that the bureau was injudicial.

The Natchez agent, for example, allowed Mangum to dismiss the laborers from his plantation, but required that their share of the crop be delivered to them at the end of the year. That ruling would, in all probability, have been a blessing to a planter who, after ordering a field hand off the plantation, was informed by the hand that he had been hired (and therefore could only be fired) by the squad leader, not the planter. The agent at Vicksburg agreed that in such cases the landlord could not fire share hands. Alternatively,

[25] BRFAL, film M757, roll 37, July 1866, pp. 695–96. Galaxy, vol. 12, no. 3, Sept. 1871, p. 332. Aiken, "What Is the Duty of the Hour?", Speech before the Barnwell County Agricultural and Mechanical Society, reprinted in Rural Carolinian, vol. 11, no. 4, Jan. 1871, p. 195 (my emphasis). Southern Cultivator, vol. 26, 1878, p. 133 (my emphasis); also ibid., p. 14 and vol. 28, 1870, p. 378. BRFAL, "Reports of Operations, Cuthbert, Georgia, November 14, 1867," in Cox and Cox, p. 341. BRFAL, film T142, roll 66, 1867. Ibid., film M1027, roll 30, 1867, E. Baton Rouge, p. 337. Ibid., Dec. 1867, p. 348. For a planter's description see Southern Cultivator, 1869, p. 208. Wiener, p. 68. Barrow, "A Georgia Plantation," p. 834. Fleming, p. 722. Joseph Reid, in a series of articles, has argued that planters and laborers both preferred the share system because untalented freedmen were clamoring for managerial direction from talented planters. This explanation, while consistent with respect to the distribution of talent (managerial), is totally ahistorical as regards the motivations and actions of the agents. Joseph D. Reid, Jr., "The Evaluation and Implications of Southern Tenancy," Agricultural History, vol. 53, Jan. 1979.

a complaint from freedwoman Celia Smith that William Hardin had "discharged her from the squad to which she belongs and of which he was headman" drew the response from the agent at Greenville that Hardin would "not be permitted to discharge those having a like interest with himself in the crop." "The discharge," it was explained, "must come from the employer," who, judging from this agent's cases, would still be required to pay the proper share at the year's end.[26]

The general intent of these decisions was of course necessary. A policy of full payment was a strong motivating force in preventing employers from fraudulently dismissing laborers prior to payment of their share. One approach to circumventing these issues was to write into the contract a monthly wage to be paid in case of dismissal. This procedure was clearly unattractive. If the planter was to avoid the risk of paying a wage greater than the ultimate value of the laborer's share, the wage must necessarily be very low. Such a solution would hardly be satisfactory to the laborers. The ability of the share contract to stabilize the labor force eroded the squad leader's authority, forcing both the squad and the employer to search for internal methods of maintaining discipline.

Share squads contracting with James A. Gillespie possessed "the right to name their own foreman" and "punish trouble-makers." John B. Burton and his four "companies," of about twenty-one hands each, "mutually agreed that should any of the hands persist in disobedience, idleness and laziness, the laborers with the concurrence of" Burton "would have the right to discharge them." How the squads may have implemented a purely economic punishment is suggested by the contract of the Baileys, two planters in the Barnwell District of South Carolina. They agreed with the laborers that any hand failing to perform the "amount of work required of" the "class he or she signed" for would "receive a portion of the crops according to the classification allowed by a majority of the squad." Peter A. Parker, who it may be recalled reserved the power to name the foremen of squads, took a more active role in the maintenance of plantation discipline by bargaining for the right to either fine recalcitrants himself or "inflict any other punishment that any three of" the laborers said was "right to be inflicted."[27]

Sometimes labor management by majority rule began prior to the decision to decentralize into the squad system. For example, A. C. Brown of Hinds County, Mississippi, contracted with a large group for one-half the crop, agreeing that any malingerer, "on the complaint of any other freedmen, shall be tried upon such charge by a jury of not less than 6 or

[26] *Southern Cultivator*, 1869, p. 54. *BRFAL*, film M826, roll 30, "Complaints, Adams County," 1867, p. 448. Ibid., "Complaints, Warren County," Aug. 1867, p. 352; also "Champion Bradley vs. Watson Lewis, 'foreman of the squad,'" p. 352, and pp. 355, 360.

[27] Davis, p. 114. *BRFAL*, film M979, roll 41, "Labor Contract—John B. Burton and Freedpeople," Jan. 2, 1867, pp. 176–77. *BRFAL*, South Carolina files, "Labor Contract G. W. and A.W. Bailey and Freedpeople," cited in Shlomowitz, "Transition," p . 112. *BRFAL*, film M826, roll 50, "Labor Contracts—Peter A. Parker and Freedpeople," pp. 198, 730, 732.

more than 12 freedmen." Juries were to be elected monthly by all the males, each jury serving a one-month term. A variant was the meting out of discipline by a "family jury of three" composed of two members selected by the complainant and the defendant, and the third by those two. Arkansas planters W. O. and S. W. Bradley reserved the right to select "a jury of three disinterested parties" themselves. On their plantation, "if the jury decide[d] to discharge" offenders "from the plantation the parties discharged" were "to be paid five dollars per month for time worked." Tennessean H. S. Williams agreed to a plan similar to that of the Baileys of South Carolina. Any hands discharged from Williams's plantation were to "receive such pay as the majority of hands left on the place think he or she is justly entitled to." Mrs. Sallie M. Parrish of Fayette County, Tennessee, agreed to delegate this authority entirely to the squad by allowing "the foreman" to "dismiss any hand by getting the consent of one half the hands under his control."[28]

Decision by majority rule could also encroach upon other areas of labor management, as is indicated by the contract of a Halifax County North Carolinian, which stipulated that the laborers would do no digging or ditching and no work in bad weather unless "the majority say so." The partners of Victor Washington, who were to work in squads that were furnished "lots of no less than 100 acres" each, were to "cease labor at noon on Saturday" but agreed "always to work during the half Saturday if the majority of the squad so decides." The general complaints of employers imply that these rules were merely formalizations of labor practices as they actually were throughout the South. Where such informal decisions were arrived at by a group process, as seems to have been the usual case, formalization through a democratic procedure provided a stabilizing influence. Within this structure, the emergence of the squad as a social organization able to exert strong social pressure against deviant behavior was a necessity if it was to provide internal modes of discipline. Both a major source of trouble and the social importance of the work group are suggested by the fact that when freedman Jackson Irving refused "to let his wife assist the other women in weighing down the lever to raise the gristhouse," his punishment was an apology "before the squad" with "promises not to interfere" thereafter "with the work assigned his wife."[29] However successful it may

[28] *BRFAL*, film T142, roll 71, "Labor Contract—A. C. Brown and Freedpeople," Jan. 11, 1866, Hinds County, Mississippi. Ibid., "Labor Contract—J. C. Weaver and Freedpeople," Jan. 9, 1866, Shelby County, Tennessee; "Labor Contract—Thomas Billups and Freedpeople," Jan. 9, 1866, Lowndes County, Mississippi. *BRFAL*, film M979, roll 41, "Labor Contract—W. O. and S. W. Bradley and Freedpeople," Jan. 11, 1867, Arkansas. *BRFAL*, film T142, roll 71, "Labor Contract—H. S. Williams and Freedpeople," Jan. 1, 1866, Shelby County, Tennessee; "Labor Contract—Mrs. Sallie M. Parrish and Henry Key (foreman) and Others," Jan. 10, 1866; "Labor Contract—Mrs. Sallie M. Parrish and Bob (foreman) and Others," Jan. 1, 1866.

[29] *BRFAL*, film M843, roll 43, "Labor Contract," North Carolina, Feb. 21, 1866, p. 679. Ibid., film M826, roll 50, "Labor Contract," Washington County, Mississippi, Jan. 25, 1867, p. 812. On the use of social pressure see, Pearson, p. 208; Benham, p. 202. *BRFAL*, film M826, roll 30, "Record of Complaints," Jackson, Mississippi, Oct. 1867, p. 441.

or may not have been, a primary value of the squad was as a maker and enforcer of rules governing the group's internal operation.

Probably the charge of being refractory was most often made when an employee dared to voice an independent opinion about management or the distribution of profits. One candid planter admitted that American planters were no more tolerant of their laborers' meddling in the bookkeeping than were French industrialists: "when a negro questions his account the majority of white men are too easy to get insulted, and knock him down or curse him out." Interference with managerial decisions would most often be concerned with investment strategies and the choice of crops. Freed laborers and their employers often disagreed about the optimal division of a given amount of land between cash and provision crops. The typical direction of this disagreement is exemplified by the complaint of a North Carolina squad leader, who "was to find nine hands and with them cultivate part of a farm and receive one fourth the crop," the landlord boarding them. The squad leader sued the landowner's overseer in the bureau courts, complaining "that defendant will not do what is just neglecting the corn for the sake of the cotton." In this case the bureau agent decided "there is no proof," only a "difference of opinion between farmers." Therefore the squad leader was "stopped from doing anything because of the contract which reads all work to be done under the superintendence and direction of" the overseer. The incident may have had a conclusion similar to that of another planter's discovery that his laborers had surreptitiously planted "corn between the cotton rows" to "an outrageous extent." After many of the freedpeople had "refused to take" the corn out, the superintendent confessed to a friend that he was "rather more in the power of the negroes than they in" his. That much friction, if not clear violation of instructions, continued over this point is demonstrated by the repeated complaint that "the freedman is unambitious of accumulation," because "he strenuously insists on a full grain crop for subsistence of his family and stock, and only a moderate cotton crop."[30]

Even when the payoff to an investment meant some benefit would accrue to the laborer, there could be disagreement if the costs were certain and the payoff uncertain. Charles Nordhoff ventured the opinion that one of the reasons Georgia field hands disliked the share system was because they were responsible for part of the large costs for fertilizers which were "an uncertain element in making the crop." The point was corroborated by a contemporary, who, in an effort to portray an unflattering caricature of the "average" Negro farmer of the 1870s, reported that the attitude of the typical Negro was that he was "not givine to buy any juanner [fertilizer] to put on other folks' lan."

On sugar and rice plantations the conflict over short-run profits versus current investment developed naturally over the issue of how much of the

---

[30] *Southern Cultivator*, 1884, p. 348, in Taylor, *Slave to Freedman*, p. 56. BRFAL, record no. 2624, vol. 96, North Carolina, p. 14. Pearson, p. 264. *DeBow's Review*, July 1869, p. 609, *Special Report No. 3, House Executive Document No. 34*, 39th Cong., 1st sess.

year's crop should be withheld from the market and used as the next year's seed crop. For example, on William Minor's sugar plantation the share laborers desired to produce as much sugar as possible and "strenuously opposed laying any aside as seed cane."[31] These conflicts over proper management decisions and worker control go a long way toward explaining the statement, incredible coming from a planter like Aiken, that white planters had not been able to "control labor" and could also profit from the example of Negroes.

On July 31, 1866, a bureau agent in King William County, Virginia, reported an event which was to be of far-reaching import. "During the past two weeks," wrote the agent, "many of the freedmen who have finished their crops of corn have positively refused to do any work on the farms where they are employed without extra pay." The agent considered the tobacco hands of his district to be the most "hot-headed" freedmen in the South. This judgment might have been somewhat tempered had he known that this work stoppage was but one instance of a common movement whose prevalence, throughout the South, approached the magnitude of a general strike. The main point at issue was explained that very day by Frank Thomasson of Polk County, Georgia, who, in his own words, had "contracted to work for" his former owner to "make and gather the crops for one-third." Thomasson continued, "I have got the crop laid by and fence in good condition and a surplus lot of rails besides. He now wants to put me to clearing land for the next year and I still have to feed my hands."[32]

The dispute was of paramount importance to both parties. After the greater part of the work required to produce the current year's crop had been finished, the established plantation routine was to begin the *investment* activities of preparing the plantation for the next year's crop. To the planter, this was a necessary activity; to the workers it was an expenditure of labor and time that added no value to the share of the crop they would receive for their year's labor.

The continuing reports throughout the South are fairly summarized by an agent in central Mississippi who, in the fall of 1867, noted the "numerous" complaints from both parties: "Complaints from the whites, are that the freedmen will not work after the crops are laid by, they refuse to repair buildings, fences, etc.; the freedmen complain that the whites are trying to defraud them." "The Negroes have pretty generally quit work," reported the governor of South Carolina, in regard to the blacks' refusal "to make rails, repair fences, open ditches, cut trees, etc.; so that the farm may be in condition for a crop the ensuing year."[33] The laborers' victory in this struggle

[31] Nordhoff, p. 107. *Southern Magazine*, March 1874, p. 305. W. F. Messner, p. 86. Brooks, p. 65.

[32] *BRFAL*, film M1048, roll 45, July 31, 1866, p. 67. *BRFAL*, film M798, roll 13, July 31, 1866, p. 147.

[33] *BRFAL*, film M826, roll 30, Sept. 30, 1867, p. 426. *BRFAL, Synopsis of Reports*, "Letter Governor Orr to General Robert K. Scott," Dec. 1866. *BRFAL*, Virginia, entry 4259, "Complaints," 1866, p. 33. Ibid., South Carolina, entry 3150, "Complaints," 1867, p. 55.

is beyond dispute. The economic solution was adopted by R. G. Oliver of St. Francis County, Arkansas, who guaranteed that "all work performed outside of crop will receive fair compensation from employer if done on his plantation." For those who could not or would not adopt this economic solution the complaints throughout the 1870s would echo the words of Samuel J. Gholson of Columbus, Mississippi: "there is no fencing now; no fencing has been done since the war scarcely; we cannot get the laborers to make fences."[34]

It is possible to argue upon economic grounds alone that these difficulties could have been overcome. There is always an optimal contractual arrangement that can be reached by reasonable and rational bargainers seeking to maximize their individual gain. But the historian cannot afford the luxury of confining inquiries to the sphere of pure economic analysis. It would be wise to admit that where the economist fails to provide a convincing explanation the sociologist may be upon firm ground. It is important to observe that for nineteenth-century employers in general, and especially for ex-slaveholders, it was not so much the points of dispute which mattered, but the very audacity of laborers so impertinent as to foment a disagreement over managerial domain. The point is amply illustrated with the details of a contractual dispute between John H. Wade, a Virginia tobacco planter, and S. Tinsley, the foreman of a six-person squad which had contracted in January 1866 to "cultivate a portion" of Wade's land for a one-half share. The squad was obliged "to clear fence and cultivate as much land as they can in tobacco; to do necessary fencing and ditching, get firewood, attend to stock, and do other necessary things about the plantation." Among these, they were to "build a tobacco house; to cure tobacco, and cultivate [Wade's] vegetables without share." Such a long list of obligations, coupled with general terms such as "other necessary things" and "work faithfully and be diligent," no doubt left Wade secure in the belief that he was master of all.

An unforeseen contingency surfaced in the laborers' view of the new order of contractual relations. Late in the year, concerned that his cattle and horses were running loose in the corn and fruit orchard, Wade wanted a fence built dividing the orchard. Tinsley refused, arguing that the squad had "run every fence they had agreed to and there was a fence around the corn." Technically the dispute revolved around the point that the contract stipulated that the share squad was "to save and if necessary house Wade's fruit," but there was no provision giving them any claim to fruit. The altercation came to an end when Wade reverted to an appeal to paternalism, telling the workers that "the land and the stuff on it was his" and promising that "as long as there was anything in the ground he would not see them want." Tinsley replied that, "they had hired the land for the year," and then Wade struck him.[35]

[34] *BRFAL*, film M979, roll 41, "Labor Contract—R. G. Oliver and Freedpeople," 1867. *KKK Hearings*, Mississippi, p. 875.

[35] *BRFAL*, Virginia, Rocky Mount District, "Complaints 1866 and 1868," entry 4259, pp. 46–47.

The strength of the employers' desire for unmitigated autonomy is revealed by the fact that late in the nineteenth century, America's most ardent salesman of profit sharing still found it necessary to convince employers that the plan was not a diabolical scheme for the introduction of industrial socialism; that

> The partnership into which the employer himself invites the men is industrial not commercial; that he surrenders in no manner or degree his absolute control over affairs; that he is just as much of an autocrat as he was before; that he keeps his books entirely free from troublesome inspection; that he himself fixes the percentage of the bonus on wages.

D. Wyatt Aiken would not have been convinced. "The share system was one so obnoxious to me, I tried it but a single year, and would have sacrificed my plantation rather than continue it," he wrote, "I could not appreciate the equity of a co-partnership between capital and intelligence on the one hand, and labor and brute force on the other." Aiken's words reveal how irksome it must have been to former masters to submit to the independence of former slaves, such as the Georgian who asserted his right to stop work and attend a political meeting in no uncertain terms: "I am not working for wages," declared the freedman, "but am part owner of the crop and as I have all the rights that you or any other man has I shall not suffer them abridged."[36]

A detailed example which indicates the management problems planters like Aiken must have experienced with share labor is provided by the narrative of Smith Powell, a veteran overseer who in 1865 was reinstated as general manager of his absent uncle's Alabama plantation. In the fall of that year Powell wrote to his uncle, John Hartwell Cocke, complaining of severe difficulties. "I can not manage *Free Negroes* to make crops and keep the plantation in good condition," Powell complained, and made it known that he intended to resign from the business of plantation management. The Negroes "feel their freedom, and I fear they will never do much more good work." Powell advised Cocke that he had two options: "Either let the freedmen have all the produce, or sell the estates." Nevertheless, Powell was persuaded to continue as manager in 1866, contracting to pay the laborers a share wage of one-third "as an incentive to production." It is at this point that his real troubles with management began. In the early spring he reported that the work was "moving on pretty well for freedmen," but he faced "a difficult matter to keep up *improvements* with them." By the month of May it became clear who was really making the managerial decisions on the plantation. Powell complained that the freedmen, desiring more corn, had planted forty acres less cotton than he had ordered. "It is very troublesome to manage Freedmen," he reiterated, and by the end of

---

[36] Nicholas Paine Gilman, *Socialism and the American Spirit* (New York, 1890), pp. 291–92. Aiken, "Labor Contracts," p. 114. Peter Kolchin, *First Freedom* (Westport, Conn., 1972), p. 42.

the year recommended that if no one in Cocke's immediate family would come out to live on the plantation it should be sold.[37]

Peter Kolchin has suggested that for many ex-masters the difficulties of adjusting to free labor may have been overwhelming. For these planters, contractual bargaining and compromise were psychologically impossible. Either the planter must have universal control over the laborers and his plantation or he must minimize the contact between himself and the lower caste laborers. In extreme cases, such as that of a Virginia planter who demanded complete deference and refused to pay Negroes wages, there could be no middle ground: "I called my people together when your army first came here, after General Lee's defeat, and told them I should not pay wages. 'You are free,' I said, 'to go where you please, but if you choose to stay here you may; you shall work for me as you have heretofore, and I will give you the same treatment you have always had, the same quantity and quality of food, and the same amount of clothing.'" "A moral principle," Charles Sumner once declared, "cannot be compromised." This slaveholder had at least one point of agreement with Sumner, for he announced that if social order among the Negroes could not be maintained, as he defined it, he would "like to send them all off the plantation"—"I should feel better to work my garden with my own hands than to endure such insolence." The disease, as in the case of Edmund Ruffin, could prove terminal. Freedwoman Anna Miller described the classic symptoms as they appeared in her ex-master after freedom had come: "De marster gets worser in de disposition and goes 'round sort of talkin' to hissel'f and den he gits to cussin' ev'rybody. In 'bout a year after freedom, Marster Loyed says he don' want to live in a country whar de niggers am free. He kills hissel'f 'bout a year after dey moves." In milder cases planters were reported to have sold their lands and left the country for Cuba, Brazil, and other places where slavery was legal.[38]

The hypothesis has been put forth that those men and women who had made the best slaves, in the sense of accepting the planter's ideal code of paternalism and mutual responsibility between slave and master, often were the least able to cope with the responsibility of freedom. Similarly, the ranks of the unsuccessful managers of free labor may well have been disproportionately composed of the best of the antebellum paternalists. The choices available to such men were few and unattractive. Especially if they were to attempt planting upon the share system they must, as one planter argued, make "the hardest, and closest bargain possible." This, he cautioned, was a procedure "more difficult than it may appear." "It is opposed to all our training. The 'barbarism of slavery' eventuated sometimes in hard words, and now and then in hard knocks, but never in *hard bargains*. This is our deficiency."[39]

[37] Coyner, pp. 583–85 (my emphasis).

[38] Kolchin, p. 47. Dennett, p. 82. *Congressional Globe*, vol. 39, part 1, p. 673. Rawick, *Texas Narratives*, vol. 5, pt. 3, pp. 83–84. M. S. Wayne, pp. 74–77.

[39] *Southern Cultivator*, vol. 29, 1871, p. 253.

For the employer who refused to yield any authority, the share system would, if it were to be profitable, often demand coercive action that went far beyond the dictates of sound business practices and mere hard bargaining. An alternative would have been to accept the share system for what it was. But that would have meant accepting the laborers or tenants as partners. An 1868 article in a Texas newspaper recognized this when it advised that successful planting on shares required that the employees' daily records be open to them on a weekly basis; that any work done outside the crops "be paid for as extra work" and that in any event such work "should be with the consent of the said laborers"; and finally that the "division [was] to be regulated and controlled by the laborers by and with the assistance and counsel of the employer."[40] This alternative was a fundamental violation of the principles of what might be termed a group of very principled men. The Virginia planter discussed above abandoned his plantation by leasing it to his overseer. But overseers who would be both experienced managers and acceptable to the laborers were in extremely short supply in the Reconstruction era.

One solution of this personal dilemma was to give up active planting by abandoning the plantation to tenants, who would become independent renters nearly free of all landland control. Hiriam Tilman, in 1866, divided his Shelby County, Tennessee, plantation into separate farms of forty to seventy acres and leased them to individual families of freedmen for a yearly rent that averaged about two 500-pound bales of ginned and packed cotton "of at least middling quality." Control of the cotton gin and press was placed with freedman Billy Griffin, who was also to board Tilman during his visits to the place and maintain "a small frame house" on the plantation for that purpose. This solution (using the word liberally) could have sad consequences. Andrew Goodman, a ninety-seven-year-old freedman, recalled the benefits of slavery in Texas under the mastership of his benevolent and paternalistic owner: "sometimes I think Morse Goodman was the bestes' man Gawd made in a long time." Goodman's description of his life in thralldom identified his master as one who lived up to the code of paternalism in every respect. But the management of free labor must have proved too much for the old master. After a short trial "the bestes' man Gawd made" abandoned his plantation, allowing his former slaves to operate as independent sharecroppers. "It seem like the war jus' plumb broke old Morse up. It wasn't long till he moved into Typer [Texas] and left my paw runnin' the farm on a half-ance with him and the niggers workers. He didn't live long, but I forgits jus' how long."[41] Allowing the laborers this much independence was an extreme solution, and while evidence indicates that the

[40] *Texas Almanac*, quoted in *American Farmer*, vol. 11, no. 7, series 6, Jan. 1868, pp. 218–19. For the same argument see D. Wyatt Aiken, "Southern Farming and Farm Labor," *Rural Carolinian*, vol. 1, 1869, p. 141.

[41] *BRFAL*, film T142, roll 72. Rawick, *Texas Narratives*, vol. 4, part 2, pp. 74, 79. Brooks, pp. 18–30.

number of planters who adopted it was substantial, it is doubtful that the fraction doing so was.

A more profit-oriented system was the early version of fixed rental tenancy recommended by D. Wyatt Aiken. This system promised all the theoretical benefits of profit sharing and allowed the landlord the opportunity to dispense with many of the distasteful aspects of day-to-day supervisory contact required by the share contract. This was a most important benefit according to one planter who, in 1867, recommended that planters divide their properties into "forty-acre farms, to be leased to families of freedmen." The benefits to be reaped were a diminution "of all the cruel vexations of overseers and personal overseeing yourself" and "an increase in the value of real estate and more clear cash in your pocket."[42] This prescription seems to accord well with the current view that the major value of rental tenancy to the landlord is the reduction in supervisory time and costs made possible by the fact that the laborer, being able to keep all output above his rent, has an incentive to work voluntarily. An ignored but important aspect of this reduction in supervisory time and costs is the distinction between economic and psychic costs. For the unredeemable slaveholder, the psychic costs were paramount. While this is suggested by the motivations of those who abandoned all active supervision, it is made clearer by the system of renting used by planters like Aiken. Aiken "apportioned off twenty acres of good land to the mule, and required a rent of sixteen hundred pounds of lint cotton if the laborer owned the mule." If the laborer had no mule, Aiken provided one at the cost of an extra 400 pounds of cotton. Laborers and their families worked their own acreage independently under Aiken's managerial superintendence. "Extra wages were paid for all work done outside each of their respective crops."

This system and the one described just before must be carefully distinguished from the contract which would allow tenants to be totally unsupervised on the property of absentee landlords. In fact, Aiken's recommendation carried with it an increased responsibility upon the landlord that a more scientific agriculture be practiced and that contracts with the tenants stipulate "the *fixed condition* that this improved system be practiced to the letter in every department of each small farm."[43] It was not a reduction in supervisory time or its economic costs which these planters sought when adopting rental tenancy, but a reduction of the psychic costs of driving free workers to deliver more labor than they desired. Landlord supervision time did not diminish but was reallocated from personal driving to managerial decision making. An account from nineteenth-century Germany describes well Aiken's view of profit sharing by closely supervised renting:

The proprietor derives, independently of the pecuniary result, many advantages from the half-profit system. He has perfectly trustworthy laborers, and

---

[42] *Southern Cultivator*, vol. 25, 1867, p. 365.

[43] Ibid. Aiken, "Labor Contracts," p. 114.

[44] Sedley Taylor, p. 95.

each piece of work is taken in hand at the proper moment. He is no longer obliged to urge and drive, while fretting internally at the many instances of neglect which he is powerless to prevent. When his back is turned, he knows that his business is as well attended to as if he were directing it himself. He can dispense with all intermediaries, as no formal overseeing is required.

But this did not allow the landlord to retire to a quiet life. On the contrary, the author continued:

Nevertheless the position of the managing head has grown in importance. He must show more than was formerly necessary that his management is sound, and that with regard to every department of his business he is firm in the saddle, for he now has a responsibility towards his associated laborers. He is more than ever bound to set them an example of diligence, economy and other virtues, on the exercise of which the success of the whole undertaking depends.[44]

Aiken's advocacy of fixed rent tenancy as a means of avoiding the laborer's claims to partnership rights appears to present a paradox. The renter, among all tenants, is usually considered the most independent and therefore vested with the most proprietory rights. The paradox is cleared up by Professor Schloss's discussion of the distinction, drawn by nineteenth-century employers, between profit sharing and industrial cooperation. The latter, which Schloss considered a "democratic working-class" ideal, was predicated upon the principle of eliminating the role of the capitalist as employer. He was restricted to the role of provider of capital to groups of workers who organized production and reaped all profit above the rental payment on borrowed capital. Profit sharing, in contradistinction, was an incentive "device adopted by middle-class employers—men who have not the smallest intention of ceding the government of their business into the hands of their employees."[45]

The democratic share squad satisfied the prerequisites of the workers' "ideal" on all counts. Planters like Aiken, similar to capitalists everywhere who were convinced of their essentiality to the process of production, fought the spread of this workers' ideal. Victory required the adoption of profit sharing in the sense just defined.

In 1869, cotton planters were advised that they would find it "greatly to their advantage to adopt a system embodying the best features of both systems" then being tried. "Paying the freedmen first with a smaller share in the crop" and also "with a limited amount of money per month" would give the planter control of the labor, allow the general work of the plantation to be kept up, and keep the laborer up to his work, since with a share of the crop "his interests are identical with those of his employer." Many planters had already adopted this system. In 1867 the agent for the central eastern counties of Mississippi, bordering Alabama, reported that contracts in his subdistrict were "principally" for "so much money and a portion of the crops." The dual motivation for this method of remuneration is illustrated nicely by the contract of a North Carolina planter. He offered, to

45 Schloss, pp. 203, 151.

a group of five, one-third the crop plus $150 if all laborers "perform all incidental labor needed on the farm before and after croping season." The labor included "ditching, clearing, and preparing land for improvements." One of the laborers was "to be foreman over the farm," managing "it by his own judgement during the cultivation of the crop." He directed "all labor outside of crop season under the instruction of the" landowners.[46]

The arrangement was in most essential details equivalent to the rental plan advocated by D. Wyatt Aiken. The system of close renting was designed by the planters as the optimal labor incentive arrangement that was to combine the comparative advantages of labor with the skillful direction of the manager. The price that the laborers had to pay in order to avoid the driving aspects of day-to-day supervision and to work their own tenancies was to recognize the planter's authority as the decision maker. The manager, in order to insure that the renter would not object to providing labor for investment activities outside the crop, had to pay extra for those services. The provision of extra pay for work outside the crop made the plan profit sharing in its strict sense. The risk to the tenants was reduced substantially, because, as Aiken stated, the extra wage "paid, fed and clothed them" for the year, making any profit above their rental payment a true bonus for individual effort.

Just how carefully the close renter was supervised varied across plantations and, within a plantation, across tenants. D. Wyatt Aiken's supervision was largely limited to managerial functions concerning investment, crop choice, and financial marketing: "I allot so many acres to the mule, and dictate what crops shall be planted." He provided "a general supervision of the various crops" and "reserved the right to direct the *mode* of cultures." A similar system was used on the Georgia plantation of David Barrow. On other plantations the amount of supervision could be much stricter. Thomas J. Edwards, a supervisor of rural colored schools in Alabama during the early twentieth century, provides a description of close renting:

> A quarter of a century ago, one kind of renter was commonly found upon large plantations where wage-hands and share-croppers were employed. He was subject to the same plantation management as other classes upon the plantation. He received the same supervision, plowed, cultivated, harvested, and received advances in the same manner as the sharecropper.

In 1875 Charles Nordhoff witnessed the practice of close renting in the cotton regions of Arkansas: "In practice furthermore, the planter finds it necessary to ride daily through his fields to see that the renters are at work, and to aid them with his advice. During winter, he hires them to chop wood for his own use, and to split rails and keep up the fences." Perhaps the most typical arrangement was that practiced by a Georgia planter from Reconstruction into the twentieth century: "The owner exercises close

---

[46] *Hunt's Merchant Magazine*, Oct. 1869, pp. 272–73. *BRFAL*, film M826, roll 30, Meridan, Mississippi, Sept. 7, 1867, p. 302. *BRFAL*, film M843, roll 34, "Labor Contract," Jan. 1, 1866, Edgecombe County, North Carolina.

supervision over them in an advisory way, visiting each place once a week. He believes this supervision to be 'absolutely necessary.' His renters are very willing to work, 'but do not know how to farm profitably.'"[47]

How adequately this explanation serves as a legitimation of the "average" planter's direction and control of the "average" laborer's day-to-day work routine is a reasonable matter of dispute. However, it cannot be denied that where the planter of ability and wealth practiced scientific agriculture with a system of crop rotation and sound business management, the laborers often succumbed to his demand for complete managerial authority. Proud, intelligent, paternalistic and fundamentally racist, such a man was David Wyatt Aiken.

II

To explain why close renting failed to capture the approbation of the majority of planters is to remember the point stressed early in this chapter, that recognition of the full range of benefits available from profit sharing was limited to the most progressive employers. In regard to nineteenth-century agricultural employers it is safe to pressume that the extent of this enlightened view was very limited. This is illustrated by one such landlord's experience in Germany. After a five-year period of operating under a system essentially identical with close renting he was forced to sell his property. "I must admit," wrote Herr Jahnke in 1877, "that by introducing this arrangement I made myself many enemies among the landed proprietors, and that it was this circumstance which induced me to part with the estate." Two years earlier he had described the opponents of profit sharing as being "proprietors who were for high prices and low wages, labourers who wanted high wages for a small quantity of bad work and persons as found their advantage in the misunderstanding existing between agricultural employers and employed."[48] Most employers lacked the imagination to grasp the profit potential of increased incentives due to high labor earnings, beyond the more rudimentary aspects of the share system that were obvious to all. Their tenacious reliance upon the share contract, partly because of this deficit of imagination and partly because of their own insufficient access to credit, was to force both their laborers and themselves to pay a terrible price for the next seventy-five years.

The *first* objection to profit sharing that Sedley Taylor thought worthy of his attention was the argument put forth by employers that "it is unjust that workmen should share in *profits* unless they are willing and able to share in *losses* when these occur." On this point the personalities among the American

---

[47] Aiken, "Labor Contracts," p. 115. Barrow, pp. 831–34. *BRFAL*, film T142, rolls 71, 72, Tennessee; Film M869, roll 42, South Carolina, "Labor Contracts—E. J. English and freedman Venus Black," p. 394; "Labor Contract—E. J. English and Toby Smalls," p. 399, Jan. 1868, St. Helena Island. *Senate Report: Relations Between Labor and Capital*, vol. 11, 1883, reprinted in Fortune, *Black and White*, p. 248. Edwards, "The Tenant System and Some Changes since Emancipation," p. 25. Nordhoff, p. 39. Brooks, p. 63.

[48] Sedley Taylor, p. 94. This was the typical response of postbellum American planters.

Society of Mechanical Engineers were paragons of consistency. John T. Hawkins was predictable: "If the workmen, so-called, are to share in the profits of a given business, it can only be done by their sharing, also, in all the responsibilities, as well as in the losses." The alternative view was well stated by Mr. Towne: "It is argued in all discussions of this kind, that profit-sharing should always include loss-sharing. I do not think that this is a correct view. The difference in the conditions of the classes of men must be taken into account. The employer, as a rule, has reserve capital. He can stand loss for awhile, and still not suffer in his domestic conditions; whereas the workman generally has but little, if any, reserve of that kind. Loss to him, therefore, means immediate suffering for himself and family." Towne concluded with his opinion "that, as society exists at the present day, better relations of labor and capital must be sought by systems of cooperation or profit-sharing, tending to identify the two interests, under which the employer gives a guarantee of fair wages to the workman, and also arranges that he may have, beyond that, a reasonable and equitable interest in the economy and increased efficiency with which he does his work, some portion of the gain."[49] In southern agriculture falling product prices, combined with fears of continued crop failures, made the prospect of forcing some of the risks of business onto the laborers through loss sharing under the share system extremely attractive. For the employers, this is an advantage in the share system which we have so far neglected.

The end of the short period of unusually high prices following the war forced those employers who were still paying standing wages to deal with the problem of reducing wages in proportion to output prices. In 1868 the editor of *DeBow's Review* argued that wage contracts entered for a year's duration must give way to the practice of making several short-term contracts sequentially throughout the season. The editor believed that planters should not take on the risk of becoming "insurers of the weight and price of the crop." Under shorter contracts, it would be "in the power of the planter to save himself from large loss by reducing or closing his labor contract." In spite of the editor's objections to the contrary, contracts of short duration were considered even riskier because of the increased exposure to labor strikes during the middle of the season or at harvest time.

One proffered remedy to this was to offer year-long contracts with the annual wage indexed to the selling price of cotton at the end of the year. Some employers had put this plan into action with the beginning of full-scale planting in 1866. For instance, William N. Waterer, a Mississippi planter, contracted in January 1866 with fifteen freedwomen and freedmen under the provision that contractual wages would be increased a specific amount "if cotton is worth 50 cents" a pound in the fall. After the federal government contracted the supply of paper money in an attempt to restore a hard money policy, the financial panic of 1873, falling prices, and economic recession throughout the country impressed upon the minds of employers

---

[49] Ibid., p. 66. *Transactions*, p. 656.

everywhere the benefits of wage indexing during economic recessions. Under share contracts the laborers' gross earnings were by definition perfectly indexed to the selling price of produce. This point was perceived clearly by the editor of the *Southern Cultivator*, who recognized the advantage of share contracts' automatic reduction of labor costs without the danger of strikes concomitant with attempts to reduce wages. In a belated conversion from advocacy of wages to shares, the editor was most circumspect: "Under existing circumstances, when labor is high compared with the products of labor, it is better to pay in part of the crop, and thus compel the laborer to share with the landlord the loss from low prices, bad seasons, etc."[50] The power of profit sharing cum loss sharing to reduce the likelihood of laborers' engaging in strikes at critical times was a much-heralded aspect of the "harmony of interests" envisioned by employers of the period.

How did the agricultural laborers view the planters' new-found love for the share contract? In the dispute among the engineers Henry Towne advised that the asset-starved laborer, if "a prudent householder or head of a family should not incur risk of that kind if he can avoid it, and rightly; therefore, with reference to their present responsibility, and to those who are dependent upon them, the workmen should decline, or at least, be very cautious about entering into relations which involve a possible sharing of losses on their part."[51] According to written history, the emancipated blacks failed to heed Towne's advice. In their thirst for freedom from control, the risks inherent in sharecropping were supposed to have been a negligible price to pay. This interpretation is at variance with the revealed behavior of other laboring classes throughout history, and therefore is forced to rely implicitly upon predicated special characteristics of emancipated blacks. This comes dangerously close to a racial theory of economic behavior. In northern industry, men of experience had learned that laborers were unwilling to accept such risks. Nicholas Gilman's discussion of the matter is worth quoting at length:

> Suppose that he [the manufacturer] then endeavors to engage workpeople by promising them simply a fixed share in the profits. What would the sensible workman have to say to such a proposition of cooperation, where he himself would invest no money capital and would receive no regular wages? Would he not answer: "How shall I live, and support my family, while the woolens are making and are not yet sold? Other workmen object to the delay of a month in getting their wages. They wish to be paid every fortnight, or every week; and such frequent payment is very advantageous to them. I should have to wait an indefinite number of weeks or months until you effect a sale of the goods I have helped to make. I might indeed manage to get along on credit, paying more, in the end, than if I bought for cash. But will the grocer and the

---

[50] *DeBow's Review*, vol. 6, 1868, p. 152. *Rural Carolinian*, Oct. 1870, p. 118. *BRFAL*, film M826, no. 49, p. 507. Somers, pp. 30–31. *Southern Cultivator*, vol. 37, no. 1, 1879, in P. S. Taylor, pp. 1–2. *Southern Cultivator*, vol. 27, 1869, p. 55.

[51] *Transactions*, p. 646.

butcher, and the tailor, and the houseowner give me credit if I am to receive no wages, and must depend entirely for my deferred recompense upon your skillful conduct of the business? For here comes the pinch. While my associates and myself may do our best in making woolens, you, with all your efforts, may reap but a small profit in selling them. Nay, who knows if there will turn out to be any profit at all? No! I cannot take such risk."

Gilman's dramatization reads like a biography, the life and times of I. A. Sharecropper. If so, it was a drama in which few laborers with alternatives cared to play the leading role. Southern immigration societies discovered that few white men, even the poorest foreigners, were willing to immigrate to work upon shares at any terms.[52] The evidence does not support the established view that freed labor willingly accepted the risks incumbent in sharecropping. The reasons for the unwilling assumption of risk were the same ones that have led laborers in general to prefer more certain incomes to less certain ones.

Based on evidence gathered during his southern tour of 1865, Carl Schurz concluded that in the only unbiased test relevant to modern economics the Negro revealed a preference for money wages. "It may," he testified, "be stated as the general rule, that whenever they are at liberty to choose between wages in money and a share in the crop, they will choose the former." This assessment was seldom contradicted. John Turner, a major in the Union Army, who had the opportunity to observe in detail, explained the preferences of Virginia blacks:

> No man likes to labor for a contingency. The negroes were necessarily com-pelled to do so, because the farmers were poor and had no money to pay them. They were consequently employed upon the contingency of receiving a portion of the crop, that was in the future, and the Negro could not see it. That com-bined with a distrust of his employer, tended to make the negro discontented; but in every instance where he has been paid his wages at the end of every month he has been contented.

Assistant Commissioner Whittlesey found that the freedpeople "labor much more cheerfully for money with prompt and frequent payments than for a share of the crop for which they must wait twelve months." The general observations of these northerners were confirmed by the specific ex-perience of tobacco planter Robert Hubard. Hubard's preference for 1866 was to hire labor for a share of the crop, but he feared an inability to find any takers. His own freedpeople said they would "only hire themselves by *the month* for 1866, as they [would] need some money from time to time to buy clothing, and if they have an interest in the crop of 1866, they [could not] get any money until the year is half over." A southerner, the Reverend C. W. Howard, reporting on conditions in Georgia, declared that freedmen "in many parts of the state" were unwilling to make contracts for farming labor to be paid at the end of the year "in kind or in money," but "were

[52] Nicholas Paine Gilman, *Profit-Sharing* (Boston, 1889), pp. 35–36. Wharton, *Negro in Mississippi during Reconstruction*, p. 102.

more willing to work when they can be paid by the week or month, as they thus obtain ready money." The Reverend Howard attributed this preference to "indolence and want of thought for the morrow." Another Georgian who was unable to obtain all the laborers he desired with an offer of half the crop explained that the reason was the laborers' preference for "something certain." [53]

Edward Philbrick, when managing the network of South Carolina plantations under his control, received indisputable evidence of these points. In January 1865 Philbrick was dubious that cotton prices would remain as high as they had been for 1864. Therefore, when bargaining with his laborers over their demand for wage increases he "told them if they wanted to share this risk with [him] [he] would give them a share of the cotton for their wages." Philbrick reported that "they all objected to this except one or two of the men." But even these one or two refused the offer, saying "they would like such an arrangement, but their families couldn't wait so long for their money." Philbrick's own opinion was that they "showed good sense" in rejecting the offer. [54] The reason for the rejection of Philbrick's offer was that he paid wages regularly throughout the season and his workers were confident they would receive their pay. As discussed in Chapter 9, where the workers were not confident of receiving their wages the proprietary value of shares was demanded as a means of lowering risk. However, the ability of shares to offer more security depended upon the laborers' having ownership rights over their share of the crop and upon high cotton prices. Throughout the 1870s political developments were eliminating the first requirement and economics the second. Careful consideration will be given to the political changes in Chapter 14.

The effect of falling cotton prices and the experience of crop failures upon the contractual preferences of blacks did not escape the attention of planters. William B. Shields, the experienced overseer of the William Mercer plantation in Adams County, Mississippi, left in his plantation reports a particularly revealing piece of evidence on this subject. The share laborers on Mercer's land were so savvy in their attention to contractual details and self-interest that they required Shields to provide evidence of the market price of cotton at settlement time. The attention to prices must have gone beyond the single objective of preventing underpayment for their share, to general concern with the movement of prices over time. For in 1869 Shields, writing to his employer about the Negroes' preference for a share in the crop over wages, reported that the laborers had become concerned with the "falling price of cotton" and that he thought this "would cause many to

---

[53] Schurz, *Report to the President*, p. 29. *Report of Joint Committee on Reconstruction*, Virginia, p. 5; ibid., pp. 189, 182, North Carolina. Thomas, p. 398. C. W. Howard, "Conditions and Resources of Georgia," *Department of Agriculture Report, 1867* (Washington, D.C., 1868), pp. 573–74. Robert Ferguson, *America during and after the War* (London, 1866), p. 230.

[54] Pearson, *Letters from Port Royal*, pp. 294–95. For similar motivations see *BRFAL*, film M1048, roll 46, Virginia, p. 294.

change their plans." This change in plans was a general response to declining prices throughout the South. An Alabama planter explained the laborers new-found desire for wages in terms which fit well with political scientist James Scott's characterization of the conservative aversion to bearing risks among modern-day Southeast Asian peasant farmers and laborers: "Just after the war they had a fancy for renting land, because the Yankees talked so much to them about it, but they are fast abandoning it for set wages, because, like regular soldiers or college boys, they don't want the trouble of balancing chances and precasting the future."

How well these examples attest to the preferences of the mass of laborers is indicated by the resolutions reached in numerous large labor conventions and meetings held by the freedmen throughout the South. The best known of these labor conventions was a mass meeting held in Montgomery, Alabama, in 1873. The convention resolved that the share system be abolished and replaced by contract wages. In addition, the squad system was to be abolished and wages *secured* by a lien on the crop. A key reason given for this preference was that the laborer would not be responsible for bad crops. Especially after the panic of 1873 there was a marked attempt upon the part of agricultural laborers to replace payment by a share in the crop with wages paid on a monthly basis.[55]

The laborer's desire for regularly paid certain wages was partly due to this desire to avoid risk and to a realization, after experience with annual contracts for wages or a share in the crop, that short-term wage contracts offered the greatest independence to workers. This point should be stressed, since most scholars, implicitly referring to the annual wage, have argued that share contracts offered the most independence and were therefore most desired. The entire range of reasons given by the field laborers is covered in an 1884 discussion by a planter complaining of the growing prevalence of the laborers' most preferred contract of all, the day wage:

> There is another class of laborers against which I think it would be well for us to use our influence, and that is day laborers; because they demoralize the balance of the hands who are hired by the year. They tell them that their employers are swindling them, and that they can make just as much clear money and not work half the time. It gets them in the habit of loafing; and not being able to get employment all the time, hence they take to pilfering. And indeed there is nothing reliable about them any way. No man can depend on them to cultivate his crops, unless he lives right in the edge of a town. And if he can get them whenever he wants them, he has to stay right in the field with them all the time; because if he does not they will slight his work and ruin his crop. They don't care how much they injure you; all they are after is pay for that day's work. And then it brings about another trouble. It is getting so a great many men are dependent on this day labor for most of their farm work. Consequently if they get in the grass, or if their crop needs gathering, they are

[55] R. L. F. Davis, "Good and Faithful Labor," p. 108. James Scott, *The Moral Economy of the Peasant* (New Haven, 1976). Powers, *Afoot and Alone*, p. 75. Rogers, *Agrarianism in Alabama*, p. 17. Fleming, p. 731. *New Orleans Daily Times Picayune*, Jan. 20, 1874, p. 1. Taylor, *Negro in Tennessee*. Wharton, pp. 63–64.

compelled to hire these hands at any price, or lose all. The man who offers the most gets them first. And then, you see, *it takes cash money all during the year* to employ this class of labor. They have the farmer at their mercy. And this day labor increases every year. Gentlemen, we ought to have some system and combined effort against such things. Let us see what has caused so much of this day labor. Of course, there always will be some, under any circumstances; but when there is so much of it there is some special reason. I have asked numbers of them why they preferred to run about by the day, when they could get plenty of regular work to do by the year. And they all tell me it is because *men will not pay them at the end of the year.* They say it doesn't matter how economical they live during the year, that it is impossible for them to clear anything; and when they *work by the day they are sure of their money.* And, gentlemen, I think they are about half right.

In the middle of the 1870s Louis Manigault attributed the desire for day labor to an attempt to prevent "being bound down to the same plantation all the year round."[56]

In areas where freedmen could obtain employment for promptly paid cash wages, planters found the recruitment of labor on other terms a difficult task. A Northampton correspondent to the *Richmond Dispatch* complained in 1873, "Labor is scarce here. Many Negroes are engaged in dredging for oysters in the sound up the bay, and others prefer to work by the day. . . . To hire a good work-hand by the month or year is difficult." Where the planters had the means to pay money wages on a regular basis, the freedmen protected themselves by refusing to contract for more than a week or month at a time. From Kentucky, early in 1867, the Freedmen's Bureau reported that the freedmen "evince no faith in the promises of employers," and that, in the Lexington subdistrict, "the freedmen are generally refusing to engage for a longer period than a week or month *at a time.*" In Florida, after the crop failures and defalcation of wages in 1866, many freedpeople refused to sign long-term contracts, preferring day labor for ready cash wages.[57] Perhaps the most compelling argument against the unsubstantiated assertion that shares were preferred to wages is the fact that throughout the postbellum period thousands of blacks working on the sugar and rice plantations of the South labored faithfully for cash wages.

To demonstrate that many freedmen preferred cash wages promptly paid to a share of the crop does not *establish* that the majority of black workers had this preference. The intention is to show that there is a problem with the traditional interpretation, and resolution of this problem has important implications. A resolution of this quandary requires that we clarify two

---

[56] *Southern Cultivator*, vol. 42, 1884, p. 348, in Taylor, p. 56. *Rural Carolinian*, Jan. 1872, p. 278. Manigault, "Journal," pp. 30–31. *BRFAL*, film M1048, roll 44, Virginia, p. 16.

[57] *Richmond Dispatch*, Jan. 21, 1873, p. 3. *BRFAL, Synopsis*, March 1867, p. 64, 68; Dec. 1866, p. 118. On the preference for wages also see *BRFAL, Synopsis*, 1866, p. 93, Louisiana, May 1866, p. 212, Virginia. Andrews, pp. 100, 212–14. "Agricultural Labor at the South," *Galaxy*, p. 335. Loring and Atkinson, pp. 5, 13–14, 16, 24, 123. Leigh, pp. 15, 226–27, 203. *House Executive Doc. 329*, 40th Cong., 2nd sess., p. 29. *Scribner's*, vol. 8, p. 138. *Dept. of Agriculture Report, 1867*, p. 416; Nordhoff, pp. 99, 107.

issues. First, it must be made clear precisely which two types of contracts are meant when the terms *wages* and *shares* are used. Second, when asking what type of contract was preferred we must be careful to indicate where and when the preference was expressed. The political and natural events which occurred between 1865 and 1880 meant that it made a great difference whether a laborer was given a choice between a standing wage contract and a share wage, a daily, weekly, or monthly money wage and a sharecropping contract, or any permutation of these. As has been shown, circumstances were such that the choice of labor contracts, by employers or employees, cannot be explained independently of the historical development of the postbellum southern economy. Share laborers or those working for annual wages, dependent upon their landlord or a merchant for long-term credit with which to feed their families, not only were subject to much closer superintendence of their everyday lives than were short-contract wage workers, but were also burdened with a direct share of the risk inherent in production. As we have seen, the very reasons which were pushing field laborers towards short-period wage contracts, coupled with the planters' inability or unwillingness to obtain risky credit, were simultaneously leading the landlords toward a strong preference for sharecropping.

The fact that most of the evidence supporting the view that the laborers preferred short-term wage contracts comes from the period covering the first ten years of freedom is important. For in that period, for a significant portion of the laborers a choice of money wages would have been a rejection not of family sharecropping but of share squads. If falling cotton prices were leading laborers and planters in opposite directions during the mid-1870s, it may be much more than a coincidence that the breakup of the squad system into family tenancy occurred simultaneously.

In accepting family-based farming, the workers relinquished much of the social cohesiveness that had been supported by the collective work group. The effect of this is strikingly portrayed by the much stronger labor solidarity of wage-working sugar hands as compared to that of family-based cotton farmers. The individualism enforced by the spread of family farming served the planters' interests insofar as that individualist ethic destroyed the social bonds on which the collective work squad had built a demand for agrarian democracy. But even granting the interpretation that holds that cotton and tobacco hands chose family farming, it appears that for roughly ten years after 1865, the field laborers were no obstacle to the institutionalization of a money-wage-based agriculture.

The putative market power of the field laborers appears to have vanished in less than ten years. The landlords won the battle over short-term money wages, but they would continue to complain of the scarcity of labor. This scarcity, as Aiken and others perceived it, was due not so much to a deficit of laboring bodies as to inefficiency caused by loose discipline and a lack of planter control. As tobacco planter Thomas S. Watson put it in a 1867 letter to his niece, "I have plenty of labourers, not labour for they work badly, at $9, $10, $11 per month." This should put us on guard not to accept without evaluation the validity of arguments which stem from the power of the laborers to dictate terms. John Hope Franklin provides an insight that is directly to the point when he states that the laborers, bereft of property, were forced to accept those terms the employers offered them. Franklin's statement has been interpreted to mean that the planters possessed monopoly power in the labor market. There is, however, another interpretation. Economic theory suggests that competition for labor will affect the *value* of those terms but is quite silent in regard to the form. It was no coincidence that laborers, unable to get short-term wage contracts in the early post–Civil War years, received shares when they demanded them from planters who also preferred shares. The same laborers could not alter the payment system when their demands conflicted with planter desires. As long as the underlying conditions of low prices, crop failures, and expensive credit prevailed, the share system would remain impregnable.

The trend would prove to be disastrous for both blacks and whites, planters and laborers. In 1873, on the eve of the return to white supremacy, the governor of South Carolina, engaging in some comparative analysis of his own, forecast the future of the southern economy in an eloquent appeal for agrarian reform:

> In the common interest of the agricultural laborer and the land owner, I earnestly recommend a speedy change in our existing agricultural lien laws. We are now working chiefly on the share system in the raising of crops. This is known as the "Italian" plan of cropping, and it has kept the agricultural laborers of Italy poor for the past three centuries.[58]

[58] Thomas, p. 225. John Hope Franklin, *From Slavery to Freedom: A History of American Negroes* (New York, 1947), p. 310. Woolfolk, *The Cotton Regency*, p. 114.

12

*Where money is due them they have to take the notes of the planters,
as the latter seem to be entirely destitute of money.*

South Carolina, Freedman's Bureau, 1868

*I have noticed that, when a cotton-planter is embarrassed, in debt,
and not making money, he is very apt to think that "the negro don't
work," . . . but wherever I have found a planter who managed well,
and had sufficient capital to carry on his operations, he was successful, and was also well satisfied with the negro. The answer I
oftenest received from planters and others whom I asked about the
negro as a laborer was this: "If you pay him regularly, cash in hand,
and do not attempt to sell him any thing, but let him trade elsewhere,
and if you deal fairly with him, he is the best laborer you can get, and
you can always keep him."*

Charles Nordhoff, 1875

*I would rather walk about and do nothing than to work all year and
get nothing.*

Georgia freedman, 1880

The Long Pay

Contemporary planters and economic historians of the late nineteenth and
early twentieth centuries placed most of the blame for the poor performance
of postbellum agriculture on the sharecropping system. This imputation
was consonant with the economic indictment of sharecropping as an inefficient mode of organization by a long line of classical economists from
Adam Smith to John S. Mill. Sharecropping was viewed by the economists
as an inherently negative incentive system for landlord and laborer. Much
of this indictment stemmed from the kinds of problems examined in the
previous two chapters. A number of contemporary economists and economic
historians have challenged this view, arguing that sharecropping is and was
just as efficient as any other organization of free labor.

The latter group has recognized the existence of strong disincentives to
efficiency under sharecropping. The main premise of their counterargument
has been that if sharecropping were inefficient, rational economic agents
would not choose to adopt it. Since sharecropping has been observed as a
stable institution and economic agents should be presumed rational, they
conclude it must be efficient.[1] This argument contains an implicit premise

[1] On the classical economists see Gerald David Jaynes, "Economic Theory and Land

that is never stated by its proponents: the economic agents must face no constraints that prevent them from making a choice other than share-cropping and all would-be borrowers must have access to a textbook-perfect capital market. However, in the absence of a perfect credit market some employers will be able to offer money wages only of the postharvest variety.

In the present chapter these questions will be reexamined in an effort to give a final analysis of the agrarian structure of early postbellum southern agriculture. In one sense this is the most purely economic chapter in the book. We shall concern ourselves with technical questions about productive efficiency and the supply of labor by the field laborers. In the process, new quantitative evidence is presented to help assess the problems. But we shall stray from these traditionally economic questions to consider closely related social issues which, in my opinion, should not be separated from their economic causes. Finally, the chapter will present a brief critique of the federal government's failure to provide an adequate agrarian policy for a prosperous Reconstruction.

"Such is the unparalleled scarcity of money here that I would advise no creditor of our people to rely upon prompt payment." This advice, from a financial agent to planter Thomas S. Watson of Virginia, uncovers the quintessence of most financial dealings in southern agriculture—waiting. From the poorest laborer to the large planter and merchant, the problem of obtaining cash and credit on the basis of crops which would not be sold until the end of the year proved paramount.

Northern creditors, often situated at the end of a long chain with many broken links, were sometimes forced to wait with the others. Merchant William Simms of Americas, Georgia, explained the elements of Keynesian multiplier theory to his northern creditors, Bitts, Nichols and Company:

> Gents I fear that I am due you an apology for not writing you before this. I had expected to have been in New York before now but have failed to collect money that was due me and had been promised. I now expect to start in about two or three weeks wither I get any money or not. . . . Since I was in N.Y. I have from six to eight thousand dollars worth of Execution Notes, etc. which I think I can collect after a while, but there is no money in the country now, nor will there be until another crop of cotton is made and I fear then the people will have so little left after paying expenses that they will not have much to pay debts with. My creditors will have to wait on me for a while I am good for what I am owing; if I can get time.[2]

At the base of this pyramid of credit were the laborers, who were forced to extend credit by waiting for their pay. Many agents of the Freedman's Bureau saw the problem in the share system. By 1867 bureau officials were strongly opposed to the use of share contracts, but found themselves

Tenure," in Hans P. Binswanger and Mark R. Rosenzweig, eds., *Contractual Arrangements, Employment, and Wages in Rural Labor Markets in Asia* (New Haven, 1983).

[2] Thomas, "Four Plantations," p. 216. *BRFAL*, Georgia files, entry 171, vol. 2, no. 138, June 30, 1866, p. 5.

unable to overcome forces leading the economy to its ultimate destination. The Washington office received a typical statement of policy from the field: "The share system, though preferred by the freedmen, has been discountenanced by the Bureau as leading to discord and disappointment." A year earlier, the same North Carolina bureau had reported precisely the same sentiments and announced that the assistant commissioner, General Robinson, would "recommend that all contracts hereafter made, shall be for a stipulated sum in money, settlements to be made every month." Again at the end of 1868, this bureau sanguinely predicted that "the system of contracting for a share of the crop is fast losing public favor; while that advocated by the Bureau, viz for a stipulated wage, periodically paid, will undoubtedly be adopted by all." This optimism was blunted by the admission that the shift to monthly money wages would occur "whenever the means in the hands of the planters prove sufficient."[3]

How unrealistic such optimism was is indicated by the actual practices of many employers who had signed contracts promising periodically paid money wages. "The greatest plague we have," wrote a friend to tobacco planter Edmund Hubard, "is to raise the money to pay" the laborers. Edmund Hubard's own financial position was extremely dire. His decision to contract for share wages was probably forced by the fact that he was a victim of credit rationing. Hubard had been advised that financiers in New York refused to grant loans on the security of land, and that "in Richmond the money is not to be had on any terms." The First National Bank of Richmond subsequently informed him that "money continues to be scarce, without any prospect of an early improvement. I have not been able to discount your note, and I fear I shall not be able to do so."

Planters in better financial condition often found themselves in the position of Thomas Watson, who, after contracting to pay wages quarterly, found himself forced to issue the hands promissory notes because of insufficient funds.[4] Kentuckian John Thompson took precautions against the inability to procure funds by making the vague promise to pay his hands money from "time to time." J. J. Powell of Wake County, North Carolina, was more specific. He promised that "if he [had] the money" he would pay Mack Hinton "fifteen ($15) dollars the fourth day of July," but "if not the sayed Hinton [agreed] to wate until the end of the year for the whole of his wages."[5]

Very frequently those working for wages were forced to seek civil action for payment. The bureau agent at Williamsburg, Virginia, evinced some ambivalence as to the reasons: "The whites are delinquent about paying the

---

[3] *BRFAL, Synopsis of Reports*, North Carolina, Sept. 30, 1867, p. 194; Nov. 12, 1866, p. 68; Sept. 30, 1868, p. 438; Nov. 1866, p. 49, Virginia; Dec. 1866, p. 121, Alabama, Georgia; Sept. 1867, p. 208, Georgia; 1867, p. 216, Arkansas; and Oct. 31, 1866, Mississippi.

[4] Thomas, pp. 504, 505, 218.

[5] *BRFAL*, Kentucky Assistant Commissioner Labor Contracts, box 42, April 13, 1868. Ibid., North Carolina files, entry 2819, box 53, "Labor Contract between J. J. Powell and Mack Hinton," Feb. 13, 1868.

freedmen for labour they do for them. Whether it is for the purpose of embarrassing them [or] on account of lack of funds to pay with I am unable to say, but I am of the opinion it is a little of both." Another agent was less ambivalent but still cautious: "The fact that there is a great scarcity of money is another cause. The whites claim to be unable to pay high wages, actual monied transactions in everyday business dealings are rare and the enormous rates of interest which money commands would seem to justify a belief in their plea of lack of money."

The argument is difficult to ignore. For the "long pay" is a familiar phenomenon in the annals of industrial development. It was a widespread and continual problem in eighteenth- and nineteenth-century England. There, as in the postbellum South, all the social evils which necessarily attend the long pay were evident. For if workpeople were to receive no money for protracted periods of time, it was still a biological weakness of the species that they must receive food to sustain life. The requirement is dealt with by the employer's delivering goods to the workers himself or arranging for a merchant to provide life's necessities through credit. In either case, the payment of wages in goods (truck) is seldom healthy for either laborers or an industry.

Both the cause and the likely consequences were commented upon early by contemporary observers. One such commentator, based in Jefferson County, Florida, was especially perspicacious: "A system of giving orders on the stores in town, to the freedmen—which is unavoidable from the scarcity of money—will tend to dissipate the earnings of the freedmen—and cause them to pay exorbitant prices for what they get. I cannot see any remedy for this evil, which will find the freedmen after selling his crop and paying for his goods—destitute of means at the end of the year."[6]

The general problem of truck and the long pay is often associated with economic recessions. The grain of truth contained within this association is merely indicative of the root causes of the problem. In industrializing England even such great manufacturers as Sir Robert Peel and Samuel Oldknow were driven to economizing on credit during trade downturns and credit crisis by issuing notes upon stores to the workmen in lieu of cash. Given the monopoly power of their factories in their local labor markets it is certain that any discounting of the value of these notes by the merchants was borne by the employees. In this sense the practice can be viewed as a reduction in wages, a common enough practice during recessions.[7]

For those working for employers of less stature and wealth, and especially within the domestic system, where small masters of little wealth were the

6 *BRFAL*, film M1048, roll 45, July 31, 1866, p. 90. Ibid., Augusta County, Oct. 1866, p. 674. Ibid., film M752, roll 37, July 1866, p. 695.

7 Samuel Oldknow experienced a personal financial crisis during 1793–94 due to his own overinvestment. He managed to survive a severe liquidity constraint by issuing notes for all his employees' wages during the period, a practice which should not be confused with Oldknow's earlier and later policies of making partial payment of some wages in kind or by notes drawn on banks. The victims of the shopkeeper's tendency to discount these crisis notes are identified by the following letter:

rule, the long pay and truck were chronic features of the workpeople's existence. There, as in the postbellum South, the people and industry suffered because of it.

Carl Schurz, in his report to President Johnson in 1865, noting the difficulties planters had securing credit and the prevalence of shares, warned the nation that "the unsatisfactory regulation of the matter of wages has certainly something to do with the instability of Negro labor."[8] Schurz's observation is related to the contraction in the supply and efficiency of agricultural labor. Every study of postbellum agriculture places great emphasis upon the withdrawal of the labor of freedwomen. These studies locate the entire motivation for household labor supply decisions in such sociological criteria as the freedpeople's desire to create for themselves the sexual division of labor putatively existing in white households. Contemporary testimony supports the conclusion that this desire played an important role, but as Carl Schurz suggested, the supply of labor might also be presumed responsive to the rate and reliability of remuneration. The question is susceptible to a quantitative answer—but only partially.

We might initially inquire into the quantitative impact of the reduction in the labor force participation of freedwomen compared to female slaves. In regard to this we are fortunate to have the richly detailed *Plantation Inspection Reports* of the Louisiana Freedmen's Bureau *BRFAL)*. These reports permit the construction of a very reliable estimate of the labor force participation rates of freedwomen in a large and important state with two primary staple crops. The estimated percentage of adult blackwomen in agriculture who were gainfully employed *full time* is exhibited in figure 12.1. The huge and immediate withdrawal of female labor from the sugar fields (47%) is consistent with a historiographical tradition which has for the most part accepted the guesstimates of contemporary planters, and with the comparative relevance of the British West Indian experience. The figures for cotton plantations are more problematical. What is meant by "large" is of course elusive and subjective, but the 17 percent reduction in the rate of participation or, put another way, the 83 percent participation rate by females in the first full year of freedom does not to my mind offer solid support for the pure sociological view. The further reduction to about two-thirds of the available female labor supply in 1867 and 1868 seems much more consonant with this view. But how should the slower response in cotton, as compared

---

Mr. Oldknow,

   Sir this is to let you know that I am Extortiant in buying Mr. Downs Shop Goods eavery Article his a penny a pound dearer than I can by at Marpor bridge and a menny of the artles three hapence and 2d worse. Sir I should be greatley obloige to you if youll give me a note go to Marpor bridge it would save me 4s a payment and then I can pay you 4s a payment to wards the ould monny if I ceep trading at Mr. Downses I can never pay you for I can hardly Geet vittles

From your
Humble Servant
Thos. Austin.

[8] Schurz, *Report to the President*, p. 29.

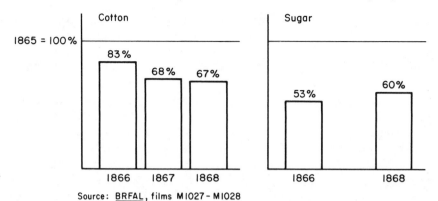

Source: BRFAL, films M1027 - M1028

Size of sample for Figure 12.1

		Female laborers	Male laborers	Total laborers
Cotton	1866	9,817	11,826	21,643
	1867	6,631	9,757	16,388
	1868	3,594	5,371	8,965
Sugar	1866	2,276	4,237	6,513
	1868	2,760	6,842	9,602

Source: BRFAL, films M1027 - M1028

Figure 12.1 Female Labor Force Participation Rates, Louisiana
(% of all Females on Plantations Working Full Time)

to sugar, be interpreted? Were black families on cotton plantations less inclined to strengthen their attachment to the traditional sexual division of labor in households?

A more probable explanation stems from the simple fact that the two crops used different production techniques. During certain periods of the year most tasks on a sugar plantation placed a premium upon physical strength. This requirement for seasonal but extremely arduous labor affected both the supply and demand for female labor. The importance of the demand factor is signified by the complaints of freedmen in the parish of Orleans that during the early months of the year their families had to subsist upon one income, because "their employer didn't want women in the field" at that time. Now that the planter had to pay for their labor, women on sugar plantations were "not always hired by the year, but often by the day."[9]

That the desire to keep women from having to perform fieldwork existed generally among both sexes seems beyond dispute. Labor force participation by freedwomen turns not so much upon the status of this desire but

[9] Dennett, *The South as It Is*, pp. 322, 326. Phillips, *American Negro Slavery*, p. 245.

upon the ability to afford the consequences of lost earning capacity. In view of the fact that the capitalistic division of labor into disjoint market and household activities is a luxury, it is not surprising that an 1869 convention of black politicians who urged the men of Georgia to "take their wives from the drudgery and exposure of plantation soil" had the savoir faire to add that the withdrawal should occur "as soon as it is in their power to do so." In the first years of freedom many freed households sacrificed a great deal to consume this luxury. Some discovered, after the fact, that they did not possess the power to do so.[10]

Bureau agents familiar with elementary principles of supply and demand attributed the fall in agricultural employment after 1866 to the fact that the fall in cotton prices induced a decline in wages and therefore a lower supply of labor. Related arguments emphasized the pessimistic expectations of planters and creditors due to the early crop failures. These conditions resulted in a downward shift in the demand for labor and, according to some bureau agents, caused involuntary unemployment when planters were refused sufficient credit. Part of the problem stemmed from the employers' pessimistic expectations concerning the connection between the supply of Negro labor and the level of wages. A South Carolinian explained that "when wages, which so far have been quite too low, shall be raised," then "more labor will be performed, especially in growing cotton, for they [blacks] exhibit their human nature in being more willing to work when they get good wages." In conclusion it was admitted that "some who raised the wages of their hands last spring [1869] got much more planted than any of us expected."[11]

This prerequisite to the withdrawal of female services from the labor market is so fundamental and elementary that it seems incredible that no source I have found discusses the supply of black labor and the movement of postbellum agricultural wages in a form which might remotely suggest that the two were in any way related. This omission is all the more incredible given the large decline in agricultural wages in the first three years following emancipation. The reductions in wages shown in table 12.1 would be expected to be accompanied by a reduction in the supply of labor if they represented a decline in purchasing power. Under the very unlikely assumption that the fall in men's agricultural wages was matched by an equivalent fall in the prices of consumer goods, so that the purchasing power of men's wages remained constant, the purchasing power of women's agricultural wages still fell from 1866 to 1868. This reduction in household purchasing power would be expected to induce a decrease in the supply of that portion of the household's labor least valued by the market. This would especially be the case if the labor valued least by the market were the most valuable in household production activities.

[10] Edmund L. Drago, "The Black Household in Dougherty County, Georgia, 1870–1900," *Journal of the National Archives*, vol. 14, no. 2, Summer 1982, p. 82. *BRFAL*, film M1048, roll 46, p. 38, Halifax County, Virginia.

[11] *BRFAL*, film M826, roll 32, Mississippi, 1867, p. 2; ibid., Feb. 1868, p. 189. *BRFAL*, film M1027, roll 30, Louisiana, Concordia Parish, Dec. 1867, p. 334. *New National Era*, Jan. 27, 1870. *BRFAL*, film M869, roll 35, South Carolina, 1868, p. 636.

Table 12.1   Male and Female Yearly Money Wages and Rate of Decrease by State, 1866–68

| | 1866 | | 1867 | | 1868 | | % Fall, 1866–68 | |
	Men	Women	Men	Women	Men	Women	Men	Women
Alabama	144	96	117	71	87	50	39.5	47.9
Florida	144	108	139	85	97	50	32.6	53.7
Georgia	132	96	125	65	83	55	37.1	42.7
Mississippi	168	114	149	93	90	66	46.4	42.1
North Carolina	120	72	104	45	89	41	25.8	43.0
Tennessee	144	96	136	67	109	51	24.3	46.8
Louisiana	157	144	150	104	90	66	42.6	54.1

SOURCES: 1866 wages estimated from labor contracts in *BRFAL*, Record Group 105, National Archives, 1867 and 1868 data taken from U.S. Department of Agriculture, *Report of the Commissioner of Agriculture, 1867* (Washington, 1868), p. 416. Louisiana, 1866, obtained by ordinary least squares regression estimates.

Bureau agent James DeGrey of East Feliciania Parish, Louisiana, recorded at the end of 1867 that "a great many women who worked last year, did not earn a cent" in 1867. DeGrey, who believed that "one great trouble among the freedpeople" was that they supported "too many consumers to the producers," took the trouble to inquire of the freedwomen why they had abondoned the labor market. The women gave "as a reason that their husbands [could] support them as well as the White folks [could], and that at any rate they would not have any more coming to them at the end of the year than if they had worked."[12] If one stresses rather than merely mentions the fact that the withdrawal of female labor from the marketplace was far from a withdrawal of females from labor, the women's reply takes on a significance that goes beyond self-rationalization. The opportunity cost of female market labor may have been highest in its effect upon the production of household consumption.

In addition to the obvious job of caring for her family and insuring the reproduction of the planter's labor force, the freedwoman was generally given primary responsibility for the maintenance of the family garden. Molly Mitchell's work activities on the Georgia farm where she cared for nine children and a husband required that she "spin the cloth" for their clothing, prepare herbal medicines to cure "worms" and "upset stomachs," and do "men's work too." "Us always kep a good gyarden full of beans, corn, onions, peas and taters, an dey warn't nobody could beat us at raisin lots of greens," reminisced Georgia Telfair: "us always made a good livin on de farm, an still raises mos of what us needs." How important these activities could be was explained by freedman Andy McAdams: "If we did other work we did not get fair wages for it, they were so low that we could not

[12] *BRFAL*, film M1027, roll 30, p. 348, Dec. 1867.

feed ourself and so we farmed. We would make more that way because when we farmed we growed near about all that we ate and made our own clothes with [a] spinning wheel."

To do these tasks in a manner which allowed the quality of home care to improve, the women naturally required more days off from market labor than did men. Under the share squad system these requirements added to the familiar problems of justice and collectivity. O. H. Violet, bureau agent in St. Landry Parish, explained in 1868: "On some of the plantations [women] will go into the field a day now and then provided they get paid by the day as soon as they desire to stop work. This does well enough on sugar plantations where all the labor is paid for monthly and most women on such places earn money enough to clothe themselves, but on cotton plantations where laborers are working for shares it makes much confusion and causes trouble to determine the exact proportion of each one's labor."[13]

To state that the labor force participation of female agricultural workers fell after emancipation is to do little more than state the obvious. An examination into the causes of this reduction forces an examination of the long pay and its effect upon the southern economy.

> The problem how to make the Negroes work cannot be satisfactorily solved until the system of prompt and frequent payments is adopted. However small the sum may be let the laborer have it every Saturday night. Holding it off to the uncertain future has been productive of most of the uncertainty and embarrassment of labor which now prevails.[14]

This common opinion among agents of the Freedmen's Bureau was occasionally accompanied by prescriptions for more religion, education, and discipline. How much the implied inadequate labor supply complained of by the employers was attributable to a deficit of these last three qualities and how much to the grossly inadequate work incentives provided by the long pay is a question of some importance for the economic history of all industrializing societies. So-called preindustrial work habits are certainly explained to some extent by preindustrial payment schemes.

Payment by a share of the crop is often lauded as a method of increasing the laborer's incentive to work. Many planters agreed with modern analysis upon this point, but just as many, and probably more, disagreed. The disagreement is basically concerned with the timing of labor payments. Invariably, when some planter or observer is found favoring wages over the share plan, it is not standing wages he refers to but wages promptly paid over short periods. Where modern analysis differs is in its common failure to distinguish between the waiting time for payment in the two systems. Therefore, modern discussions implicitly assume standing year wages as the alternative to shares. Because of the preferences of employers and workers for shares over year wages, the long pay and, more precisely, the credit

[13] Rawick, vol. 13, part 3, *Georgia Narratives*, p. 134; ibid., part 4, p. 39; ibid., vol. 7, part 6, *Texas Narratives*, p. 2456. *BRFAL*, film M1027, roll 30, March 31, 1868, p. 666.
[14] *BRFAL*, film M809, roll 18, Alabama, 1868, p. 537.

problems behind it led to the decentralization of production. The continuing claims that the share system was generally adopted because of the planters' inability to pay "prompt wages" specifies this as a major problem for the postbellum South. It was no accident that when discussing the problem of "waiting" in reference to the twentieth-century "laborer's willingness to work for *less* wages when he is paid daily or weekly than when he is paid semi-monthly or monthly," economist John R. Commons chose as an illustration of the point his personal observation of southern "Negroes who consider[ed] *daily* payment of wages more important than the *rate* of wages." [15]

With respect to the supply of labour during Reconstruction, bureau agents and planter-reformers were in effect chronicling the same consequences of the long and uncertain pay which earlier social critics and reformers found so detrimental in other regions of the world. The Reverend Thomas Gisborne's "observations" upon the social condition of colliers near Manchester, England, at the turn of the nineteenth century, like those of mid-century American critics of southern agricultural laborers, ascribed much of the problem to the effect of the long and uncertain pay:

> Their gains are large and uncertain; and their employment is a species of task-work, the profit of which can very rarely be ascertained. This circumstance gives them the wasteful habits of a gamester, leading them to trust, without forethought to apprehension, to the extraordinary success of tomorrow, for the support of their families.

The Reverend Gisborne reported with approval that an exception of this behavior could be found among the colliers of the duke of Bridgewater, who were paid monthly. The implied advice would be given innumerable times in the postbellum South. In 1868 agent James Gillette in Mobile County, Alabama, offered this advice and then made a claim that runs counter to the dominant view of Reconstruction historiography:

> *Agricultural affairs* are becoming more settled than they were wherever adoption to the system of *prompt payments* has taken place. The system of retaining half wages to the end of the crop breeds distrust, laziness and difficulties in settlement. The few planters who pay weekly or monthly in full, are doing well. They did well last year as they had the choice of labor. . . . There is plenty of labor of the best kind to be had by those who convince laborers of their ability and intention to pay by frequent and full payments.

The point was made too often to be ignored.[16] At its heart the issue involves the efficiency of sharecropping as an agricultural institution. Economic theory suggests that sharecropping will result in an underallo-

---

[15] John R. Commons, *Institutional Economics*, vol. 2 (1934, reprint Madison, 1961), p. 507.

[16] *Society for the Poor Reports*, p. 191. BRFAL, film M809, roll 18, Alabama, May 21, 1868, p. 523; ibid., p. 14. Charles Nordhoff, *The Cotton States in the Spring and Summer* (1875), North Carolina, p. 99, Georgia, p. 102. Campbell, *White and Black*, pp. 304, 310, North Carolina.

cation of labor and other productive inputs to the land because tenants only receive a portion of the return to their inputs. The problem of labor control under the long pay is consistent with this conclusion.

As early as 1862 Edward Philbrick had complained that "long-delayed payments" prevented him from exercising proper "control" over his work force. By lack of control Philbrick meant the inability to prevent absenteeism and inefficient work habits. Unable to speed up the government payroll, Philbrick adopted the same second-best solution as the planters and merchants would later—but with an important variation. Instead of withholding daily rations from nonperforming workers, he gave ration bonuses to those who worked well. Elsewhere even this minor improvement was insufficient to circumscribe the exorbitant difficulties of managing labor under share payments.[17]

Much space has been devoted to an examination of the work behavior of the freedpeople. In Chapter 5 this issue was discussed in terms of the evidence offered by various contemporaries who observed the freedpeople at work. There were two standards of comparison: how much work the freedpeople performed relative to their forced performance as slaves, and how their work habits compared with those of other groups of free workers. Taking into account the different presumptions and biases of those offering testimony, the conclusion of Chapter 5 was that the work habits of the freedpeople compared very favorably with those of contemporary free laborers elsewhere. Quantitative evidence on rates of absenteeism among field laborers supports this conclusion well. Work records compiled by planters in the state of Arkansas during the initial full year of freedom are

Figure 12.2 Comparative Absenteeism Rates
Black Agricultural Laborers, Arkansas, 1866, and Southern and Northern
Textile Weavers, 1922.)

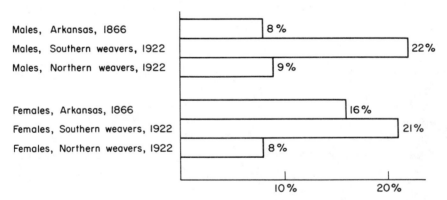

Sources: <u>BRFAL</u>, film M979, rolls 49–51, Arkansas. Abraham Berglund, George T. Starnes, and Frank T. DeVyver, <u>Labor in the Industrial South</u> (Charlottesville, Va., 1930), p. 85. Estimation methods and data base are described in Appendix A.

[17] Pearson, pp. 111, 57, 208.

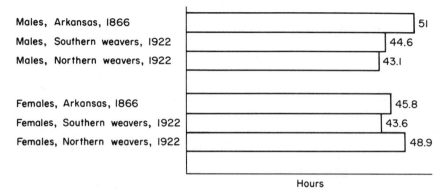

Males, Arkansas, 1866 — 51
Males, Southern weavers, 1922 — 44.6
Males, Northern weavers, 1922 — 43.1

Females, Arkansas, 1866 — 45.8
Females, Southern weavers, 1922 — 43.6
Females, Northern weavers, 1922 — 48.9

Hours

Sources: BRFAL, film M979, rolls 49–51, Arkansas. Abraham Berglund, George T. Starnes, and Frank T. DeVyver, Labor in the Industrial South (Charlottesville, Va., 1930), p. 85. Arkansas figures are calculated on the basis of a 52-week work year and ten hours per day as required by contract.

Figure 12.3   Average Hours Worked per Week
(Black Agricultural Laborers, Arkansas, 1866, and Southern and Northern
Textile Weavers, 1922.)

illuminating. Arkansas freedpeople worked very well during that first important year. The sample by the Freedmen's Bureau, constructed from employers' reports of time lost by employees, indicates that black males worked an average of 5 full days a week throughout the year. Freedwomen averaged just over 4.5 days a week. On the basis of a contractual 5.5-day work week, normal after emancipation, this gives average rates of absenteeism of 8 percent for males and 16 percent for women.

These figures are most useful as an indicator of work habits if they can be compared with similar data on the absenteeism of other workers. The absentee rates of freedpeople in agriculture probably compared well with those of contemporary white textile operatives in southern mills, if the complaints of high absenteeism in that industry are reliable. A more precise comparison can be made with data collected for cotton weavers in southern and northern mills during 1922, an average year during the period 1918 through 1928. As shown in figures 12.2 and 12.3, the freedpeople's delivery of steady labor on a day-to-day basis compares very well with this standard. These numbers overestimate the number of days voluntarily lost by field hands, because inclement weather or other uncontrollable forces frequently halted fieldwork.

Based upon L. C. Gray's estimates of the number of days worked in a year by slaves, we find that the postemancipation reduction in days worked for males and females was about 5 and 12 percent respectively. The rates of absenteeism shown in figure 12.2 aggregate share hands and wage laborers. According to the data, share hands were less reliable on average than those paid a money wage (see figure 12.4). For men this difference is not great, about one-half day less work a month, but for women the much higher rate of absenteeism among share laborers supports the view that a primary cause

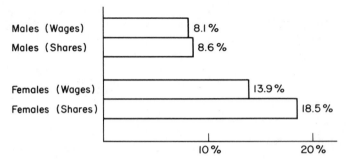

Source: BRFAL, film M979, rolls 49-51, Arkansas.

Figure 12.4   Comparative Absentee Rates
(Share laborers and Wage Hands, Arkansas, 1866)

of erratic work habits was the disincentive effects of the long pay and the "free rider".

While their choice of the phrase "Protestant work ethic," with all that it connotes, as a characterization of the work habits the slaves brought into freedom seems excessive, the evidence does support a modified version of Stanley Engerman and Robert Fogel's prognosis of early black work patterns after slavery. The blacks were not Calvinists but a group well endowed with attitudes toward *work for wages* consonant with the characteristics of any contemporary working class.[18]

After 1866, the increase in share payments and the failure of a significant percentage of planters to meet their payrolls should have been accompanied by reductions in both the demand and supply of labor. The theoretical argument that share payments provide incentives fails to take into account human psychology. When the crop promises to be bountiful and prices high, the share laborer may be expected to exert himself due to the spur of increasing profit. The rational economic calculating organism created by neoclassical economics makes little distinction between the work effort induced by the prospects of maximizing a positive profit and that put forth in minimizing a negative loss. Bureau agent James DeGrey, like Reverend Gisborne, knew better:

> At the beginning of the season the laborers seemed to be very cheerful and contented. On the 15th of January [1867] not one was out of employment, but

[18] See Ransom and Sutch, *One Kind of Freedom*, pp. 44–47, and Appendix A. The aggregate reduction in days worked, combining males and females, would then be toward the low end of the pioneering estimate of 8 to 11 percent made by Ransom and Sutch. There is a caveat, however. My estimate includes children and therefore may overestimate the aggregate for adults. However, my estimate is intended to account for 1866 only, a year that probably had lower rates of reduction in labor than later. Fogel and Engerman, *Time on the Cross*, pp. 208, 260. Lewis Cecil Gray, *History of Agriculture in the Southern United States to 1860*, 2 vols. (Washington, D.C., 1933).

as the season advanced and the rainy weather continued, and the grass, out-
growing the cotton plant; they became discouraged, and finally when the
worm made its appearance, they became indifferent as to the results of the
crops, and remained so till now. On some plantations they could hardly be
induced to gather what little cotton they did make. The reason for this
dissatisfaction, was that they saw no prospect to make anything; they said they
had worked two years and got nothing; and that the more they worked the less
they had.

Alternating seasons of relative prosperity and despair proved to be a
countervailing power to the supposed incentive effects of share labor. By
the middle 1870s a satirist, with more insight than he realized, claimed that
the favorite song of the black farmer began

> Nigger work hard all de year
> White man tote de money.[19]

The inability to control labor was a primary reason for the failure of the
share system to develop in rice and sugar culture. It has been argued that the
large capital requirements of these two industries, coupled with their
factorylike grinding and milling requirements, precluded the spread of
share payments. While there is much to this assessment, it is weakened by
the fact that the cultivating process in the fields could be separated in time
and space from the factory requirements of grinding cane or milling rice
during the harvest.[20] It is more likely that the specific production techniques
of American rice and sugar planting had more to do with the failure of
shares.

The production of rice and sugar in the United States during the nineteenth
century was most distinct from cotton and tobacco culture in the relationship of the first two to water. Rice and sugar plantations, with their land
areas divided by intricate networks of ditches and irrigation canals, might
profitably be described as huge asymmetrical chessboards. The distinction
was greater in the cultivation of rice, a well-regulated rice plantation has
been aptly described as a gigantic hydraulic system. To bring the water in
from the river, rice plantations were divided into small sections by crossing
canals so that various fields could be alternately drained and flooded as the
crops required. "Often, in the case of fields lying side by side, one must be
under water and another be perfectly dry." The timing and control of these

---

[19] *BRFAL*, film M1027, roll 30, E. Feliciania Parish, Louisiana, Dec. 1867, p. 248. Edwin
DeLeon, "Ruin and Reconstruction of the Southern States," *Southern Magazine*, vol. 14,
1874, p. 306. *BRFAL*, film M1027, roll 28, July 31, 1867, St. Landry Parish, p. 665.
Montgomery Advertiser, Sept. 19, 1866, in Rogers, *Agrarianism in Alabama*.

[20] This assessment was adopted by the sugar planters, who, prior to capitalization and the
emergence of a stable wage system, were increasingly advocating the leasing of small farms to
tenants who would independently harvest and deliver cane to a central sugar factory controlled
by the landlord or an independent corporation: Sitterson, *Sugar Country*, pp. 258-60, *New
Orleans Daily Picayune*, Sept. 4, 1873, p. 2.

activities required a coordinated labor force willing to devote major portions of its working time to the preparation and maintenance of levees, irrigation systems, and plantation infrastructure. These requirements were exactly the activities that share laborers refused to perform.

Agricultural expert James Killebrew, referring to Tennessee tobacco hands in the middle 1870s, voiced the planters' chief objection to the share system: "It is found that the amount required to procure extra labor to do what should legitimately be done by 'croppers' consumes by far the largest share of the profits of the farm." The problem for rice and sugar plantations was that work outside the crop on those farms demanded a larger share of the total year's work and was more important to successful farming. These planters simply could not allow their drainage ditches to go unattended for any length of time. In addition, the intricate systems of levees and irrigation ditches requiring regular attendance and coordination of group labor did not mesh with the noncooperative nature of share groups, who often refused to work upon one another's land or in concert for supposedly common interests.[21] For example, in the lowlands there was "a general disposition on the part of the laborers not to work in the ditches of the rice fields." On the sugar plantations the problems were similar. The leading student of the industry explained that under shares "teams were abused, farm implements destroyed, drainage ditches allowed to fill up, and the crops from time to time so badly neglected that the harvest was endangered." In 1874 one commentator concluded that "successful culture of sugar cane and the share system [were] incompatible."[22]

Actually the share system was never numerically significant in sugar culture. The successful defense against the encroachment of the share system is explained by the higher costs incurred by the lack of complete central control over the organization of work and by the significant influx of credit capital made possible by the reorganization of the ownership structure of sugar plantations. The process was initiated by a large turnover of proprietorship to northerners, which one historian has estimated reached as high as 50 percent by 1870. The production of sugar became increasingly capitalized under the corporate form of ownership.[23]

Postbellum problems confronting southeastern rice planters were largely caused by increasing relative costs of production due to technology advances favorable to their competitors in the southwest, and to insufficient

---

[21] James Killebrew, *Resources of Tennessee* (Knoxville, 1874), pp. 350–51. Somers, pp. 60, 147.

[22] BRFAL, film M869, roll 35, Georgetown, S.C., Oct. 1867, p. 490. Sitterson, *Sugar Country*, pp. 240, 241. In the rice fields much of the ditching and preparation of soil was given over to white labor.

[23] Sitterson, pp. 262–68. Roger Shugg, *Origins of Class Struggle in Louisiana* (Baton Rouge, 1939), pp. 248–49. Nordhoff, p. 69. A hidden aspect of this ability to avoid the share system was the large percentage of plantations left inoperative during Reconstruction. Less than a quarter of the prewar sugar mills were in operation in 1870. By 1879 this number had climbed to 85 percent. Mark Schmitz, "The Transformation of the Southern Cane Sugar Sector, 1866–1930," *Agricultural History*, vol. 53, Jan. 1979, pp. 272–73.

credit.[24] The failure of rice production in this region is often blamed on a collapse in the quantity and quality of labor supply. Again the charge has some credibility, but the major influence of the long pay is seldom discussed. From Georgetown, late in 1865, Lieutenant Colonel A. J. Williard pinpointed the difficulties that short and expensive credit would cause coastal rice planters throughout the rest of the century:

> The planters generally are unwilling to contract on the basis of a division of the crops for another year. The result this year has been unsatisfactory. They are disposed to pay for labor either by the month or by the task. They have not got the means to pay cash, and hesitate in regard to substituting any other mode of payment.

These planters, irrespective of their wishes, continued share payments for the next two years. The source of their dissatisfaction was communicated by bureau agent Allan Rutherford, who, in the fall of 1867, had just completed an inspection tour of North Carolina rice plantations on the Cape Fear and North East rivers. Rutherford reported to his superiors in Wilmington that although the wage system was working well, "the share system [was] a failure and unprofitable to all parties." A detailed explanation shows that the central problem was one of labor control:

> This arises from the fact that the landowners have but little control or direction in the working or management of the crop, and altho in many instances of the share contracts it is stipulated that the work shall be done under the supervision of the owner of the land, the laborers feel quite independent of such contracts, and have worked as suited themselves.

Agent Rutherford believed "that few landowners or laborers [would] be willing to work on shares next season."[25] If North Carolina rice planters were reacting in a fashion similar to that of their counterparts in Georgia and South Carolina, Rutherford was probably correct in his belief.

Near the end of the summer of 1867 rice planter Benjamin Allston, deeply disturbed about the prospects of securing labor for the heavy investment required to prepare his plantation for the coming year, explained to a friend his reasons for abandoning the share system: "I shall be unwilling to pay now for preparing land, and then in January to contract for a portion of the crop. If we pay now, which seems the only way in which the work can be secured, we must pay throughout the year." Allston, along with most other rice planters, believed that variations upon the ticket wage system of W.

---

[24] The demise of the South Carolina rice planter is beautifully chronicled by two of the last planters to continue the operation of their ancestral antebellum plantations into the early twentieth Century. See Heyward, *Seed from Madagascar*, and Elizabeth Waties Allston Pringle, writing under the pseudonym Patience Pennington, *A Woman Rice Planter* (Cambridge, Mass., 1961). Heyward, pp. 33–34, attributed much of the loss of rice production to Louisiana, Texas, and Arkansas to the inability of southeastern planters to use newly invented rice harvesting machinery on the boggy soil necessitated by the flow-tide planting system. Thus the technique of production which required intensive use of expensive labor became outmoded.

[25] *BRFAL*, film M869, roll 34, South Carolina, Nov. 13, 1865, p. 323. Ibid., film M843, roll 22, North Carolina, Sept. 1867, pp. 47–48.

Miles Hazzard would be cheaper and offer "the best prospects for getting work from the Negro." [26]

The near universal demand for controllable wage labor and the inability to command the credit to make prompt wage payments led the rice industry into the worst abuses of the long pay. A widespread attempt to circumvent the share system and the credit requirements of wages was a combination of truck and labor renting. The origin of labor renting was explained by South Carolinian Harry Hammond:

> A widely adopted system is one proposed as early as 1866, by a negro laborer in Silverton township. The laborer works five days in the week for the land-owner and has a house, rations, three acres of land, and a mule and plow every other Saturday to work it when necessary, with sixteen dollars in money at the end of the year.

Hammond, who believed that the system was successful, did not mention some of the defects inherent to the practice. A bureau agent reporting in 1867 provides a somewhat different interpretation of the movement to labor renting:

> Many freedmen who last year found that their expenses exceeded their earnings insist this season upon having lands rented to them. Planters take advantage of this by renting them lots containing three or four acres each, and requiring as rent two days labor each week. This at fifty cents per day would make $52 a year for three acres—this is more than the land is generally worth.

The agent then exposed the primary defect of labor renting: "Considering that the freedmen must work for other parties the greater part of the re-maining four days" in order "to obtain subsistence, their prospects for the year are gloomy." [27]

The primary problem was the timing of labor. As under European manorial serfdom, to which this system is analogous, conflicts generally developed over the landlord's right to the tenants' labor at crucial moments when the tenants valued their time more highly on their rented plot, or for work for immediate subsistence elsewhere. That this problem existed in the postbellum south is made clear by the scathing critique of the system given by agent James M. Johnston:

> That miserable form of contract wherein the laborers give 3 days service each week for use of as much land as he can cultivate the remainder of the times with half rations: which system can never prosper and especially this year,

[26] James Easterby, ed., *The South Carolina Rice Plantation as Revealed in the Papers of Robert F. W. Allston* (Chicago, 1945), p. 231. Benjamin Allston was an older brother of Elizabeth W. Allston Pringle (see note 24). Leigh, *Ten Years on a Georgia Plantation*, pp. 127, 226.

[27] Harry Hammond, *Handbook of South Carolina* (Charleston, 1882), pp. 83, 84. *BRFAL, Synopsis of Reports*, 1867, p. 91. Julius Fleming, Sumter correspondent of the *Charleston Courier*, endorsed labor renting as an improvement over shares, because it avoided the dif-ficulties of division and gave the landlord full control over his entire crop: Moore, *The Juhl Letters*, p. 252.

when the freedmen generally commenced the year with no means of buying provisions, and consequently the greater portion of their [remaining] (3) days is occupied in laboring elsewhere to obtain means to feed themselves and families. And it naturally follows, they soon become dissatisfied and disheartened, and the labor he gives the owner of the lands is an unwilling service, and many excuses are framed for violation of their agreements by feigned sickness, etc.

These specific complaints about this system, we will recall, had already been voiced by northern planters in the earliest days of Reconstruction. Wherever labor renting was prevalent, the charges of inferior labor were most pronounced. Inherent in the method of remuneration, the problems would show no respect for time, continuing into the twentieth century.[28]

If the conflict resulting from the allocation of labor between the landlord's crops and the tenant's was most transparent under labor renting, that is not to say that this conflict was nonexistent under other nonwage forms of remuneration. The cause of these same problems under the European domestic system is often ascribed to the decentralization and dispersal of production which lessened the employer's ability to adequately supervise the domestic laborer's work rhythm. Relatively loose supervision over the employee's time was certainly a necessary condition for the existence of semiautonomous choice over work rhythms, but again the long pay must have played a significant role.

If the domestic outworker received no immediate payment for work finished quickly, then there existed no benefits to doing so, and in any case the allocation of time toward the employer's business had to be balanced with work in the employee's garden or the chance to earn extra income or subsistance elsewhere. "When I get out of corn and out of meat both," explained freedman John Lewis, "and anybody has got corn and meat, I jump out and work for a bushel of corn and a piece of meat, and work until I get it." This behavior was castigated by employers throughout the South.[29] It explains a great deal about the conflicts inherent to sharecropping. Once the share contract was entered into, the cost to the employer of any increment of labor that could be induced from the laborer was zero. Without a wage rate neither party had any consistent and common value by which to measure the benefits of additional labor. The landlord's incentive is to coerce labor from the tenants, who, although they will not necessarily seek to do as little as possible, will attempt to perform less labor than the employer demands. Ultimately the situation, while it may show some sensitivity to abnormal periods of extreme excess supply or demand in the labor

---

[28] *BRFAL*, film M869, roll 35, p. 137, South Carolina. Williamson, *After Slavery*, p. 101. Heyward, *Seed from Madagascar*, pp. 184–85. J. C. Hemphill, "Agriculture in South Carolina and Georgia," *Department of Agriculture, Special Report no. 47*, 1882, p. 12. The reference to serfdom is not intended to imply social equivalence, only similarities in the "purely" economic relations.

[29] *KKK Investigations*, South Carolina, p. 436. *Report of the Department of Agriculture, 1867*, p. 421. *BRFAL*, film M1027, roll 29, Louisiana, p. 572.

market, will generally be resolved by social custom, due to the bargaining power of labor as a whole versus the landlords as a whole.

Wealthy planters and agricultural experts who favored the wage system over shares ascribed their preference to the relative inefficiency of the share system, under which the employer lacked control over the quantity, quality, and timing of the laborer's supply of labor. To contemporaries, this inefficiency was due to the inherent inferiority of the laborers, but, as mentioned earlier, economic theory predicts that just these problems will inevitably exist under a share contract. To substantiate their claim that the wage system was superior, its supporters often offered quantitative evidence that output per acre was lower on share-organized plantations.[30] This line of argument was well conceived, for the relative efficiency of share and wage plantations is a question that can only be answered, if at all, by quantitative methods. With a given endowment of land, labor, and capital, which organization of the labor force would produce a greater output? More precisely, taking the ratio of output to input, which organization has the higher ratio? This the partial productivity index, output per acre, does, but it has the insurmountable problem that it ignores the productivity of other inputs. Wage plantations may have produced more output per acre, but this may have been accomplished by using more capital and labor per acre, so that the wage plantation may not have been more efficient at all. From both a conceptual and empirical framework, the best approach to a comparison of efficiency is to compute a geometric index of total factor productivity for the two organizations. This index measures productive efficiency by taking the ratio of total output (revenues) produced to a measure of an average amount of inputs used. Equivalently, the measure is a nonlinear average of the three partial productivity indexes: output per unit acre, labor, and capital.

The usefulness of this method depends upon the reliability of the data base and estimation techniques used to construct the index, and upon the validity of the method of averaging the index of inputs. The usual method of making this kind of efficiency analysis is to construct an aggregate index

[30] It is this unavoidable clash between labor and management under share contracts which no doubt led Professor Mandle to conclude that sharecropping is a nonmarket mechanism for allocating labor. In a sense he is correct, but in the economist's parlance there exists an incomplete set of markets, i.e., there exists no "price" for a marginal unit of labor. The importance of this missing price is clearly shown in the work of economists like Steven Cheung who attempt to prove that sharecropping is efficient. Cheung and others are forced to assume that all sharecroppers could choose to labor in a competitive wage market if they so desired. This wage market supplies the price for labor that sharecropping cannot. See Steven N. S. Cheung, *The Theory of Share Tenancy* (Chicago, 1969), chapt. 2; but also see Pranab K. Bardhan and T. N. Srinivasan, "Cropsharing Tenancy in Agriculture: A Theoretical and Emprirical Analysis," *American Economic Review*, vol. 61, 1971; Gerald D. Jaynes, "Production and Distribution in Agrarian Economies," *Oxford Economic Papers*, Aug. 1982; and Joseph E. Stiglitz, "Incentives and Risk Sharing in Sharecropping," *Review of Economic Studies*, 1974, for alternative views. J. C. Hemphill, "Agriculture in South Carolina and Georgia," passim. John Caldwell Calhoun, "Testimony—Agriculture in Arkansas," *Report on Labor and Capital*, vol. 2, p. 158.

for each class of organization to be compared. This is done by summing outputs and inputs over each farm within a class. The indexes for each class are then ranked in order of magnitude. The approach used here departs from this procedure by constructing an index of productivity for each individual plantation. This more microeconomic construction permits a statistical analysis of comparative organizational productivities through the technique of simple regression analysis. In this way the independent quantitative effect of wage and share organizations upon plantation efficiency can be tested by controlling for other influences, such as increasing returns to scale, and allocative decisions, such as the choice of crops produced. The regression analysis also produces results which allow a determination of the statistical significance of the quantitative relationships. The data base used is a sample of plantations from the 1880 census manuscripts of cotton farms. I believe the methods used in measuring inputs and in constructing the weights used to average input use are a considerable improvement upon previous work in this area. Since these methods do diverge in considerable detail from other approaches, a thorough description of methods and data is presented in Appendix B.

The results of the analysis of comparative organizational efficiency can be summarized in two statements:

1. Wage organized plantations were about 35 percent more efficient than share organized plantations. The statistical test formulates this result in the following precise way: holding constant the total acreage cultivated in all crops and the percentage of those acres planted in cotton, a 1 percent increase in acres worked by wage labor (1 percent reduction of acreage worked by share hands) would on average increase the productive efficiency of a *given* bundle of inputs by about .35 times 1 percent.
2. Under free labor there existed no productive economies of scale on cotton plantations. Again in the precise sense of the statistical test, we must *reject* the hypothesis that, holding constant the crop mix and the organization of the labor force (wage or share), an increase in acres cultivated would produce more output per unit of inputs.[31]

An equivalent way of expressing the first statistical result is that conversion of a plantation worked completely under the share system to a complete wage organization would result, on average, in a 35 percent gain in productive efficiency. As explained in full detail in Appendix B, the construction of the index of total factor productivity implies, in the logical meaning of that word, that the discrepancy in efficiency is due to the undersupply of labor by share laborers. This implication is consistent with the known existence of incentives for share laborers to reduce their labor effort below levels which under a different form of payment would result in an improvement in the welfare of themselves and their employers. Part of the apparent voluntary reduction in income may perhaps be explained by an

---

[31] It would be a gross error to infer from the nonexistence of returns to scale with free labor that no returns to scale existed with slave labor. That issue remains an open question.

attempt to spread market risks by organizing part of the plantation under the share system. In the sample used here, the average proportion of plantation acreage worked under shares was 70 percent for those plantations that did not use a complete wage organization. The implied reduction in efficiency of 25 percent seems far too large to be justified as a risk premium.

Why would planters and laborers voluntarily forsake such a large percentage of their income? A reasonable answer, consistent with the structure of the economy, is that the choice was not voluntary. The wage plantation required that the planter have access to enough credit to finance operations and guarantee the wage payment. Without that guarantee the laborers refused to engage in wage contracts. As a bureau agent explained:

> The great source of distrust that freedmen have of the planters is because of this failure to pay them for their labor, many make a scanty living from a little piece of ground rented as a garden, that would gladly work steadily on a plantation, if they felt sure of their wages for this service.[32]

The true legacy of federal agrarian policy in 1865 was the failure to provide the financial assistance to southern agriculture that would have allowed a restructuring of a sounder credit system. The irony of this failure is that the end result could have been achieved by a variety of redistribution schemes, each of which favored a different group. A plan which discriminately granted or loaned capital and land to freed slaves would obviously have improved their welfare. The lack of increasing returns in cotton production ensures that this decentralization of the plantation by creating a class of independent owners–not sharecroppers—would have resulted in no efficiency losses. Perhaps more feasible would have been a policy of extending federal credit to southern planters. Such a policy would have allowed employers to avoid the initial problems that enveloped them in the credit cycle that eventually emerged. The freed laborers, we have seen, presented no initial obstacle to the establishment of a money wage labor market if the wages were secure.

Even the all-important crop failures of 1866 and 1867 were largely a consequence of ineffectual federal policy. For the spring floods which drowned the crops were due to a failure to repair and maintain river levees. This was an investment few planters had the time or funds to undertake without aid. Even the destruction of the cotton crop by the cotton worm in the fall could have been largely avoided. Planters unable to obtain loans early were forced to begin their cleaning and preparation of the land for planting much later than the antebellum dates for such procedures. The uncertain government policy led many Negroes to abstain from entering labor contracts early, on the chance that they might receive federal grants or assistance. If the lands had been cleared and planted early, the cotton worm larvae might have been destroyed in their nests, and in any case an early crop might have been harvested before the worms took it. The cotton worm did not appear new to the South in 1866, and the occurrence of two, in many places three, suc-

---

[32] *BRFAL*, film M809, roll 18, p. 0772, Alabama.

cessive crop failures was not an event due solely to an unfortunate roll of the dice.

But in taking this view I leave myself open to the charge of political naiveté. Any well-conceived plan of agrarian reconstruction had to face the fact that it would cost money. The issues of a large federal war debt and an inherited price inflation have been discussed in Chapters 1 and 2, and there is no need to repeat that discussion. Even without those two obstacles, it was not an age that practiced Keynesian-type deficit financing to spur economic growth, and northerners were not reconciled to granting aid to southern planters who had just laid down their weapons. Nonetheless, it is difficult to ignore the cogency of prognostications of the long-term development of southern agriculture made by such people as Edward Pierce. Pierce, it may be recalled from Chapter 2, felt that the policy of limited federal involvement and provision of capital would ultimately result in failure.

Gavin Wright's work demonstrating the insufficient growth of world demand for southern cotton provides the background for this view. With world cotton demand growing at a pace that did not keep up with the postbellum growth of supply, United States cotton producers, collectively forming a major factor in world supply and therefore facing an inelastic demand curve, needed to retard their supply at or below the growth in demand if profits were not to stagnate. This the postbellum economy did not do. A major reason for this was the increased specialization in cotton as a percentage of acreage cultivated. It is important to note in this regard that a major factor in this specialization was the great increase in cotton acreage made by small farms and especially by sharecropped and rental tenancies, whose producers were most tied to the credit-debt mechanism identified with cotton specialization in the work of Wright and Roger Ransom and Richard Sutch.

Considering the relative productive inefficiency of the sharecropping system and its contribution to the growth of specialization in cotton, an early agrarian policy of providing federal capital to the South could have produced a more efficient, diversified, and prosperous agriculture. Even a limited plan of confiscation and loans of capital and land to the freedpeople would have set the long-term possibilities for prosperity upon a sounder basis. The black landholder, once free of high market rates of credit, evinced no propensity to specialize in cotton in lieu of food crops. Alternatively, a well-financed money wage system might have allowed the accumulation of savings for the purchase of homesteads, as many Republicans believed. In any event, we may agree with three northerners who, while inspecting southern Georgia and northern Florida in 1865, despaired of the widespread use of share payments: "we regard regular cash payments as better calculated to promote the industry of the freedmen as they will by that means see an immediate result from their labor."[33]

[33] Gavin Wright, *The Political Economy of the Cotton South* (New York, 1978). *BRFAL*, film M752, roll 20, June 12, 1865, p. 196. I am indebted to Gavin Wright for raising the question of the likelihood of three successive crop failures.

II

The use of the term "wages" for the method of payment practiced by most planters claiming the term is one of the true misnomers of the period. Probably no example of the impact of the long pay rivals that of the rice industry during the middle 1870s. We have had occasion to mention the "Combahee strike" of 1876. The reason for this strike was the workers' demand that planters stop paying their laborers with personal scrip redeemable at plantation stores at a high discount. Planter J. B. Bissell, it may be recalled, was prominently singled out by the strikers as deserving special condemnation. This may be understood when we learn that Bissell was paying his hands with scrip that was redeemable in cash only after a wait in excess of three years. The discount rate on such notes was infinite. No merchant would accept the tickets in exchange for goods at any interest rate. The consequence was that such notes were tradable only at plantation stores.[34] This seems like a perfect example of planter-merchant monopoly power. The problem is that it is too perfect. Below the surface we find the lurking presence of the main source of trouble.

In 1876, rice planters found themselves in the midst of a period of falling rice prices exacerbated by "the loss of their crops (by the June Freshet)" of 1876. In a situation where credit had already been tightened, the fact "that factors were losing money" converted a difficult situation into a credit crisis. Gabriel Manigault recorded in his journal that the factors were refusing to make advances. Manigault "sought in vain to obtain advances in Savannah but failed." In the end, he was forced to regress to the business option he had taken ten years earlier. Gowrie plantation was leased to James B. Heyward, who contracted to pay Manigault not the money rental of earlier years but one-half the net proceeds of the crop. This ironic situation and the conditions that caused it seem a much more plausible explanation of the rice planter's recourse to scrip payment than either monopoly power or exceptional cupidity.[35]

With the failure of the cotton crop in 1867, the impact of the long pay on the cotton belt came to rival its impact on the rice industry. Freedmen's Bureau records are replete with complaints such as the one made by freedman Floyd Bunkly of Bolivar, Mississippi, who in September 1867 charged that his employer "gave him a due-bill for $56.30, payable June 1, 1868." Mr. Bunkly's response may be taken as generic. "The amount is too small and the time too long," complained Bunkly, who added "that he [had] agreed to neither." Where the primary industry was so depressed it is not surprising that other industries were also affected. A classic example of the influence of poor trade conditions upon the form of payment received by laborers is the deal promulgated by the owner of a brickyard who, "unable to pay" his workers "the money (has not got it)," suggested that since he had "plenty of

---

[34] Sterling, *The Trouble They Seen*, quoting the *Savannah Tribune*, Sept. 2, 1876, p. 287. See above, Chapter 6.

[35] *Plantation Records of Gabriel Manigault*, pp. 36, 39.

Brick" he would "turn over to [them a] sufficient amount to pay the debt at the market price." [36]

The claim that these conditions were isolated or confined to a few short periods of recession seems weak. As we have seen, the poor conditions in the late 1860s due to crop failures and disorganization were general throughout the South. Unfortunately, the nationwide recession of the middle 1870s was not restricted to the Georgia-Carolina rice country. The condition of the period was well represented by an agent of the R. G. Dun Mercantile Agency in an 1874 report on the credit-worthiness of a plantation in once-wealthy Jefferson County, Mississippi: "I could not rec. them to unlim. cr. In fact I would advise great caution in dealing with any planters in this country as labor is scarce & nearly every one plants all cotton & during the past yr. but very few made their expenses." [37]

Arkansan Ambus Gray probably summed things up pretty thoroughly:

> The Reconstruction time was like this. You go up to a man and tell him you and your family want to hire for next year on his place. He say I'm broke, the war broke me. Move down there in the best empty house you find. You can get your provisions furnished at [a] certain little store in the closest town about. You say yesser. When the crop made bout all you got was a little money to take to give the man what run you and you have to stay on or starve or go get somebody else let you share crop wid them.

In 1880 James A. Stokes would cite the long pay as one of the major reasons for the mass emigration from North Carolina to Indiana:

> I could only get from five to seven dollars per month for labor, and was paid in orders at the store, and had to pay from ten to fifteen per cent above the regular prices for goods and groceries, because, as was said, the orders were "time orders"—that is, not payable for some months, they being paid in the fall and spring. But living with the most stringent economy, on the plainest fare, and working all the time, I could hardly keep out of debt. [38]

Even as conservative an assigner of social causes as British historian T. S. Ashton was driven to the conclusion that in eighteenth-century Britain "a close connection between the 'long pay' and embezzlement is evident." In the postbellum South many would-be social scientists were willing to go further in the assignment of cause and effect. "Criminal Cases," reported a bureau agent in 1868, "were principally confined to stealing and killing stock and to stealing anything which may sustain life." A planter writing concurrently from the same general region argued for the institution of cash wages and an end of the contract system:

> It is the fruitful source of half the stealing in the country. Does any one suppose that the freedmen, if found by their employers in rations and paid small

---

[36] *BRFAL*, film M826, roll 30, pp. 354, p. 378. The offer of brick was accepted by a portion of the men and refused by others. If the market price had meant anything concrete the employer would have sold the brick himself and paid the cash wages due.

[37] M. S. Wayne, pp. 106, 94, 106–7. R. L. F. Davis, p. 130.

[38] Rawick, vol. 9, part 3, *Arkansas Narratives*, pp. 78–79. "The Removal of the Negroes from the Southern States to the Northern States," *Senate Report no. 693, 46th Cong., 2nd sess., 1880*, p. 98 (hereinafter cited as *Exodusters*).

sums of money through the year will indulge in as many burglaries and stock killing scrapes as is now the case? By no means.

Ten years later, testifying before a congressional committee, the black editor of an Alabama newspaper offered a pithy statement of this socioeconomic cause of crime: "where a man has got a wife and child and gets $6 a month he cannot live and be honest." Israel Jackson, recalling his childhood during Reconstruction in Yalobusha County, Mississippi, told how his parents had taken care of their family. "No maam, dey did not pay em. I'se old but I ain't forgot dat. Dey fed theirselves by stealin and gettin things in the woods." [39]

That much crime, perhaps most petty larceny, was due to the long pay and hunger seems too likely to be reasonably doubted. This in turn undercuts the long-standing myth that rural peoples, because they are on the land, do not suffer from the spells of unemployment and lack of subsistence that an urban proletariat must periodically undergo. The reports of mass unemployment due to the landowners' inability to obtain the credit to support a workforce are too numerous throughout the Reconstruction period to accept such a view when a large percentage of the rural population are landless employees. At times, such as in 1874 Alabama, when large numbers of laborers were thrown out of employment by a severe crop failure due to the ravages of the army worm, the field laborers were driven to extraordinary measures, such as the unsuccessful sortie by an "armed band of blacks" who rode into Birmingham starting fires in one section of town "with the design of diverting attention" until "the provision stores were robbed." [40]

All crime, however, was not directly a cause of hunger. A serious problem for employers, common wherever the long pay and decentralized production were found, was the increase in embezzlement consequent upon their weakened ability to control the distribution of produced goods. If the nineteenth-century English weaver or journeyman could always find an obliging publican or small merchant who was willing to carry out a nocturnal trade, the nineteenth-century American sharecropper could do the same, locating a buyer for rice or tobacco surreptitiously harvested from the common fields. Complaints about such practices were made early in the Carolina cotton districts, where planters desired "to give to the freedpeople a share of the net proceeds of cotton instead of the cotton itself intending thereby to prevent dishonest Country and River traders to play their games like last year by inducing the freedpeople to steal the cotton and trade it to them." By the early 1870s the traffic in stolen seed cotton is supposed to have reached such heights that legislation was proposed to make it "penal to trade with freedmen after dark." [41]

[39] Ashton, *Economic History of England*, p. 209. *BRFAL*, film M869, roll 36, July 1, 1868, Sumter, South Carolina, p. 28. Chartock, p. 95, quoting *Anderson Intelligencer*, Aug. 5, 1868. *Ku Klux Klan Hearings*, vol. 20. Rawick, vol. 9, part 4, *Arkansas Narratives*, p. 6.

[40] King, *The Great South*, pp. 331, 337.

[41] *BRFAL*, film M869, roll 34, Monck's Corner, South Carolina, May 31, 1866, p. 531. *Galaxy*, p. 333. Fleming, pp. 769–70.

The problems of labor discipline and embezzlement, both traceable to the long pay, formed much of the basis for the planters' legitimation of their involvement in the Ku Klux Klan. For if these problems were not structural, not due to the failure or nonexistence of social institutions, but to the personal biosociological defects of black people, it was not difficult for ex-slaveholders to accept the view that justice and the natural order of society sanctioned the whites' violent means of addressing these problems. It will be more convenient to discuss these matters carefully in the final two chapters. For the present it should simply be noted that the declarations of just provocation made by the planter class were not wholly hypocritical. The planters we have seen were well aware of the close connection between industrial organization and social discipline. In sanctioning various groups of "regulators," planters perceived these organizations as variants of William Wilberforce's Society for the Suppression of Vice or the ill-famed Worsted Committee of master clothiers who banded together to protect themselves from the transgressions of those perceived as lazy and larcenous outworkers in late eighteenth- and early nineteenth-century Britain. The question which needs addressing is not the personal character of planters or master clothiers but the moral legitimacy of external interference in the affairs of individuals in a putatively free society.[42]

Reconstruction historiography has generally treated the material contained in this chapter separately. This chapter has sought to show that this trichotomy is a false one. The decrease in agricultural output, the fall in the supply of black labor, and much (though not all) of the crime and interracial violence were direct consequences of inadequate cotton prices, insufficient credit, and the long pay.

[42] Davis, *The Problem of Slavery in the Age of Revolution*, p. 461. Herbert Heaton, *The Yorkshire Woolen and Worsted Industries: From the Earliest Times Up to the Industrial Revolution* (Oxford, 1920), pp. 418–37.

Part IV
Economic Origins
of the Color Line

Jus as de niggers was branchin out and startin to live lak free folks, dem nightriders came long beatin, cuttin, and slashin em up.
William McWhorter

They hung Wyatt, Outlaw, who was a man of sufficient influence to be elected mayor of Alamance.
R. C. Badger

13

They want niggers to stay bad enough; and most of em haint got no mo use for we po men than a coon has fur Sunday. Thats what makes niggers sech a cuss to us. And any furriner as comes hyur in reggard of benefitin of hisself, he's a comin to a goat fur to git wool.

Tammany Jones (white)

We have got the best mechanics there is on the globe in regards to carpenters and bricklayers and blacksmiths, and all such things. As far as concerns the machine shop, the veil of darkness hangs there; they don't allow us to learn the machine trade

Robert Williams (black)

The Free-Born American

In the South, white laborers had few friends outside their class. This was so much the case that one is hard pressed to find a contemporary who had anything complimentary to say about them. Poor whites were invariably described as "ignorant" and "unreliable,"—"lazy and dissolute plebeians," in short (as a tobacco planter characterized them), "the most offensive class of society." Even the slaves, and their progeny, were wont to sing, "I'd rather be a nigger than a poor white man."[1]

Given such a democratic recommendation, it would be expected that white laborers had a difficult time obtaining employment. Consistently, during the near-famine following the war, reports on the condition of the people concurred with Tennessean William Spence's observation that "the poorer classes of whites are not getting along so well" as the blacks. Colonel Spence explained that the whites could not "get employment as readily as the colored men can. The richer men will not employ them, for the truth is, they are not as valuable for laboring," being "inclined to be idle and lazy, and think[ing] it degrading to work." Over a decade later, Louisiana cotton

---

[1] Reid, *After the War*, p. 348. Andrews, *The South since the War*, p. 177. King, *The Great South*, p. 346. Thomas, "Plantations in Transition," p. 245. Hammond, *The Cotton Industry*, pp. 96–97.

planter R. T. Vinson stated flatly that he "would not give one colored laborer for five white men." Vinson must surely have exaggerated the differential in productivity, but his belief that one existed was nearly universal. Consequently, "most of the large planters" would "not have white men on their plantations if negroes" were "to be had."[2]

In response to the charge that white southern laborers were inclined to be too idle, one of them, Mississippian Tammany Jones, would not frame a denial but did offer an explanation: it was due to a closed shop composed of blacks and their best friends. "Secessioners has all the land," explained Jones, and "the niggers gits all the work." This gave "a po man *cassion* fur meditatin a good deal in a settin postur." The laboring whites would have stressed that they associated the degradation not with work per se but with work relations between employer and employee that were akin to slavery. "When a nigger is hired," Jones continued, exaggerating somewhat, "it's mighty nigh as if he wuz a slave agin. They knows they is onpleasant to white men, and that ar makes em sorter meek like." Jones's explanation that the planters' preference for ex-slaves was based upon the white worker's independence and willingness to do "his own cussin back" at a former slave owner who was "used to slingin his orders round" must be allowed some merit.

The antebellum planter's complaint that an employer "never could depend on white men, and you couldn't *drive* them any" because "they wouldn't stand it" is strong support for Jones's explanation. And indeed, the post-bellum planter's valuation of white productivity seems to have been dominated by constraints upon possible social relations. According to one planter, the "difficulty" obstructing the hiring of white laborers was due to their unwillingness to exchange "their independence for dependence upon a *master*." Contempt for manual labor, and especially hired labor, is common to all postslavery agricultural societies. In India, where agricultural fieldwork was once done by slaves and then associated with the lowest castes, to walk behind a plow was to perform, in a religious sense, "unclean labor."[3]

In 1865, Confederate veterans returned home with empty pockets to find their farms run-down from neglect, their livestock greatly diminished, and prospects for an immediate resumption of independent farming dim. Many white laborers in this position were in no condition to indulge in the luxury of avoiding hired labor. From the start, white and black labor were thrown into intense competition for employment.

With the large plantation areas being the exclusive province of black labor, most of this competition occurred in the upland areas dominated by smaller planters and farmers. A portent of what the outcome would be if it was left to be decided by unbridled competition, was given by a bureau

---

[2] *Report of the Joint Committee on Reconstruction*, Tennessee, pt. 1, p. 117. *Exodusters*, Louisiana, pt. 2, p. 342. *Rural Carolinian*, vol. 10, July 1871, p. 637.

[3] Powers, *Afoot and Alone*, pp. 91, 90. Olmsted, *Seaboard Slave States*, p. 84. *American Farmer*, vol. 2, no. 8, Feb. 1868, pp. 250–51.

agent near Harper's Ferry, West Virginia. He reported that because "colored laborers" had done "as well" or better than whites who had worked on shares before the war, "some farmers [were] taking away land from the whites and giving it to the colored people to work." This outcome was indicative of the economic fact that wherever freed labor was supplied in quantities abundant enough to be a significant factor in the market, white labor would be forced to exit from that market or submit to the industrial relations and rate of remuneration conferred upon the blacks.

The presumed greater profitability of the freedpeople's labor was a point of neither pride nor shame. For it was an inheritance of their former servitude, and in many cases due to their lack of any alternative means of support. They often acquiesced to social relations which, according to one planter, would have been "regarded as insulting by the [white] free-born American whose duty" should have been "to hear and obey."[4]

The debasement of the white workers' social existence due to the presence of the blacks was a source of great bitterness. But while two and a half centuries of unsuccessful competition with slave labor must have been a festering catalyst to the development of a population of Negrophobes, it is untenable to argue that racial antagonism was solely a product of capitalism and black slavery. Much recent scholarship on the origins and growth of racism and racial ideology has confirmed the judgment of nineteenth-century historian George Bancroft that to white colonists "the negro race was from the first regarded with disgust, and its union with the whites was forbidden under ignominious penalties."[5]

With racism so deeply imbedded in society, the mass of laborers, white and black, were, in 1865, probably incapable of taking a stance toward one another other than one of conflict. Tammany Jones reasoned that as "long as niggers is round, us po men's not gwine to git any work." Implicit in his statement was a logical course of action that had been attempted before the war and would become the major strategy of white labor for over a century after the war's end. The elimination of black competition through colonization schemes and legal restrictions upon slave occupations had been an antebellum issue. In the postwar era a long and intensive struggle to segregate black labor spatially and by occupation was initiated anew. Major General Clinton B. Fisk testified at the beginning of 1866 that the small farmers and laborers of Tennessee, many Union soldiers, and "the bitterest opponents of the negro" desired that they "be entirely removed from the state." In contrast, the "largest and the wealthiest planters" cooperated

---

4 *BRFAL*, film M1048, roll 46, Virginia, 1867, p. 0557. *American Farmer*, p. 251.

5 George Bancroft, *A History of the United States* (Boston, 1834), vol. 1, p. 190. The social origins of racism in Western culture have received a good deal of attention in the last decade: see Davis, *The Problem of Slavery in Western Culture*; George M. Fredrickson, *The Black Image in the White Mind: The Debate on Afro-American Character and Destiny, 1817–1914* (New York, 1972); Winthrop D. Jordan, *White Over Black: American Attitudes Towards the Negro, 1550–1812* (Chapel Hill, N.C., 1968).

with him in giving protection to the freedpeople. Throughout the South this was the case.[6]

The most effective means of curtailing the Negro presence was demonstrated by workers in the North in 1863, when white dockworkers, incensed by the military draft and the use of black labor imported from the South to break strikes, precipitated a series of race riots that drove much of the black population from the docks and from their homes. Immediately upon the war's end race riots erupted in Memphis and New Orleans. Throughout the South, the thousands of atrocities and altercations between blacks and whites during the next ten years were replications of these situations.[7]

The reign of terror presided over by bands of Regulators and Ku Klux Klanners were frequently organized for the purpose of improving the economic prospects of whites. In southern Kentucky organized "Regulators," late in 1867, "began to notify freedmen that they must leave on or before the 20th day of February." White men were notified "that all tenements rented to freedmen would be burned." There resulted an exodus of freedpeople into the cities. The Regulators' motive was reported to be a "desire to drive the negro away that wages may advance." Throughout the South freedpeople were beaten and driven out of employment that was sought by whites. Black renters or purchasers of preferential properties were often burned out or forced to abandon their homes by some other means.

Ridgeley Powers, lieutenant-governor of Mississippi, explained that many "prominent planters" who were opposed to "driving the labor" from their country informed him that the violence "was made by the poor unreliable nonproperty holders with a view of getting rid of black labor" in "order to command their own prices for labor." The explanation, it should be recognized, applies generally only to nonplantation counties, like those in northeastcentral Mississippi to which Powers referred. As shall be seen, in counties dominated by large resident planters Klan activity served a much different purpose. In the counties composed chiefly of small farms the large planters were often victims. An attorney practicing in Winston County, Mississippi, presented the "agrarian idea" as the "one idea" that most thoroughly explained the outrages directed at "driving off negroes." The small landowners were saying to the large planters, "We do our own work, and our families have to do their own work, and you must do it too; we are not going to let you have servants." This Jacksonian philosophy would prove explosive later in the century, when small farmers and laborers would turn their energies and enmity upon merchants, financiers, and government policies in a period of agrarian populism that sometimes showed signs of

[6] Powers, p. 91. Sterling D. Spero and Abram L. Harris, *The Black Worker* (1931, reprint New York, 1968), chap. 1, pt. 1. *Report of the Joint Committee of Reconstruction*, Tennessee, p. 112.

[7] Du Bois, *Black Reconstruction*, pp. 103–4. Spero and Harris, *The Black Worker*, p. 17. Herbert Gutman, *The Black Family in Slavery and Freedom, 1750–1925* (New York, 1976), pp. 24–28.

ignoring the race question. In the 1870s these issues lay largely dormant for white farmers, who devoted their political activity to driving out large planters and their black servants.[8]

There exists in northwestern Alabama, well north of the plantation belt, a namesake of Winston County, Mississippi. There all the ironies existing in eastern Tennessee were also prevalent. James H. Clanton, an antebellum whig legislator and chairman of the Democratic Executive Committee during the period of redemption, described Winston County, Alabama, as "the only white county in the state that voted for General Grant" and "passed resolutions that no negro shall settle in the county."

The collective efforts of small white farmers and laborers to spatially separate themselves from black competition were frequently successful. The 1871 convention of the African Methodist Episcopal (AME) Church in Florida "resolved" that since the governor of the state admitted that "he could not protect the [colored] people," the "colored people should move out of Jackson County forthwith." A North Carolinian claimed that most Ku Klux Klan activity in his state was confined to counties close to or bordering white majority counties in South Carolina. Essic Harris, a black resident of Chatham County, agreed that the blacks in that region had been so brutally dealt with that a "heap of them have gone off." In western Texas, and especially the Panhandle area, black renters and homesteaders were "run out" after having their cotton houses and cribs "burned" by "cow-boys" and other whites.

Reverend Charles Pearce of the AME Church in Florida gave about as clear a geographical description of this phenomenon as anyone. Pearce thought that in the "large counties" blacks were very welcome. "But in the small and sparsely settled counties they are very much opposed to" blacks "having lands and settling upon them." In those areas freedpeople were "threatened and whipped, and driven off their places."[9]

Elimination of black competition through a policy of spatial segregation was most readily implemented in areas where most whites agreed with that action. Where the large planters were sufficiently numerous, they could not only declare that they "hated" lower-class-controlled Ku Klux "as bad as" they did the "devil himself," but could protect themselves and their property. In such areas the white laborers' attempts to eliminate black competition was less than successful. An understanding of this failure is crucial for a proper comprehension of emerging race relations and the mechanics of economic discrimination. A great quantity of work has been devoted to proving that white racism did not result in effective economic discrimination against blacks, although it did relegate them to an inferior social position.

---

[8] *House Executive Document No. 329,*, 40th Cong., 2nd sess., pp. 2–3. *Senate Report no. 41,* 42nd Cong. 2nd sess., *The Condition of Affairs in the Late Insurrectionary States* (Washington, 1872), "Mississippi Testimony," p. 590. The hearings contained in this Senate Report are in 14 volumes, and are hereafter cited as *KKK Hearings*. Ibid., pp. 659, 661.

[9] *KKK Hearings*, Alabama, p. 230; Florida, p. 166; North Carolina, pp. 62, 52, 99. *Exodusters*, Texas, pp. 416–17. *KKK Hearings*, Florida, p. 167.

Much of this work has been sorely misspecified, relying as it does upon a conceptual framework that has white employers favoring the hiring of white labor until the competitive pressure of wage cutting by blacks made the favoritism too costly for employers to withstand temptation. This model would make some sense if the bulk of southern employers had originated from the nonslaveholding population. But for obvious reasons, southern employers had absolutely no compunctions against the use of black labor. As long as race relations observed the caste system the employers actively sought black servants.

In 1883 Mobile businessman W. H. Gardner spoke truthfully for an entire class of southern gentlemen: "I would not have any other servant than a negro even if I could get him." Gardner's generic use of the term "servant" for "laborer" reveals much. Any white laborer or small farmer with a superior grasp of southern labor markets, like Tammany Jones, would have been quick to claim that white labor was the class discriminated against.[10]

Geography, demographics, even, if you will, statistics have a large amount of explanatory power in regard to the structure of racial competition for jobs in a southern county. Remaining in Alabama for the moment, consider a pair of neighboring counties like Coosa and Tallapoosa, situated in the east central region of the state. Whites represented 72 and 75 percent of the counties' respective populations in 1870. In 1860 the most frequent farm size for both was between twenty and fifty acres, with 23 and 24 percent above one hundred acres. Consistent with these landholdings were the slaveholdings. Twenty-two and 16 percent, respectively, of all slaveholders owned one slave—the most frequent size holding. Twenty-four and 30 percent of the slave owners had ten or more slaves, and in each county only eleven, or 1.7 and 1.4 percent respectively, held fifty or more slaves. These were hardly plantation counties but did contain enough large property holders in 1860 to form the basis of an employing class capable of affording themselves some protection against laboring whites in 1870.

Daniel Taylor, owner of a large plantation in Tallapoosa, told a familiar story. The Regulators began when "some of the negroes" became "saucy and impudent, and hard to put up with." In time the organization came to be controlled by poorer men of the "meanest characters" who tried "running the negroes out of the neighborhood" to "make labor scarce," so "they could choose a choice of the land" as renters. In these two counties the terrorism was partially effective, forcing the Negroes onto the larger places and out of portions of the country. Taylor, who thought the Klan was "injuring the men who wanted to have their farms cultivated by freedmen," made it clear that he "wanted [his] negroes let alone as long as they behaved themselves."

The tendency of the terrorism to concentrate the blacks on the large plantations is shown by the racial composition of Taylor's labor force. Working "sixty to seventy hands," he claimed to "like the negroes best."

---

[10] *KKK Hearings*, Alabama, "Testimony of Daniel Taylor," p. 1126. *Senate Hearings, Labor and Capital*, vol. 4, p. 67. Campbell, *White and Black*, p. 118.

However, in 1871 he "had about half of [his] labor negroes and the other half white." Given his expressed preference' the mixed labor force may have been due to a temporary absence of black labor, because in two previous years he claimed to have "worked white labor altogether."[11]

In the river parishes of Louisiana, where the laboring population was predominately black, the role of the white laborer was often that of a scab hired to break a Negro strike. In a true plantation county like Tensas Parish, with 52 percent of the 1860 landholdings over 500 acres, the individual white laborer in the area was virtually shut out of employment by a social and economic process that was destined to work in his favor outside agriculture. When black and white labor was worked together under close relations, their mutual antagonism often led to physical violence that disrupted the productive process. On sugar plantations, where the gang system was the norm, a planter explained that large plantations had to depend "almost entirely upon negro labor," because the two groups could not work together any more than it was possible to "mix oil and water." The choice of black labor was based upon the fact that there were "plenty of negroes and but few whites in the county seeking employment."[12]

Within any geographical area wherein all the laborers and employers were in close enough proximity to compete with one another directly there is little evidence of wage discrimination. Almost no source, white or black, employer or employee, visitor or resident, was aware of wage differentials between black and white agricultural workers.[13] The one difference was that white laborers often demanded greater social equality with their employers. This demand sometimes extended to material compensations of a special nature. W. W. Arrington, a black planter owning a thousand acres in Nash County, North Carolina, stated that the one "difference in the hiring" was that a "white man" was "taken into the house" of the employer and fed "a little better." This social prerequisite was apparently expected by white workmen as a natural right due to their superior social position as Caucasians. It often backfired, because many employers rightly viewed it as equivalent to a demand for higher wages, and some disliked the presumption of social equality. Arrington admitted that the distinction in treatment was "generally" of no consequence, because there were "not many white men who hire out," but this was partly due to the whites' demands, as when Virginian Robert Hubard advised his son to employ a "negro carpenter" rather than a white in order to escape "the trouble of entertaining him in your house."[14]

[11] *Agriculture of the United States in 1860, Eighth Census* (Washington, D.C., 1864), pp. 193, 223. *Nineth Census of the United States: Statistics of Population*, pp. 11–12. *KKK Hearings*, Alabama, pp. 1126–27, 1129.

[12] *Agriculture of the U.S. in 1860, Eighth Census*, p. 202. *New Orleans Picayune*, Aug. 3, 1877. BRFAL, film M1027, roll 28, Louisiana, p. 186.

[13] *Exodusters*, North Carolina, pp. 260–63, 253, 305. Louisiana, p. 572. Campbell, *White and Black*, passim, pp. 297, 303. *Senate Hearings on Labor and Capital*, vol. 4, Tennessee, p. 144; Alabama, p. 23. *Report of Joint Committee on Reconstruction*, Arkansas, p. 58.

[14] *Exodusters*, North Carolina, p. 253. Thomas, "Four Plantations in Transition," p. 309. *Senate Hearings on Labor and Capital*, vol. 4, p. 165. Nordhoff, *The Cotton States*, pp. 38, 39.

As important as these motivations were, they were probably less so than the fact that the presence of free-born Americans could have a negative effect upon the productivity and cost of black labor. George Benham, planting cotton in northeast Louisiana, decided in early 1866 that he would in the future have no use for white workers. Short of labor, he had hired a "squad" of whites to work alongside his black squad under the same contract. The whites turned out to be inexperienced cotton hands, but even so when put at tasks which they found "harder work than they cared to engage in" they demanded that the "negroes" be given "that kind of work" so the whites could "have something easier." The overseer acquiesced. "Impossible to please" at the dinner table, the whites "twice" dashed "what was set before them on the ground." They proved to be the "veriest eye-servants," rising late and performing "little work," and that "shabby." The black squad soon began to take the whites' "labor for the standard." When rebuked they remarked that they were "doing more than the white folk." Blacks and whites ate their "meals in separate rooms," the whites getting a different menu. Benham as a matter of course "joined the white squad" when eating with the men. Within three weeks the freedmen were resenting the "little favors" given to the whites. The black squad, remarking that the whites had "dashed out the grub," threw out theirs.

The day after the first payday the employers were relieved to find that all but two of the Caucasians had deserted the plantation. Convincing themselves that it was no "loss," because "more labor" could be "accomplished" with the blacks "alone," the partners fired the remaining two whites. George Benham's self-described lesson is worth quoting:

> Never was there a more forcible illustration of the utter fallacy or impossibility of successfully attempting to feed and lodge two men, who are grinding the same grist, in different stalls, and on different diet, simply because one is black and the other white. What is flesh for one must be flesh for the other and where there is equality in labor, perfect equality must run through the whole government.[15]

Benham's lesson might be accepted as a social theorem. If so, it was one which white laborers perceived well. In Chatham County, North Carolina, the "laboring class of white people" were reported by a white of higher status to be "unanimously" in favor of "colonizing" the blacks "or anything else to get them away." The people felt that

> universal political and civil equality will finally bring about social equality; in fact it is commencing to do it already in that country; there are already instances in the county of Cleveland in which poor white girls are having negro children. . . . The white laboring people feel that it is not safe for them to be thus working in close contact with the negroes.[16]

[15] Benham, *Year of Wreck*, pp. 98–99, 103, 99, 121–22.
[16] *KKK Hearings*, North Carolina, "Testimony of Plato Durham," p. 318.

The poor whites, ignorant though they were, in their attempt to drive the black population from the small farm counties demonstrated a thorough understanding of Benham's Theorem. That they could not delve deeper and see that the two laboring classes must rise together or remain depressed was a serious failure.

A corollary of Benham's Theorem that economic equality must result in social equality was the basic tenet underlying Booker Washington's philosophy of separate but equal. Aware of the validity of the theorem but unwilling to accept the realization of the consequent, the New South had to devise a workable alternative that could utilize black and white labor in a reasonably efficient manner and preserve white supremacy, especially for the disadvantaged whites. The process began during Reconstruction with the "whiting" of the upland counties. In the next thirty years the evolution of new social and industrial relations would be closely tied to the growth of nonagricultural industry in a complicated system that was patently separate and unequal.

II

Throughout the 1870s and certainly by the end of Reconstruction, the evidence related to the question of color discrimination in wages and employment opportunities outside agriculture is very different from that found for farm work. At this early period clear signs of the developing racial division of labor are evident. In unskilled labor, even in factories, the abundance of efficient and reliable black laborers made it impossible for whites to displace the indispensable pool of black workers. The refusal of unskilled whites to work with blacks under terms of equality was at best a poor bluff when they knew the employer could easily replace an entire force of striking whites with an equally productive number of blacks. The consequence was that few sources report the existence of discriminatory wage policies for unskilled labor.

Where the labor was skilled many sources claimed that black labor was confronted with discriminatory hiring and wage practices. Because of the lower education and often specific but poor training given to slave mechanics, a confident valuation of some of this evidence is problematical. For example, in cases where black carpenters were basically illiterate and had only received a rudimentary training to enable them to perform simple work on a plantation it would not be correct to say that they were discriminated against if paid less than white or black carpenters with better-developed skills. In this period, before the standardization of wages by strong trade unions, we would expect to find dispersion in wage rates commensurate with the dispersion in skill and training between white and black artisans and factory hands. If the generally agreed-to verdict of historians that slave mechanics often received an inferior training is correct, then the "average" black mechanic would have received lower wages and employment opportunities even in the absence of discrimination.[17]

[17] Lorenzo Green and Carter G. Woodsen. *The Negro Wage Earner* (Washington, 1930), pp. 4, 17, 35. Spero and Harris, *The Black Worker*, p. 6. Genovese, *Roll Jordan Roll*, pp. 389–90.

In an economic environment where workers competed as individuals on the basis of individual skills, neither black nor white workers would have gained any advantage from discrimination. Any two workers with identical skills would have received identical earnings in reasonably competitive markets. In the actual environment, where competition was not merely between individuals but also between the two color groups, the economic opportunity available to an individual was also a function of the relative market power of the group. This relative group power varied by trade and geographical location. However, in general, the presumed inferior distribution of skills among black artisans in most occupations decreased the opportunities open to even the most skilled of their number. Striking white craftsmen might be replaceable, but often at the cost of a reduction in the average productivity of an employer's work force. Unable to completely eliminate the competition of the blacks, white workers often compromised by demanding a wage differential based on color.

A black bricklayer making a success as an independent contractor in Birmingham, Alabama, during the early 1880s testified that cases of color discrimination among "common labor" were rare, but that in the building trades white mechanics "always" received "50 cents or $1 a day more than colored men," even if "the colored man be a better workman." In contrast, a white manufacturer of Columbus, Georgia, testified that there "the great bulk" of "masons and carpenters are negroes" who were employed "indiscriminately with whites." Implying that this was not the usual circumstance, he added that when "they pursue those businesses they do very well indeed, because where their capacity is equal to that of white men we pay them equally well; they command the same wages."

Descended from a long line of free blacks, Wiley Lowerey, owner of a livery stable in Kinston, North Carolina, made it clear that he was aware of the fact that different grades of labor receive unequal wages in competitive markets. According to Lowerey, among skilled workmen only whites enjoyed a "preference sometimes in wages." He believed that in such cases the employer actually had a preference for the lower-paid "colored men." [18] If true, this would only be consistent with the wage differential's being due to social rather than strictly economic factors. This preference for the cheaper but commensurately skilled black laborers was the major weakness in a supposed social compact among whites that espoused preferential treatment for Caucasion labor. If the individual enterpreneur could make greater profits with black labor, how could the social compact be enforced?

III

Of the numerous labor strikes occurring in the South during 1865, two, each upon the docks of one of the region's most populous and industrial cities, were of momentous social import. Black workers had played a long and im-

---

[18] *Senate Report: Relations between Capital and Labor*, vol. 4, pp. 403, 514. *Exodusters*, North Carolina, p. 305. Campbell, *White and Black*, pp. 285, 360.

portant role in the development of the shipping industry in both Baltimore and New Orleans as sailors, skilled artisans, and longshoremen. For years the craft of ship caulking in Baltimore had been dominated by slaves and free blacks. Frederick Douglass's account of the physical attack upon his person by white craftsmen during his tenure as an apprentice caulker in a Baltimore shipyard is a stirring depiction of the animus between white and black workers on the Baltimore docks prior to the war. Another omen of what was to come occurred in the summers of 1858 and 1859, when white caulkers attacked blacks in the south Baltimore shipyards with the objective of driving them from their employment.[19] This effort was resumed in the late summer of 1865 when a committee of men, representing the unemployed white caulkers returning from the war, demanded that the black work force of an east Baltimore shipyard be fired and replaced by whites. The owner of the shipyard rejected the proposal. What followed was a strike by white workers in the Baltimore shipyards. A major debate, seeking public support for the opposing positions of white and black caulkers, raged in the city newspapers. White labor, appealing to employers and the white population at large, moderated their demands to the desire that whites be given preference in hiring and be appointed foremen.

Newspaper accounts indicate that shipbuilders and merchants in general aligned themselves with the continuance of black employment, on the grounds that employers had the "right to enter the market and buy labor of the best at the cheapest rates. They may prefer to have their work done by the colored caulkers, who were at one time the only caulkers in Baltimore." This confrontation is of extreme theoretical as well as practical importance. There was no question about the competence of black caulkers, who were deemed at least as skilled as the whites. Black caulkers appear to have had a numerical majority, and certainly had the support engendered by their employers' profit motive. Despite all this, after only one month of confrontation the white caulkers emerged victorious. The *Baltimore American* regretfully reported that the whites had "gained all or nearly all that they contended for, namely the exclusion of all negro caulkers, excepting such as are too old or infirm to do work of another kind."

The appeal to preference based upon color probably played some role, inasmuch as it can be assumed that most whites in the city who did not employ labor probably supported the white caulkers. But there is no evidence of any special social pressure being applied to the shipbuilders, so this probably played a fairly minor role at this time. The more important pressure was economic, and was alluded to by the black caulkers when they publicly attacked white "mechanical organizations" outside the trade of ship caulking for discontinuing work and putting a virtual freeze upon the commercial interests of Baltimore. *All* white craftsmen in the shipyards had joined their brother caulkers by discontinuing work in a sympathy strike. The

[19] Douglass, *My Bondage and My Freedom*, pp. 308–15. Eric Arnesen, "A Little Corner of Creation: Black Workers and the 1865 Baltimore Shipyard Strike" (typescript, Yale University, 1981). The following account draws from Arnesen's paper.

Journeymen Carpenters' Association of East Baltimore enforced its strike agreement by a fine of twenty dollars on "each and every member" working in a shipyard where blacks were employed. This incredible victory, achieved mostly by nonviolent means, served as a lesson to all concerned. It would be some time before a Baltimore businessman would preach with confidence that any combination "among workmen" was "bound to defeat its own object, because it is simply impossible to overcome the constant tendency to establish an equation between supply and demand." For the workers, this important first demonstration that the "natural laws of supply and demand" were sometimes amenable to human forces proved prophetic.[20]

Two months after the racial contest for control of employment in the Baltimore shipyards a very different strike erupted in New Orleans. Black and white longshoremen, in unison, struck for higher wages, forced a cessation of all work upon the riverfront, and marched "up the levee in a long procession, white and black together." Racial solidarity and the strike were broken when the city authorities ordered the arrest of "colored longshoremen on the charge of rioting and interfering with nonstrikers."

The presence of the nonstrikers is the key to understanding the unusual solidarity of white and black labor. In the latter half of 1865 the New Orleans labor market was becoming inundated with cheap black labor from the plantations more than willing to underbid the relatively high wages earned on the Mississippi levee. Eight years later the recession of 1873 and the continual flood of workers into the area had so increased the competition for jobs that all semblance of racial solidarity was erased. Blacks and whites engaged in riotous armed combat. The blacks, in an attempt to drive the whites off the levee, attacked the police, who wounded several of them. White longshoremen, overwhelmed by the greater number of freedmen, realized they were threatened with near extinction in longshore work. To protect their position they appealed to the federal army to arrest as vagrants the "low, ignorant negroes, who slept under tarpaulins and in barrel houses" and were willing to work for half the prevailing wage.[21]

The long-run solution to this situation, and probably the only one that could have preserved white employment, was a political agreement between the two groups. A quota system was devised to allocate jobs among white and black longshoremen, giving each group an equal percentage at a uniform wage. This plan actually favored the whites, because their guarantee of one-half the jobs ensured them a higher employment rate, due to their smaller numbers. As an economic treaty this quota system spread through the South and in the early twentieth century was in force in each major Atlantic seaboard port from Galveston to New York. As

[20] Arnesen, "A Little Corner," pp. 5, 8 (quoting the *Baltimore American and Commercial Advertiser*, Oct. 3, 1865); p. 15 (*American*, Oct. 28, 1865); p. 11 (*Baltimore Sun*, Oct. 2, 1865; *American*, Oct. 2, 1865); pp. 4, 8 (*American*, Oct. 3, 1865).

[21] Trowbridge, p. 405. John W. Blassingame, *Black New Orleans: 1860–1880* (Chicago, 1974), pp. 64–65. Shugg, *Class Struggle in Louisiana*, pp. 301–2.

a social practice it was closely tied to an analogous procedure in the political arena.[22]

In the battle for political control of southern states, the Democrats used a variety of strategies to nullify the Republican vote. In counties or legislative districts with clear white majorities it was a fairly simple procedure to maintain political superiority over the outnumbered black voters. In districts where the two races were more evenly matched elections during the 1870s were often carried by the use of the so-called first Mississippi plan, which emphasized ballot stuffing, terrorism, and economic coercion of dependent black voters. During Reconstruction, and in some cases into the 1880s, neither of these techniques were totally successful in many regions where the blacks enjoyed a large majority over the whites. In the interest of preventing a full-scale race war that might result in a Pyrrhic victory at best for the successful group, a more peaceful solution called political "fusion" was often negotiated among politicians of opposing parties. Senator W. B. Roberts recalled that in Bolivar County, Mississippi, where the black voters outnumbered the whites sixteen to one, "we used what was known as the fusion system, and that was to agree for the Negroes to have some of the offices, and the whites of course the best ones." In Georgetown County, South Carolina, where the Negroes held a five-to-one population advantage in 1880, the political fusion system lasted until late in the 1890s. In 1880 the fusion plan that was adopted "provided that there would be no contest in the general election." The Democrats were allocated the offices of sheriff, clerk of the county court, coroner, two county commissioners, and a representative. The Republicans received a state senate seat, one representative, one county commissioner, a school commissioner—all black—and a white probate judge.[23]

The parallel between the quota system and racial division of political offices and patronage, giving more and better positions to the whites, and the admitted objective of white labor to institute a racial hierarchy in the labor market is more than mere analogy. Many contemporaries considered the struggle for white supremacy in politics to be fundamentally connected to the need for a similar hierarchy in the labor market. During the tumultuous seventies this relationship was most easy to see. The editor of the most circulated Democratic newspaper in Louisiana held that the "principle involved" was "identical" in both cases. Employers "choosing their employees" or, equivalently, "individuals with patronage to bestow" had the same "right as public officials to favor their friends in preference to

---

[22] Shugg, *Class Struggle in Louisiana*, p. 303. Spero and Harris, *The Black Worker*, pp. 184–86.

[23] W. B. Roberts, "After the War Between the States," *History of Bolivar County* (Jackson, Miss., 1948), ed. Florence Warfield Sillers, p. 163. George Brown Tindall, *South Carolina Negroes*, p. 63. *Compendium of the Tenth Census* (Washington, 1883), p. 596. Woodward, *Origins of the New South*, pp. 217–18, 80–81. See Wharton, *The Negro in Mississippi*, pp. 202–4, and Williamson, *After Slavery*, pp. 397, 408, for discussions of political fusion.

their enemies." During the political campaigns ending in Democratic redemption, opinion among whites ran high in favor of ousting Negroes and especially Republicans from preferred occupations. Even merchants, understanding the "exigencies" of the broader social problem, were advocating that "preference" be given to "*white* over *black* labor." Replications of the victory of the white caulkers in Baltimore were numerous. Merchants in Vicksburg, Mississippi, hired white porters for the first time ever in 1874. The Vicksburg & Meridian Railroad fired all of its black mail agents and hired whites. Earlier, Virginia's *Lynchburg News* had been "gratified to learn that one hundred and fifty negroes employed at the Wythe Iron Mines, all of whom voted the straight out radical ticket, were discharged."[24]

This policy was only effective where the employment being offered was acceptable to whites. In general, low-paying day labor and domestic service were left uncontested to the Negroes. The domination of higher-paying, more attractive jobs was not a simple operation. In many cases it simply failed, victim of the competitive process and entrepreneurial drive for profit. While market competition and old-fashioned individual greed should not be underestimated as strong factors tending to eliminate or ameliorate racial differences in market earnings, neither should they be overemphasized. The difficulties of holding together a coalition that had formed a political fusion were no less than the problems presented by independent employers who might attempt to steal away differentially paid Negro laborers. As long as the Negro retained the franchise, political disagreements among the whites could always be exacerbated by competition for the black vote. Independent party movements among the less-wealthy whites often vied for the black vote and threatened property with a coalition among the lower classes. The Negroes were admittingly accepting the politics of fusion with the redeemer Democrats only until a better opportunity was available. Black Republican congressman John R. Lynch's instructions to his Mississippi constituents demonstrate the plasticity inherent in fusion strategies and the extreme competitiveness of the political system. The Negroes were told that it was their duty to follow their instructions, whether it was "to fuse with the Independents instead of the Democrats, or with the Democrats instead of the Independents, or to make straight [Republican] party nominations instead of fusing with either."

Third-party movements, whether they actually sought to fuse with the black vote or not, were always accused of doing so and of threatening the sanctity of white supremacy. Where public condemnation and the dreaded punishment of social ostracism from the white world failed, the last resort, violence, was seldom shirked. Political fusion served conservative business and planter interests by keeping the laboring and small farmer white and black votes from coalescing. While this blunted the political power of the workers, it guaranteed to the poor whites that social equality for blacks could not be legislated.

[24] *New Orleans Picayune*, Aug. 1, Aug. 4, 1874. Wharton, p. 127. Taylor, *Negro in Virginia during Reconstruction*, p. 223.

The white laborer's ideal racial division of labor was to designate agricultural and unskilled menial labor as Negro jobs. Where the blacks were too numerous and competent in a specific occupation, the maintenance of white superiority depended upon the success of an economic fusion that delivered unequal pay to the two colors of labor. Where successful, the relationship to broader social concerns was divulged by the common defense that it took "more for the white man to live on than the colored man, and that, consequently, they pay him higher wages."[25]

The first comprehensive data on the occupational distribution of the black labor force were compiled in the United States census of 1890. The low status of the entire black labor force in 1880 is attested to by the fact that ten years later 86 percent of all black workers earned their living as agricultural laborers or domestic and personal servants. Of those remaining, approximately 7 percent were skilled artisans or engaged in manufacturing industries. Given the close relationship between iron and steel production and mining, where a high degree of vertical integration was characterized by the holding of all three activities under a single ownership, mining and iron producers were the largest industrial employers of black labor, with a combined total of 27,378 workers. The least significant employer, in proportion to its total employment and present and future economic importance, was cotton textiles, where 1,077 blacks represented three percent of 1890 employment. The divergent opportunities open to black labor in these two areas are indicative of the problems encountered in an attempt to give a simple description of southern labor markets as either highly competitive, with minimally effective discrimination, or totally discriminatory. In fact the opportunities open to black laborers outside agriculture depended upon a variety of factors, some social and others economic. Study of these two important industries reveals much about these factors everywhere. Differences in the postbellum utilization of black labor by these two industries cannot be explained by unfamiliarity with black labor, since both had made liberal use of slave labor prior to the Civil War.[26]

Black workers were the primary source of labor supply to southern iron mills and related mining operations. Virtually every firm in the industry employed a large percentage of freedmen. A partial explanation of this was the hot, dirty, and extremely strenuous nature of the work. Forced to select hardy and vigorous laborers, the industry was in no position to exclude what Albert C. Danner, president of the Mobile Coal Company and the Bank of Mobile, considered the "only reliable" supply of labor available. Immediately after the war the freedmen were largely excluded from skilled work "because if you" placed them in such positions "the white men would strike." By the early 1880s the white union's monopoly of skilled positions was broken. A survey of iron mills in east Tennessee and Alabama reveals

[25] Woodward, *Origins of the New South*, pp. 217, 104–5. *Senate Report on Labor and Capital*, vol. 4, p. 404.

[26] Charles Dew, *Ironmaker to the Confederacy* (New Haven, 1966). Robert Starobin, *Industrial Slavery in the Old South* (New York, 1970). Tom Terrill, "Eager Hands—Labor for Southern Textiles, 1850–1860," *Journal of Economic History*, vol. 36, no. 1, 1976.

that five of the nine mills for which information about the racial mix of skilled workmen was found employed blacks in those positions. Two of the firms, the Knoxville and Chattanooga iron companies, had nearly 100 percent black labor in both skilled and unskilled jobs.

The penetration of skilled positions by freedmen was effected by the opposite of the later tradition of replacing whites in the midst of a strike. The management of the two Tennessee companies provoked a strike by forcing blacks into the skilled positions, in a move to lower labor costs. The subsequent strike by skilled whites, largely northerners and men of European origin, failed, resulting in the complete domination of both plants by black labor. Joseph Reid Anderson, manager of the Tredegar Iron Works in Richmond, Virginia, the leading producer of iron in the South and one of the largest in the nation, had essentially pioneered this technique when he "broke the opposition of his free workers" before the war "and introduced large numbers of Negro slaves into skilled rolling mill positions." After the war blacks continued to fill skilled jobs in the Tredegar Works. In Alabama blacks trained as slaves to perform skilled labor in iron works found places to use their skills to advantage.[27]

The major factor leading to management's decision to place Negro labor into skilled occupations was the inferior competitive position of the southern iron industry. Faced with severe competition from more efficient English and Scottish producers, southern manufacturers were finding it difficult to maintain a market position, even with the advantage of a substantial tariff. In the early eighties Lothian Bell, an English authority on the iron industry, after completing a tour of the entire southern industry reported that in "all the southern iron districts we are confronted with a list of dismal failures, and cheap iron seems a myth." Of Georgia iron producers Mr. Bell wrote that "if there is a single furnace in the state that for the past twelve years has not sunk the original owners all the money they put in and not changed hands, we do not know it." Tennessee producers had "not fared much, if any, better." Willard Warner, president and manager of Alabama's Tecumseh Iron Company, testified to the distress of the companies in his state. Of seventeen iron companies operating since the war, by 1883 twelve had gone bankrupt. Of the remaining companies the Eureka plant's stock was selling below "par" value, the Selma Works had shut down for a year, and the Anneston furnaces had "been successful," but Warner did not know if they had made any money. Warner, whose own company had not paid a stock dividend in ten years, declared that the industry "could only reduce the cost of iron by reducing the cost of labor."[28]

[27] Senate Report, Labor and Capital, vol. 4, pp. 102, 133. Survey extracted from testimony of iron manufacturers, ibid., pp. 133–34. Dew, Ironmaker to the Confederacy, p. 22. Richmond, Virginia, City Directory, 1872. Ethel Armes, The Story of Coal and Iron in Alabama (Birmingham, 1910), p. 77.

[28] Senate Report, Labor and Capital, vol. 4, pp. 263, 187, 233. Lothian Bell quoted in Ethel Armes, The Story of Coal and Iron in Alabama (Birmingham, 1910), pp. 301–3. Senate Report, Labor and Capital, vol. 4, pp. 261, 255–60. Armes, p. 253.

Social and racial ideology generally played a secondary role in industries characterized by low profit rates and dependent upon cheap male labor for their survival. Industries like iron and mining, which required corporate organization to raise huge sums of investment capital, were producing a new kind of capitalist for a new South. Enoch Ensley of Memphis, Tennessee, a young planter who had inherited "one of the largest plantations on the river between Cairo [Illinois] and New Orleans," was a fine specimen of the new breed of planter-capitalist emerging to take the lead in southern industrialization. President of the Pratt Coal and Iron Company, the largest producer of coal in the South, Ensley expressed a peculiar concatenation of corporate economic philosophy and the political demands of white supremacy that could have been regarded as consistent only in the New South. A self-proclaimed Democrat "without endorsing hardly a single tenet or principle of the party," Ensley considered the "hard money" policy of his party a "humbug" and favored protective tariffs and government expenditures for internal improvements. Men like this were concerned much more with the low wages and efficiency of their labor than its color.[29]

The philosophical conversion, evolution, or mere continuance of Whiggism, as the case may be, was a necessary requirement if aspiring southern industrialists were to form financial combinations with the flood of northern capital and capitalists penetrating the postwar economy. The southern iron industry had been demolished by Union troops during the last years of the war. All sixteen of Alabama's blast furnaces, each of its six rolling mills, and "practically every forge and foundry" that had operated during the war were destroyed by 1865. The first company put back into operation, the Irondale Works in Jefferson County, was seeded by capital obtained from Crane and Breed, a Cincinnati firm, in 1865. The Shelby Iron Company, idle for two years after the war, was reopened by a consortium of fourteen New England capitalists. Writing in 1882 about the prospects of the Republican Party in Alabama, a northern immigrant in Talladega pointed to the influence of the manufacturing interests in and near Birmingham: "Alabama Furnace; Anderson Furnace, controlled by General Tyler of Pennsylvania, Tecumpsa [sic] Furnace controlled by Ohio people, Shelby *Iron* Furnace, the largest in the state—some millions of Capital, controlled by *Hartford Connecticut people*—The Stonewall Furnace, the Aetna Furnace, Cedar Creek Furnace and other large manufacturing interest—*all controlled by Northern men* who will cooperate if the proper influences are brought to bear."[30] The influence of northern capital and capitalists on the industrial and employment philosophy of the developing southern mineral industries was knotted in a complicated web of profit seeking that meshed railroads, iron, and coal with the politics of state bond issues to help finance private industry and the utilization of cheap Negro labor.

[29] Armes, p. 425.
[30] Armes, pp. 186, 196–97. Woodward, *Origins*, p. 150. King, *The Great South*, p. 533.

The Tennessee Coal, Iron, and Railroad Company, whose name proudly advertised the tripartite vertical integration that dominated the early industry, became the leading producer of coal in the southern states with its acquisition of the Pratt Coal and Iron Company of Alabama in the mid-eighties. The company got its start in the 1850s with Tennessee ingenuity and New York capital, obtaining a charter to build a railroad for the purpose of hauling newly discovered coal deposits out of the Cumberland Mountains. Much coal mining in the antebellum South, short of capital and often undertaken by inexperienced men, was of the type described by the "head of a company of northern men" who had purchased large coalfields twelve miles above Richmond, Virginia, in 1865: "these Virginians would dig a little pit and take out coal until water came in and interfered with their work; then they would go somewhere else and dig another little pit." These methods, achieved with mule power, left most of the coal "undisturbed" and readily extractable in deep shafts kept free of water by the use of expensive "steam-pumps." This capitalist, like others in the industry, worked "negro labor," which he preferred "to any other class of laborers."[31]

During a period when white southerners were supposed to have been committed to but one cause, the "divided mind" of the South was nurtured by the southern capitalist's need for profit and, therefore, Northern capital and black labor. John T. Milner, railroad magnate, iron maker, coal supplier, and one of the leaders in the founding of Birmingham, wrote in the midst of Reconstruction that there was "nothing wanting" in Alabama "but capital and labor, and now negro labor does better in coal and iron business than in farming."[32]

Mining, with its rugged physical requirements and need for *skilled* workmen who could only become highly proficient with time, was one of the first southern industries to face competition for its supply of black labor from northern and border states. In the 1870s large numbers of black miners were being imported as strikebreakers and through more regular channels into the mines of the North and West. By 1877 the *Richmond Enquirer* was concerned enough about the outflow to issue a warning to southern employers to prevent the "exodus" of the best class of colored workers:

> The invariable testimony North, West, and South credits the black man as a more efficient and tractable laborer than the white man. Mining and milling companies and large farmers writing for laborers to the agency of J. P. Justis in this city always make it a condition that no whites be sent.

The growth of the mining industry in the South, coupled with the tendency for experienced miners to migrate to the North for the higher wages available there, should have been a much more significant source of nonagricultural employment for white and black southern labor than it actually proved to be. Two obstacles developed to slow the creation of employment

---

[31] Armes, pp. 364–67. Trowbridge, *The South*, p. 192. On antebellum mining practices see Armes, pp. 49–50, 53–55, 78–79, 151–56.

[32] Armes, p. 269. Woodward, *Origins*, chap. 6.

in southern mining. First the flow of black labor to northern mines was greatly checked by the violent opposition to the influx of low-wage competition, white or black, but especially black.[33] Perhaps even more important was the widespread use of convict labor in southern mines as a source of cheap and necessarily reliable labor, which served to depress wages and prevent changes in laboring conditions.

In Alabama, General Wager Swayne described the convict labor system as a "savage mode of punishment" as early as 1867. Ridiculously long prison sentences, often for trivial misdemeanors, provided contractors with a continuous supply of fresh laborers forced to serve an apprenticeship during which, according to the warden of the state penitentiary, in a defense of the system reminiscent of proslavery arguments, they could become thrifty and honest and learn the skills of "an expert miner."

In 1882 the Pratt Coal and Iron Company employed a total of 2,408 men, women, and children. Its subsidiary, the Comer and McCurdy Mining Company, did all of Pratt's coal mining with a force of 1,212 men from the above total. Among these men, 520, or 43 percent were convicts leased from the state. Joseph E. Brown, Redeemer governor of Georgia and later one of the state's U.S. senators, was also president of a railroad, a railway, and a steamship company. Brown's coal mines held a twenty-year lease guaranteeing "three hundred able-bodied, long-term men" a year at a cost of about eight cents per laborer a day. Southern capitalists, speaking also for their silent stockholders in the North, made little attempt at this time to conceal the motivation for the use of convict labor. The United States commissioner of labor reported that Alabama "mine owners say they could not work at a profit without the lowering effect in wages of convict-labor competition." Florida contractors of state convicts claimed that "stone-cutting could not be done at all at the ordinary rates of pay to free labor." Profit calculations and cost estimates often stressed the benefits of low wages induced by the presence of convict labor. Colonel Arthur Colyar, a Tennessee Democrat bigwig and general counsel of the Tennessee Coal, Iron, and Railroad Company, explained in 1892 that the company "found that we were right in calculating that free laborers would be loath to enter upon strikes when they saw that the company was amply provided with convict labor."[34]

Drawing dividends from an industry which, in the three iron- and coal-producing states of Alabama, Georgia, and Tennessee, utilized over 2,000 convict laborers in the middle 1880s, holders of mining and iron stock were the primary beneficiaries of the convict lease system. The benefits were

[33] *New York Times*, July 8, 1877, quoting *Richmond Enquirer.* Herbert G. Gutman, "Reconstruction in Ohio: Negroes in the Hocking Valley Coal Mines in 1873 and 1874," *Labor History* vol. 3, Fall 1962, pp. 244–64. *Senate Hearings on Labor and Capital*, vol. 4, p. 360. Rawick, *Alabama and Indiana Narratives*, vol. 6, Indiana, pp. 23–24.

[34] BRFAL, film M809, roll 2, "Annual Report Wager Swayne, October 1867," p. 130. *Senate Report, Labor and Capital*, vol. 4, pp. 437, 426, 441. Woodward, pp. 15, 215. *Second Annual Report of the Commissioner of Labor, 1886: Convict Labor* (Washington, 1887), p. 301. Woodward, p. 232. Armes, p. 305.

appreciated in other lines of endeavor, however. The highest-paying and greatest employment-generating industrial activities available to black laborers were most conspicuous for their abundant reception of northern capital and use of convict labor. The lumber industry, inundated with northern capital and speculators, became a user. The related turpentine industry, offering seasonal employment under rough, dirty conditions and forced to rely upon migratory labor became infamous for the forced work pace it impressed upon leased convicts.

Second to mining, railroad construction and maintenance were the most important employment opportunities curtailed by the competition of unfree labor. With its high wages, often exceeding a dollar a day, for unskilled men in rural as well as urban regions, railroading promised early to be a tremendous source of earning power for whites and blacks seeking an escape from agriculture. Even though he considered the wages too low, a bureau agent at Newport News, Virginia, registered 700 applicants for construction crews to build a railroad in 1866. The market rate for "able bodied Negro hands" at a compensation of "$20 per month and board" by the Mississippi Central Railroad Company in 1866 was no match for the supply of convicts by the state at a cost of four dollars per month plus board two decades later.

Reported by the U.S. commissioner of labor to be "worked to the utmost and barbarously treated" in Georgia, considered "more reliable and productive than free labor" in Alabama, performing "30 percent more work than free laborers, being worked long, hard, and steadily" in Mississippi, the convict lease, wretched and productive of an "appalling death rate" as it was, remained a source of labor supply in Negro-dominated industries for many decades.[35]

Compared to the heavy industries producing iron, steel, and coal, the two most striking characteristics of the early southern textile industry were the absences of northern capital and black labor. The two marched hand in hand, although not in the most obvious way. While the negligible role of northern capital in the textile industry has been exaggerated, it does appear that the bulk of initial investors were from the region and were very frequently local residents.[36] The connection between the dominance of local financing and the exclusion of Negro labor was indirectly made by economic historian Broadus Mitchell when, in a related discussion, he explained William Gregg's stated preference for slave labor for the development of the antebellum textile industry. Gregg, generally considered the father of the southern industry, was, as Mitchell observed, endeavoring to convince the slaveholding class of his time of the pecuniary gain to be made

[35] James E. Fickle, *The New South and the "New Competition": Trade Association Development in the Southern Pine Industry* (Urbana, Illinois, 1980), pp. ix, 4–8. *Exodusters*, North Carolina, p. 251. Woodward, p. 215. *BRFAL*, film M1048, roll 45, Virginia, Aug. 8, 1866, p. 276; July 31, 1866, p. 95. Ibid., film M869, roll 50, "Letter, Superintendent of Mississippi Central Railroad Company," p. 352. *Second Annual Report Commissioner of Labor, 1886*, pp. 204, 301, 303.

[36] Broadus Mitchell, *The Rise of Cotton Mills in the South* (Baltimore, 1921), pp. 102, 128–35, chap. 4. *Senate Report on Labor and Capital*, vol. 4, pp. 512, 686.

by diverting part of the South's wealth from agriculture to industry. After emancipation this appeal was lost. A major sales pitch used by early mill entrepreneurs to attract considerable numbers of small investors was the purported philanthropic service rendered by providing employment for Confederate widows and orphaned children.[37] While we may be comfortably skeptical of the self-serving character of this preaching on the part of those who were to reap the profits, we must not underestimate its power, for it was as potent an appeal as the finest ad campaigns ever concocted on Madison Avenue. It is difficult to believe that local capital would have been raised for the construction of manufacturing jobs for black agricultural laborers.

Despite Broadus Mitchell's insistence in his pioneering study that 1880 was the beginning of the industrial revolution in southern textiles, the data arrayed in figure 13.1 clearly show that while rapid acceleration did begin in 1880, the turning point was 1870. This supports the view that the crucial event spurring growth was not the spontaneous awakening to the need for industrial diversification by southern leaders about 1880, but rather the much more mundane introduction of a technological innovation in cotton spinning. Shortly before 1870 the invention of an automatic ring spindle, adapted to spinning coarse and medium yarns, allowed a great reduction in the amount of skilled labor required to produce low- and medium-quality cotton goods. Southern firms were thus enabled to compete with New England mills by utilizing their available supply of unskilled labor.[38]

Figure 13.1   Growth of Employment in Southern Cotton Textiles, 1850–1900

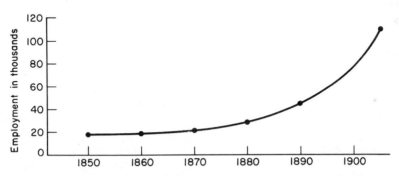

Source: Ben F. Lemert, The Cotton Textile Industry of the Southern Appalachian Piedmont (Chapel Hill, 1933), p. 33.

[37] Broadus Mitchell, *William Gregg: Factory Master of the Old South* (Chapel Hill, 1928), p. 23. Idem, *Rise of Cotton Mills*, chap. 2.

[38] Mitchell, *Rise of Cotton Mills*, chap. 2. Ben F. Lemert, *The Cotton Textile Industry of the Southern Appalachian Piedmont* (Chapel Hill, 1933), p. 33.

The exclusion of Negroes from this rapidly expanding industry can hardly be explained by ideological cant such as that the soft humming of machinery would put them to sleep or that their fingers were "too stiff" to adapt to work requiring manual dexterity. The adaptability of racial ideology to local circumstances has often been commented upon by students of race relations. Where blacks were heavily engaged in occupations requiring manual skill, such as in tobacco factories of the South Atlantic, this was cited as proof of their intellectual inferiority. Where they were excluded from such industries, it was explained that their mental deficiencies incapacitated them for the work. In the cotton fields the exceptional agility and manual dexterity of black women and children, especially as cotton pickers, was lauded. More importantly, the charade being practiced is exposed by the fact that two late-antebellum factories, the Saluda Mill near Columbia, South Carolina, and the Bell factory in Alabama, often referred to by postwar textile producers as early models for the industry, were operated with black labor. In 1880 the Saluda Mill was still in operation, turning a profit with the only racially mixed force of textile operatives in South Carolina. The Bell factory, according to Robert M. Patton, governor of Alabama in 1866 and son of the factory's founder, had worked "several hundred" operatives, all black. "The machinists even were black," the "only two whitemen about the factory" being "the superintendent and one of the partners." Former slave owners like Patton, who had witnessed the black contribution to the industrial development of the Old South, found it difficult to parrot the deceptions of the New South: "You can make anything of the black people if they are properly raised; they make the best servants, the best carpenters, and blacksmiths; they can do anything."[39]

The rising generation of mill managers and owners, often stemming from the non-slave-owning classes and evincing a great deal of antipathy toward the Negro, found it relatively easy to blend business principles with a professed philsophy of white philanthropy. The seemingly limitless supply of female and child white labor at extremely low wages was responsible for profit rates high enough to leave almost any employer's commitment to white supremacy in the mills unchallenged. The four mills under the control of the Eagle and Phoenix Company in Columbus, Georgia, where "all the labor inside the mills [was] white" and the few "outside" all "colored," made "average" earnings of 17 percent from 1870 to 1882. During that period the lowest dividend payment had been 6 percent. This experience, drawing the envy of southern iron manufacturers, was typical. The seventeen mills existing in 1880 South Carolina averaged profits above 20 percent, ranging from a low of 18 to a high of 25.5 percent. These figures were exclusive of an eighteenth mill, placed in a special category, which earned 50 percent.[40]

---

[39] King, *Great South*, p. 373. *Senate Hearing on Labor and Capital*, vol. 4, p. 47.

[40] *Senate Hearings on Labor and Capital*, vol. 4, pp. 508–9. *Charleston News and Courier.*

The brunt of the exclusion of Negroes was born by black women and children. The Eagle and Phoenix Mill, employer of 1,883 hands in 1882, was high on the list of desirable employers. Its employment of 739 adult women and 312 children of both sexes ranging in age from ten to sixteen, making 56 percent of its labor force, was about the industry's average. Women and children were 46 and 18 percent of the labor force in Columbus's textile mills, respectively. The entire southern industry in 1879 employed 36 percent adult women and 24 percent children. Children employed as spinners earned the lowest wages in the mills. At the Eagle and Phoenix the lowest wages, 25¢ a day, went to children ten to thirteen years of age. Those fourteen to sixteen earned 50¢ a day. Most of the weaving was done by older girls and women, generally 80¢ to $1.25 a day. Most importantly, the mill paid not truck but "cash promptly." A high estimate of the mill superintendent's figures for housing and food costs comes to 10¢ a day, so that the lowest wages for women gave a net daily wage of 70¢.

The Eagle and Phoenix probably paid close to the top of the pay scale of the industry. However, among more than two dozen mills, with the exception of a few very young children working as sweepers for about 17¢ a day, no mill reported a wage for children of below 25¢ or for women of below 60¢. Hidden reductions in wages no doubt occurred in rural mills that dealt in truck. But no matter how low these wages appear in absolute terms, relative to farm wages available to women and children they were large. Adult women performing farm work earned from 25¢ to 30¢ a day for "about five months" of the year. During another three or four months they were unemployed at market labor, there being no demand for their services. During cotton picking, working at a piece rate that varied with the price of cotton from roughly 30¢ to 50¢ per one hundred pounds picked, an average hand would seldom exceed 50¢ a day. The problem of in-kind payments, such as a garden plot, or rations of salt pork and corn meal (for working hands only) make a precise comparison difficult. However, a woman's weekly allowance of two to three pounds of meat and three or three and a half pounds of meal could not have made up the difference. Such comparisons seldom take account of the fact that garden plots also required time and labor.[41]

A better basis for comparison are the wages paid to urban black women. There, as explained by the Reverend E. P. Holmes of Columbus, Georgia, "everybody knows that among the female sex the mass of our people are washer-women, washers, and ironers, cooks and chambermaids." In these occupations the monthly pay was often in the $5 to $6 range, and seldom above $10. The daily rate, then, was less than one-half the cash payments available to women working in the mills.

The social prohibition against mixed labor forces in factories employing white women and children worked to the detriment of the most qualified

[41] *Senate Hearings on Labor and Capital*, vol. 4, p. 496. *Compendium of Tenth Census*, p. 1125. *Senate Hearings on Labor and Capital*, vol. 4, pp. 508, 509, 530, 535, 661.

black women, who were restricted to domestic service. For if integration had been permissible these women would have found employment in the mills. As conditions were, with whites refusing to work with blacks, the only path to mill employment for these women would have been the operation of all-black mills. Since as a group white industrial labor was probably of higher average productivity and was in plentiful supply, and since the industry was making profits, there existed no motivation for prejudiced employers to resort to black labor. Later in the century the fragility of the social forces excluding blacks would be demonstrated when a Charleston mill that had gone bankrupt with white labor made an unsuccessful fourth attempt with cheaper black workers.

With this threat of black competition always present, the separation of the workers served the industrialists well. Cognizant of the fact that white working people were against their employment in remunerative occupations, blacks often drew the conclusion that it was the white working people who "oppressed" them in their labor.[42] For the most skilled, usually victims of the severest forms of economic discrimination, the experience often led to severe antipathy against white labor organizations and unions in general.

The ousting of the black ship caulkers in 1865 Baltimore is the most dramatic example of this. Isaac Myers, a leader of the colored caulkers, was deeply affected by that event. Myers responded by organizing a black shipyard company with $10,000 capital raised by selling shares of $5 each, providing employment for hundreds of black caulkers and carpenters. Elected president of the Colored National Labor Union, Myers saw black trade unions not as working-class organizations formed to bargain against employers, but as competitive weapons to prevent the "extermination of colored labor" by white trade unions that would "not permit colored men to work in their shops." The philosophy of the mechanics' cooperative shipyard came to rule the Colored National Labor Union. Necessarily wedded to the Republican Party, entrepreneurial artisans like Myers foresaw no confrontations with employers. Opposed to strikes, they subscribed to a belief in the harmony of interest between labor and capital.[43]

The developing political coalition between black leaders and American industrialists often attributed to a lingering tradition of slave paternalism was more often due to a host of other factors. Excluded by the great majority of white laborers, the black worker was forced to look to the "self-interest" of profit-seeking capitalists for economic salvation. Bound together by a racial caste system, class differences within the community were never absent but were generally unvocalized. Under these conditions black leaders who gained national attention from the white community were often professionals or businessmen whose industrial philosophy melded easily with business interests. Southern employers, aware of the situation,

---

[42] *Senate Hearings on Labor and Capital*, vol. 4, p. 607. Tindall, *South Carolina Negroes*, pp. 133–34. *Senate Hearings on Labor and Capital*, vol. 4, pp. 399, 400, 403.

[43] Sterling, *The Trouble They Seen*, p. 283.

advertised their black workers as a "conservative element" in the labor market, "free from trades-unionism and communism," two isms that were often indistinguishable in the minds of contemporaries. This absence of a working-class presence, even as early as Reconstruction, was in many ways a myth.

In the first place, all black leaders were not anti-strike opponents of labor organizations. Men like state congressman Henry M. Turner of Georgia and Robert B. Elliott of South Carolina were at the forefront of labor organization in their states. James T. Rapier, congressman from Alabama and a large plantation owner, was clearly a man whose personal interest was tied to cheap labor. A leading figure in Alabama's black labor organizations, he endorsed a nine-hour day and a minimum wage. The political influence of these men, like that of Frederick Douglass, waned in a Republican Party that was far from pro-union in the nineteenth century. In the cities, union activities were far more common than the exhortations of contemporary employers would lead one to believe. A long and successful tradition of working-class agitation for better wages and conditions began with the chartering in 1869 of the Longshoremen's Protective Union Association of Charleston. This union, whose legal representative in 1880 was a white Democratic member of the South Carolina legislature, became a model for others all along the south Atlantic coast.[44]

The largest employer of black factory hands was the tobacco industry based in Virginia and North Carolina. Wasting little time, the factory freedpeople of Richmond engaged in their first organized effort as free laborers in the fall of 1865 with an appeal to the city at large that made their demand for a pay increase known:

> They say we will starve through laziness that is not so. But it is true we will starve at our present wages. They say we will steal we can say for ourselves we had rather work for our living. give us a chance. We are compeled to work for them at low wages and pay high Rents and make $5 per week and sometimes les. And paying $18 or $20 per month Rent. It is impossible to feed ourselves and family—starvation is cirten unles a change is brought about.
> [Signed,]
> Tobacco Factory Mechanicks of Richmond and Manchester

This effort probably failed. The employers had cut wages, having found that "labor was much cheaper than formerly," a fact no doubt explained by the influx of laborers from the country.

The difficulty of the work was described by Trowbridge, who "never saw more rapid labor performed with hands than the doing up of the tobacco in rolls for the presses; nor harder labor with the muscles of the whole body than the working of the presses." This difficult work, paid by piece rates,

[44] Campbell, *White and Black*, pp. 287, 143. Sterling, *The Trouble They Seen*, p. 286. Loren Schweninger, *James T. Rapier and Reconstruction* (Chicago, 1978), pp. 87–91. Tindall, *South Carolina Negroes*, p. 137. John W. Blassingame, *Black New Orleans*, pp. 64–66. Spero and Harris, *Black Worker*, pp. 182–83.

was only available nine months of the year, forcing the laborers to scramble for alternative labor during the off-season. Wages were kept low here, as in the white-employing cigarette factories and textile mills, by the employment of large numbers of women and children to do the light work requiring manual dexterity. Despite these difficulties the operatives in the tobacco factories were to develop a tradition of strike activity.[45]

The now legendary Richard L. Davis, Negro coal miner twice elected member of the National Executive Board of the United Mine Workers in the late 1890s, received his early industrial training in the tobacco factories of Roanoke, Virginia, which he entered at the tender age of eight. At seventeen, "disgusted with the very low wage rate and other unfavorable conditions of a Southern tobacco factory," Davis relocated to West Virginia's Kanawha and New River coal fields. The conditions in the salt and coal mines of this region were described as they existed in 1865 by Booker Washington:

> Our new house was no better than the one we had left on the old plantation in Virginia. . . . Our new home was in the midst of a cluster of cabins crowded closely together, and as there were no sanitary regulations, the filth about the cabins was often intolerable. Some of our neighbors were coloured people, and some were the poorest and most ignorant and degraded white people. It was a motley mixture. Drinking, gambling, quarrels, fights, and shockingly immoral practices were frequent. All who lived in the little town were in one way or another connected with the salt business. Though I was a mere child, my stepfather put me and my brother at work in one of the furnaces.[46]

In the company town, be it around a textile mill, a sugar plantation, or a coal mine, the low plane of living and the racially divided labor force spawned conditions that were hardly conducive to the development of a working-class consciousness. That Richard L. Davis, who quickly left the Kanawha Valley for Ohio, would soon become an instrumental force in the struggle to better the lives of white and black working people is a high tribute to the courage of a number of unknown Americans. These conditions would prove insurmountable for most black and white workers. The still largely undocumented history of southern workers' organized activity in the late nineteenth century will be a history of economic failures, but of much cultural value. The poor conditions stemming from a heritage of poverty, racism, and child labor would constrain the workers' activities to defensive measures and allow their employers and political leaders to praise the lack of solidarity among the working classes. The black labor move-

[45] Trowbridge, *The South*, pp. 230–31. *Senate Hearing on Labor and Capital*, vol. 4, pp. 3, 8. Nannie May Tilley, *The Bright-Tobacco Industry, 1860–1929* (Chapel Hill, 1948), pp. 515–21.

[46] Herbert G. Gutman, "The Negro and the United Mine Workers of America: The Career and Letters of Richard L. Davis and Something of Their Meaning, 1890–1900," in Herbert G. Gutman, *Work, Culture, and Society in Industrializing America* (New York, 1977), p. 125. Booker T. Washington, *Up From Slavery*, in Louis R. Harlan and John W. Blassingame, eds., *The Booker T. Washington Papers* (Urbana, Ill., 1972), pp. 227–28.

ment, such as it was during Reconstruction, was tied to the labor philosophy of the Republican Party and the freedpeople's own desire for land ownership. This we take up in the next chapter.

Kept undereducated, and barred from apprenticeships in industrial trades, the black worker would develop a heightened sense of racial injustice. The white laborer, soon to aid in the political disfranchisement of the blacks, would also aid in preventing the establishment of an adequate system of education for themselves as well as their competitors. In the southern press, during elections or when it was opportune to advertise the free-born American's industrious and docile qualities, the white laborer was to undergo a remarkable metamorphosis, becoming the finest worker available in a period of time so short as to defy belief.

14

We congratulate our people upon the rapid progress they have made in the past six years, and upon the increase of mixed industry, homesteads, and small farms in opposition to the ruinous plantation system.

Resolution, the AME Church of Florida

Emancipation implies that the laborers no longer have any "masters" to rule over them. But actual freedom is only attained when these laborers can actually dispense with such masters.

New Orleans Tribune

Negro Agrarianism

At the height of the political agitation that has come to dominate all subsequent images of Reconstruction, a perplexed band of white Regulators demanded that the black Mississippian whom they were whipping tell them "what the radicals had ever done for him, how much money they had ever given him, or how much meat and bread they had ever given him, and why it was that he was damned fool enough to be controlled by the radical party?" Ostensibly the answer would have been built upon the Republicans' position as the party of emancipation and the deliverers of the Thirteenth and Fourteenth Amendments to the Constitution. These acts, which earned the gratitude of generations of black Americans, can hardly be discounted as unimportant. Nevertheless, the Mississippi Regulators were not total fools for asking the question. Years later Booker T. Washington would reiterate much the same point when he suggested that "coloured men" should seek the advice of southern white people in reference to political matters. In the political sphere, Washington thought Negroes could not "afford to act in a manner that [would] alienate [their] Southern white neighbors." Washington was merely enunciating a position that had been staked out early during Reconstruction. Gilbert Myers explained that he voted Democratic because "I sympathized with my own self." A hardworking black Louisiana landowner, Myers voted with the Democrats because he "could not expect to accumulate property" unless he went "with the majority of the country whom [he] expected to prosper with." Myers, who testified that in a choice between his freedom or his property he "would rather have [his] property," would probably have agreed with an Alabama contractor who, having been set up in business by his former owner, declared that "if my vote affected my business I would not vote at all." These two represent perfect examples of the rational "economic being" so often constructed by economic historians. Their decision calculus was expected by those masters who prognosticated a continuing relationship of total dependency for their laborers.

One planter, well versed in the mechanics of market paternalism discussed in Chapter 7, explained the principles of nineteenth-century social democracy to a group of skeptical fellow travelers:

> The class of people there [North] that represent our niggers, the laborers, have a right to vote. My father-in-law employs thirty-five Irishmen. They always vote the right ticket, and he tells them which is the right one. Now, major, if you hire thirty-five niggers, work em well, pay em well, and feed em well—they don't know William H. Seward from a foreign war.[1]

In the same vein, James L. Pugh, an antebellum U.S. congressman based in Barbour County, Alabama, testified in 1871 that earlier "there was a good deal of surprise that the negroes all went to the republican meetings, and would not attend the democratic meetings." A planter who was able to extend his influence into the postwar period as a leading Redeemer and U.S. senator for Alabama, Pugh expressed the outlook of the wealthiest conservatives in the South. Favoring Negro suffrage, Pugh had been convinced that black voters would support the paternalistic policies of his class.

Some twenty years later, just prior to the complete disfranchisement of the Negro, sharecropper Nate Shaw's description of the purchasing of black votes by Alabama planters would nearly rival Francis Place's portrait of election buying by servants of the Duke of Northumberland in early nineteenth-century Britain: "Niggers just fell in there like pigs around their mammy suckin: votin the white man's way. Come time to vote, white man runnin all about the settlement buyin the niggers' votes. Give him meat, flour, sugar, coffee, anything the nigger wanted."[2] Shaw's merciless condemnation can be partly explained by the limited sense of history often found among those lacking formal education. Born in 1885, Shaw did not know that in an earlier decade the great mass of black people chose a different political course. Were the freedpeople in the days of Reconstruction

---

[1] *KKK Hearings*, vol. 1, "Testimony of W. W. Chisolm," Kemper County, Miss., p. 256. Washington, *Up From Slavery*, pp. 236, 86, 201. *Exodusters*, vol. 2, Louisiana, pp. 584, 584, 587. *Relations between Labor and Capital*, vol. 4, p. 116. Dennett, *South as It Is*, p. 31. Easterby, *South Carolina Rice Planter*, "Oliver H. Kelly to Benjamin Allston," p. 232.

[2] *KKK Hearings*, Alabama, vol. 1, 2nd sess., "Testimony of James L. Pugh," pp. 405, 406. Theodore Rosengarten, *All God's Dangers: The Life of Nate Shaw* (New York, 1975), pp. 34–35. Francis Place's account is quoted in Thompson, *Making of the English Working Class*, p. 77.

so grateful to the Republican Party that they abandoned the most obvious chance for the relative material comfort offered by their ex-masters in favor of an ideological principle? That conclusion may have been drawn by Alabama Democrats who, after telling the Negroes that the low price of cotton was due to "their voting for *Radicals*," received the reply that the blacks would "vote for their *friends*, even though it bring cotton to *five cents.*" A more circumspect analysis must consider Max Weber's evaluation that a socially dominated agricultural labor force will seldom coalesce in an effort to confront their employers as a group except during "times of extreme political turmoil." The freedpeople saw in their emancipation a real chance to seize the time.

In 1868 the southern branch of the Republican Party, expounding the gospel of "free labor," public education, and a homestead law, all solidly based on a philosophical rendition of the classical labor theory of value, was the only political home for a group of emancipated agricultural slaves. We need only recall the role of the free soil movement in the formation of the Republican Party to agree with John R. Commons about the social significance of the "views on the natural right to land" which "appeared in the Republican party wherever that party sprang into being." The intuitive appeal of a philosophical tradition rooted at least as far back as John Locke and based upon the primal claim of labor to its product is exemplified by the freedpeople's own independent and hardly surprising development of a labor theory of value.[3]

Few historians have failed to discuss the freedpeople's tremendous desire for land and the bitter disappointment which followed the federal government's decision to restore confiscated lands to previous owners. Leon Litwack's poignant presentation describes the argument of Cyrus, a Virginia freedman, who explained to his ex-mistress what was wrong with a sharecontract:

> Seems lak we'uns do all the wuck and gits a part. Der ain't goin'ter be no more Master and Mistress, Miss Emma. All is equal. I done hear it from de cotehouse steps. . . . All de land belongs to de Yankees now, and dey gwine to divide it out mong de colored people. Besides, de kitchen ob de big house is my share. I help built hit.

Cyrus was expressing more than a desire for social equality and property. His words were indicative of a strong tradition in Afro-American culture. The last generation of slaves and their free descendants were acutely aware that much of American civilization and prosperity was built with the labor and skills of generations of black Americans. For blacks there has always been a mixture of pride and a continuing sense of collective exploitation in this knowledge. Texan Andy McAdams recalled the disappointment of 1865: "I'se expected lots different from freedom than what we got yes sir, I

[3] *BRFAL*, film M809, roll 18, Oct. 1867, p. 157 (emphasis in original). John R. Commons, "Horace Greeley and the Working Class Origins of the Republican Party," *Political Science Quarterly*, no. 24 (1909), p. 484.

don't hardly know exactly what we did expect. I think they ought to have given us old slaves some mules and land too, because everything that our white people had we made for them." Up on the Carolina coast, rice planter Edward B. Heyward had received a personal lesson in the elements of agrarian justice from a former slave who told him that, "the land ought to belong to the man who (alone) *could work it*" and not those who "sit in the house." Of property, Locke had written that "as much as any one can make use of to any advantage of life before it spoils, so much he may by his labour fix a property in. Whatever is beyond this is more than his share, and belongs to others." Frederick Law Olmsted was told in the 1850's "that everywhere on the plantations, the agrarian notion has become a fixed point of the negro system of ethics: that the result of labor belongs of right to the laborer."

Cyrus's specific claim to "de kitchen ob de big house" was at once an acknowledgement of the slaves' agrarian ethic and perhaps the first direct communication to his former mistress of a strong sense of collective exploitation within the slave community. Two generations and more than one hundred years afterwards Nate Shaw would reiterate:

> I feel a certain loyalty to the state of Alabama: I was born and raised here and I have sowed my labor into the earth and lived to reap only a part of it, not all that was mine by human right. It's too late for me to realize it now, all that I put into this state. I stays on if it gives em satisfaction for me to leave and I stays on because it's mine.[4]

From precepts such as these the freedpeople constructed their claim for a distribution of the lands. One of the finest was delivered by freedman Bayley Wyatt of Yorktown, Virginia, during a public meeting in 1866:

> I may state to all our friends, and to all our enemies, that we has a right to the land where we are located. For why? I tell you. Our wives, our children, our husbands, has been sold over and over again to purchase the lands we now locate upon; for that reason we have a divine right to the land.

After this imaginative argument, Wyatt offered what Edward Magdol aptly termed a "lesson in political economy," an eloquent statement of the slave's theory of value and sense of collective exploitation:

> And den didn't we clear the lands and raise de crops of corn, ob cotton, ob tobacco, ob rice, ob sugar, ob eberything? And den didn't (dem) large cities in de North grow up on de cotton and de sugars and de rice dat we made? Yes! I appeal to de South and to de North if I hasn't spoken de words of truth. I say dey have grown rich and my people is poor.

It is well known that from the perspective of a general economic theory of value, as opposed to an ethical one, Locke's labor theory of value requires that all other factors of production be so abundant that they are

⁴ Leon Litwack, *Been in the Storm So Long*, p. 399, quoting Rembert W. Patrick, *The Fall of Richmond*, p. 125. Rawick, *Texas Narratives*, vol. 7, pt. 6, p. 2455. Litwack, p. 401. John Locke, *Two Treatises on Civil Government* (London, 1884, Morley's Universal Library ed.), book 2, p. 206. Olmsted, *Seaboard Slave States*, p. 117. Rosengarten, p. 500.

redundant—free goods. If this is not the case, part of the value of the product of the worker's labor must be attributed to other nonscarce factors, such as land and capital. This the blacks understood. "We don't want de whole valler of de cotton," explained Sea Islanders to a government plantation superintendent. "De land belongs to de Government, de mule and ting on de place belong to de Government, and we have to spect to pay somefn for um." Based upon the same axiom of justice subscribed to by radical Republicans, the blacks' labor theory of value was purely philosophical. "Labor," declared an 1871 convention of ministers and laymen of the African Methodist Episcopal Church in Florida, "is the basis of all wealth." [5] Such a clear perception of the distinction between ownership and distribution suggests a predilection for an individualistic conception of property. This, it shall be seen, was true, but its transparency is clouded by the fact that the freedmen's notion of a *right* to the land was communal, being based upon a history of group exploitation that, as Bayley Wyatt explained, encompassed far more than the expropriation of the result of their labor. The freedpeople could make the subtle distinction between a communal right shared by all descendants of enslaved Africans and an individual right based upon the laborer's claim to his or her individual productivity. Their notion of property was private; let "each one work his own and do as well as he kin," advised the Sea Islanders.

Revealingly, a favorite song of both slaves and antebellum whites contained the lines "All I want in dis creation Is a pretty little wife and big plantation." [6] This desire overrides all other considerations in explaining the political behavior of the freedpeople during Reconstruction. It is a crucial element in a complete understanding not only of Negro republicanism and planter reaction but also the "squad system" of labor organization studied in a previous chapter. The freedpeople's proverbial thirst for land is seldom discussed in terms that go beyond an individualist desire for land. During Reconstruction, this desire led to considerable collective action. We must be careful not to read too much into the recourse to collective modes of organization, for this approach seems to have been adopted more as a means to an end than as part of a philosophical orientation which viewed collective organization as a way of life. For the freedpeople, collective action was decidedly chosen as a means to advance individual gain. However, as is often the case in collective undertakings, a fervent esprit de corps developed within the collective. Something of this we have seen in the earlier discussion of the collective work squad. In the present chapter a view of Negro agrarianism during Reconstruction provides a final perspective on the social function of the collective work squad and the breakup of the antebellum plantation.

[5] Edward Magdol, *A Right to the Land: Essays in the Freedmen's Community* (Westport, Conn., 1977), p. 172. Pearson, *Freedmen at Port Royal*, p. 113. *KKK Hearings*, "Testimony of Charles H. Pearce," Florida, vol. 1, p. 171.

[6] Pearson, p. 113. John Blassingame, *The Slave Community* (New York, 1979), p. 120.

Given that radical confiscation schemes failed to materialize, it is not difficult to conclude that after emancipation the Republicans did little for the
freedpeople. That perception formed the basis for the planters' initial belief
that they could control the black vote. The logic fails to assess adequately
the revolutionary spirit of the period. If the Republicans' free labor ideology of open opportunity and upward mobility to independent proprietorship
for the masses was becoming an anachronism in the urban North, it was far
from being so on the expanding western frontier. The South, and especially
the Southwest, with its gigantic agrarian base, was, in the absence of a
structure capitalized upon slave property, theoretically extremely compatible with the tenets of the "free labor" ideology. In addition, regardless of
the sincerity or the ability of congressional Republicans to deliver on political promises to the freedpeople, the words and actions of local Republicans
in the South were far more important in mobilizing the people.

If the doctrines of self-help and individual hard work appear less than
sufficient to modern readers, it is perhaps partially due to anachronistic
thinking on our own part. With hindsight we may easily scoff at the rant-
ings of carpetbaggers and black politicians who promised the people so
much without delivering. But in the euphoric days of early Reconstruction no
one could foresee precisely what the socioeconomic structure of the South
would be in twenty years. Anyone who has studied letters of the planting class
during this period will realize that their fear of Republican prognostications
of an agrarian revolution cannot be fully appreciated by the modern mind.
Beyond this, one who would criticize must ask, as Booker Washington
might have, what alternatives existed? The federal government was not
inclined to deliver land or monetary compensation. The miniscule black
bourgoisie, even if it had had the inclination (and it certainly did not), was
too undercapitalized to attempt organization of an alternative to capitalism.
The people had, in the main, to look to themselves. This much Booker
Washington surely understood.

We have seen in earlier chapters something of the Freedmen's Bureau's
positive involvement in the laborers' first attempts to secure a livelihood.
Any skepticism concerning the validity of local agent activity on behalf of
the laborers should be dispelled by the intense vilification of the bureau by
the planter class. P. T. Sayre, an attorney from Montgomery, Alabama,
enumerated a typical list of charges, and unwittingly verified most of what
has been attributed to the bureau in previous chapters:

> They notified everybody that they must employ their freedmen, and that all
> their contracts must be submitted to the inspection of the Freedmen's Bureau;
> that no man would be allowed to employ freedmen unless their contracts were
> submitted to and approved by that Bureau. Well, they listened to every sort of
> tale that any dissatisfied negro might choose to tell; they would send out and
> arrest white men, bring them in under guard, try them and put them in jail.
> They got hold of plantations there, what they call refuges for freedmen. . . . In
> that way a large number of negroes were enticed away from plantations where
> they had been living, and they flocked to these places. Hundreds of them died

from neglect. The impression was produced upon the negro that the white man who had been his master was his enemy, and that these men were his peculiar friends; that they had nothing to expect from and through their old masters. They then commenced the establishment of these Loyal Leagues, into which they got almost every negro in the country. They would send their agents all through there, from plantation to plantation, until I expect there was hardly a negro in the whole country who did not belong to the League. In that way a want of confidence was produced between the negro and the white man, and a feeling of confidence between the negro and the agents of this Bureau.

Local bureau agents, after emphasizing a different interpretation of the basic facts, would have disagreed little with Sayre's account. Even planter-historian Walter L. Fleming admitted that much of the bureau's activity could have been overlooked if not for its political role. Alabama agent W. E. Connelly explained why in an 1867 report:

Great pains have been taken to instruct the freedmen in the principles of *justice* on which contracts should be based,—the *binding force* of contracts, and the importance of a faithful compliance with all their lawful obligations.

But Connelly continued, providing a clear picture of just what sorts of ideas he may have been promulgating in the Republican political clubs known as the Loyal Leagues:

The evils attending the present system of farming—the working of a great number of laborers on large plantations under one contract, (referred to in the *July Report* from this office) are daily becoming more apparent; while the plan heretofore suggested of dividing the large plantations into small farms, is gaining favor with farmers. Indeed, it is evident that the day is not far distant, when the present land-holders will find it to their interest to reduce their extensive tracts of land down to a comfortable *homestead*, and invest their capital in other branches of business. It is to be hoped they will turn their attention to the building of railroads, the erection of manufacturing establishments, and the development of the various resources of the state.

   Then every citizen will have an opportunity to secure a *homestead*, and the *latent wealth* of the state will be brought to the surface and distributed among the people. Then we will have a *practical* application of the principles of the *Republican Union Party*, resulting in "*the greatest good to the greatest number.*"[7]

No social philosophy was more attuned to the freedpeople's "agrarian ethics." In addition, in a world of self-help and hard work the ex-slaves considered themselves in an enviable position of comparative advantage relative to their white competitors. "Our old mistrus," recalled Tempie

   [7] *KKK Hearings*, Alabama, vol. 2, pt. 8, p. 357. Fleming, *Civil War and Reconstruction in Alabama*, pp. 443–44, 533–38. *BRFAL*, film M809, roll 18, Eufala, Alabama, Oct. 1, 1867, pp. 149–50 (emphasis in original). See *KKK Hearings*, Florida, "Testimony of Joseph John Williams," pp. 235, 237; ibid., Mississippi, "Testimony of Charles, Baskerville," p. 413; "Testimony of Samuel Gholson," pp. 852, 883.

Cummins, "she was turrible lazy." Austin Grant remembered that his master "had a nice rock house," and "the white women didn't do any work a-tall." This contempt for the work ethic of white women of the planting class manifested itself in many versions of the slave song "Mistress in the Big House," one example of which contained the following stanza:

> Missus in de big house,
> Mammy in de yard.
> Missus holdin her white hands,
> Mammy workin hard.

The slaves' assessment of their masters fared little better:

> Old Marse ridin all time,
> Niggers workin roun.
> Marse sleepin day time,
> Niggers diggin in de groun.

A postbellum ditty accused a white man sitting under a "shade tree" in a starched "white shirt" of being the "laziest man that god ever made." The freedpeople, like most people who make their living by getting their hands dirty, did not regard as work any activity that failed to give the same result. For this reason alone they would never recognize as legitimate a manager's claim to a distributive share of the proceeds of productive labor. When the blacks of Liberty County, Georgia, proclaimed themselves to be "a working class of people," we may also detect a subtle but sharp jab at their former masters. Performing no necessary labor in the eyes of the laborers, the landlords were a redundant class.

For the freedpeople during the Reconstruction period, Republicanism and Negro agrarianism seemed to be synonymous. It is not too much to say that, in 1866, the slaves of the old South were born-again Jacobins. We may both sympathize with and understand Felix Haywood's recollection of the beginning he and his father made with seventy head of cattle and their "own brand" as Texas ranchers:

> We knowed freedom was on us, but we didn't know what was to come with it. We thought we was goin to get rich like the white folks. We thought we was goin to get richer than the white folks, cause we was stronger and knowed how to work, and the whites didn't and they didn't have us to work for anymore. But it didn't turn out that way. We soon found out that freedom could make folks proud but it didn't make em rich.[8]

These ideological beliefs of the freedpeople must be invoked if Republican solidarity among blacks is to be appreciated. Even the most bourgois-oriented leaders, such as Richard T. Greener, considered the formation of industrial cooperatives among workmen for the purpose of competing in the free

[8] Rawick, *Texas Narratives*, vol. 4, pt. 3, p. 1007. Ibid., vol. 5, pt. 4, p. 1539. Howard W. Odum and Guy B. Johnson, *Negro Workaday Songs* (Durham, 1926, reprint New York, 1969), p. 117. Ibid., p. 116. Herbert Gutman, *Slavery and the Numbers Game* (Urbana, Ill., 1975), p. 172. Rawick, vol. 4, pt. 2, p. 134.

enterprise system legitimate and desirable. Isaac Myers and the leaders of the Colored National Labor Union meant by "labor organization" not the formation of trade unions but cooperative firms willing to compete as capitalists against white capital:

> Every effort should be made to make your labor more remunerative, and less dependent upon the capitalist; . . . organize cooperative mechanical associations. Let each one lay by a small sum weekly for the purchase of the necessary tools, then take his labor as capital, and go out and build houses, forge iron, make bricks, run factories, work plantations &c.

Essential to the political success of this strategy was its ability to attract disparate elements among the population needed to form a winning political coalition. The *New Orleans Tribune*, published by free blacks and an intellectual organ of Negro agrarianism, highlighted the flexibility contained within the competitive cooperative concept. The *Tribune*, arguing early during Reconstruction for "new models" of labor organization, argued that the "best way was to get rid of the old planter personnel and overseers." The plan which the *Tribune* seemed to prefer called for the replacement of antebellum planters and overseers with "excellent managers" from "among the old free colored people," who, joining with people willing to provide capital, would form a "partnership" with the freedmen. The attitude of paternalism which northern whites held toward the freedpeople was probably stronger among free blacks of the South and North. The free blacks' call for the organization of "labor-colonies" to develop agriculture and "elevate" their "emancipated brethren" says much. The elite among the free blacks, business-oriented and possessing an accumulative ethic, viewed the problem of economic reconstruction from a perspective essentially similar to that of radical Republicans. They believed that the mass of freedpeople were not of sufficient intelligence at the present time to be given unguarded control over the allocation of resources in the economy. Of necessity, the position of free coloreds on this question placed no weight upon racial inferiority. The question of class was paramount. The freedpeople, having been given no educational grounding to function as free citizens, needed a type of guidance which would not be provided by their previous owners. "Associations among men willing to work the plantations, on the plan of really free labor," were required. Free blacks, too, had nowhere to rest outside the Republican Party.

The charge that their exhortations merely camouflaged their intentions to exploit freed labor as capitalists seem very weak. Outside of any natural feelings of humanity, their own self-interest and desire for an integrated society of social equality between the races could not be achieved as long as a great pool of black labor—unskilled, uneducated, and poor—served to anchor a biracial caste system.[9]

---

[9] Isaac Myers, *New National Era*, Feb. 17, 1870. *New Orleans Tribune*, Jan. 28, 29, 1865. Ibid., Nov. 30, 1864. Ibid., Jan. 28, 1865.

It will be recalled that David F. Schloss, in his nineteenth-century classic, *Methods of Industrial Remuneration*, distinguished between profit sharing and other systems of industrial cooperation by referring to the former as a device of "middle class employers" thoroughly convinced of the economic necessity of their "class." The *Tribune*, when insisting that managers "associate" with labor rather than simply supervise them, drew up a first-rate plan of profit sharing. The broad outline of this scheme, which sought to make laborers "partners in the yield of the crops," made it imperative that they also "receive monthly or weekly pay." The payment of wages would allow them to become "accustomed to self-reliance" because they would be "free to go where they please" and "to purchase freely and at their own will," being "bound to plantations by their interest only." But the plan suggested by Isaac Myers in 1871, described six years earlier by the *Tribune* as recommending associations of families who choose their manager "from among the associates," became by far the most prevalent, manifesting itself as the squad system.

Far more dangerous to the plantation economy was the emergence of numerous independent farming colonies for the purpose of purchasing homesteads. Most of these companies appear to have been structured upon an institution of a type known throughout the world and referred to generically as the "rotating credit association." Found, for example, in West Africa among the Yoruba, who call it *esusu*; in China, where it is termed *hui*; and among the Japanese, where it is known as *ko, tanomoshi,* or *mujin*, the rotating credit association is an innovative socioeconomic structure in which member-investors pool their finances, each making small periodic installment payments, so as to finance projects for the members of the association. Use of the funds in the investment pool rotates through the membership until all participants have received a pool credit.[10]

A paradigm of the rotating credit association was approved at a large meeting of freedpeople in Norfolk, Virginia, in the fall of 1865. "The National Homestead, Settlement and Labor Agency" was created to help laborers obtain farms and employment. This "joint-stock" company, which attached itself as a branch of John A. Andrew's Boston-based American Land Association and Agency, was to cooperate with local agencies. Local groups were to "form Land Associations, in which, by regular payment of small installments, a fund may be created for the purchase of land sales, of land on behalf of any investing members."

The extent of related activity was far more extensive than has been supposed. In one way or another many black political leaders, such as U.S. congressman James Rapier of Alabama, the Reverend Tunis G. Campbell in Georgia, or Elijah Marrs of Tennessee, were connected with some kind

[10] Schloss, p. 151. *New Orleans Tribune*, Jan. 29, 1865, Feb. 2, 4, 1865. Shirley Ardener, "The Comparative Study of Rotating Credit Associations," *Journal of the Royal Anthropological Institute*, 1964, pp. 201–21. Clifford Geertz, "The Rotating Credit Association: A Middle Rung in Development," *Economic Development and Cultural Change*, April 1962, pp. 241–63. Ivan H. Light, *Ethnic Enterprise in America* (Berkeley, 1972).

of homestead association during the Reconstruction era. Even the military was active in assisting such groups, in some cases apparently initiating land colonization plans for the freedpeople. General David Tillson of the Georgia Freedmen's Bureau lent active support to a number of "societies of freedpeople seeking to purchase and settle lands in Southwest Georgia with funds they had raised through subscriptions." Colonel Manley Preston, commander of the 58th Colored Infantry, stationed in Mississippi, succeeded in getting a leave of absence from military duty for himself and a couple of his best officers in order to organize a homestead colony from which his troops would be able to obtain farms. Behind many of these efforts was the same paternalistic missionary zeal that formed part of the ideological apparatus of many northern immigrants. Colonel Preston was very concerned that his troops, who were only paid twice a year, were in the habit of quickly dissipating the relatively large sums they received on consumption of little permanent value.[11]

If the philosophy and practice of cooperative agrarianism had been limited to northerners and middle-class blacks with business and artisan backgrounds, its importance to our comprehension of Reconstruction would be limited to an interesting curiosity. Such an interpretation would seriously flaw our understanding of postbellum southern history. In the early summer of 1865 journalist Whitelaw Reid reported from Alabama that "in the country they were already talking of clubbing together and working plantations." Freedman Henry Baker, of the same state, confirmed Reid's report seventy-three years later. Baker's advocacy of this same development strategy in 1938 demonstrates its ties to Reconstruction, its bourgeois philosophy, and one of the reasons for its failure:

Ef de people would come togedder say now, we get er thousand people en come togedder en say we put up five dollars a year, en eny thousand people could put up dat much money in a yeah's time, which would be five thousand dollars, we could go en buy five thousand dollars worth er lan. Hit would be er private enterprise en we could begin settlin our people on dis lan en de next yeah raise five thousand dollars agin en in de coase uv time all uv de people could be placed on de lan. Dis would not be a Govment project, hit would be sumpin de Race done fer hitself. We nevah will be spected till we do sumpin fer ourselves. All dis must be done by honest folks. I membah back oftuh surrender, dere wuz sumpin lack dis started, en we got togedder er good deal uv money hopin tuh buy lan fer our folks, but hit wuzn't long fore de leaders run off wid de money.

Bakers plan, while probably too grandiose in the magnitude of subscribers, is clearly another instance of the rotating credit association. The "society" formed by Anthony and Ezekiel Blunt and James Hargate of Lenoir County,

[11] *BRFAL*, film M752, roll 23, pp. 670-71. Ibid., film M798, roll 20, Georgia pp. 766-69. Ibid., film M752, roll 23, Washington, D.C., pp. 788-802. See Lawrence N. Powell, "The American Land Company and Agency: John A. Andrew and the Northernization of the South," *Civil War History*, Dec. 1975, pp. 293-308.

North Carolina, to purchase "homesteads" by "joint stock" accumulated over a three-year period with payments of $48 "per annum" from each of 250 members, was more in line with practical possibilities.[12]

When in 1880 attorney James E. Ohara, a black New Yorker who had emigrated to Halifax County, North Carolina, in 1862 as an eighteen-year-old missionary-schoolteacher, testified to a belief in the philosophy of self-help for the Negro, he enunciated a conviction that recognized no class dimension within the black community. Given this, few freedpeople were as fortunate as those in the vicinity of Hilton Head, South Carolina. There the head of the bureau was Major Martin R. Delaney, the nineteenth-century's leading advocate of black self-determination. In an act typical of the man, Dr. Delaney inflamed both southern and northern whites and anticipated the great William E. Burghardt Du Bois by seventy years when in an 1865 speech he informed an assemblage of blacks and whites that the freedpeople "owed their liberty to their own strong right arms" and should "beware of all whites, northern and southern."

In a series of newspaper articles Delaney preached to the southeastern Atlantic coast what the *New Orleans Tribune* had advocated in the Southwest. "Restoration of the industrial prosperity of the South" required a "triple alliance" among the owners of "capital, land, and labor" on the basis of a "copartnership." The "net profits" were to be "equally shared" between northern capitalists, southern landlords, and black labor. Delaney's plan envisioned family tenancy, with the Freedmen's Bureau exercising a major and paternalistic role in seeing that the shares were justly divided.[13]

By the end of 1866 Martin Delaney had become disillusioned with the share system, which promised perpetual poverty and a "balance due" to the "stores of the speculators." Delaney, like the Farmer's Alliance and Grangers, who would later in the century set the South afire with the rise of populism, saw much of the problem in a credit and merchandising system which made small farmers "victims of the brokers and petty cotton traders." For the father of black self-determination there was little question of the proper course of action. It is fitting and perhaps logically inevitable that the earliest attempts at cooperative credit and merchandising in the postbellum South should have stemmed from Negro agrarianism during Reconstruction. Delaney "recommended the establishment of a Freedmens Cotton Agency." Here the small farmers were able to have their deposits of cotton "culled, ginned, bagged and sold" in order to realize the profits "themselves, instead of the speculators." The bypassing of merchants engendered by this operation and the obtaining of provisions at prices far below the credit terms offered by commercial stores

[12] Reid, p. 222. Blassingame, *Slave Testimony*, p. 668. BRFAL, film M752, roll 23, pp. 86–87.

[13] *Exodusters*, North Carolina, pp. 56, 66. BRFAL, film M869, roll 34, pp. 202–3. Also Victor Ullman, *Martin R. Delaney: The Beginnings of Black Nationalism* (Boston, 1971), pp. 332, 338, 339. Delaney argued that the Civil War would not have been brought to a successful close without the emancipation of the slaves and the utilization of black troops. W. E. B. Du Bois, *Black Reconstruction*, chap. 4, "The General Strike."

did not sit well with competing business interests, who were largely responsible for General Robert Scott's order that Delaney and the bureau disconnect
themselves from the enterprise. The Freedmen's Cotton Agency was closed
after a short period of operation.[14]

The Freedmen's exchange store in Oktibbeha County, in east central
Mississippi, seems to have survived for a longer period. This cooperative
was financed by a hundred black "stockholders" holding shares "of from
$5 to $20 each." Little is known about the operations of this enterprise. The
store was placed under the direct management of the Scot named McLachlan
who may or may not have presided over stockholder meetings, which are
known to have occurred. McLachlan, who said he had been sent from the
North by the Methodist Church to open a colored church and school, was
chased out of town by whites who feared his influence over the Negroes.
Afterwards the management of the exchange was put in the hands of freedman Tom Woody, the largest stockholder. There are few known instances
of organizational structures of these two types, but this should not be taken
as proof of the insignificance of cooperative efforts in this area. Toward the
end of the century the Farmer's Alliance generally effected cooperative
credit and buying through the device of clustering trade with a chosen
merchant, and there is reason to believe that similar activities were used by
freedpeople and northern merchants during Reconstruction.[15]

This agrarian movement, like so many others, failed for a variety of
causes. Problems connected to collectivity, as with a group of forty
freedmen who in 1866 formed the National Union Colony, leasing a large
plantation near Baton Rouge and paying one-fourth the crop as rent, must
have played some role. The colony obtained credit by pledging the crop to a
factorage house. The arrangement led to the courts and internal friction
because the factors desired to settle with the group as a body "and not with
each individual personally." Those who had managed their affairs more
frugally "objected to paying the debts of the others." These problems of
collectivity and the "free rider" were dealt with by organizing labor on the
basis of individual families and using a democratic arbitration to settle
disputes. A reporter explained the mechanics of what he termed "colored
communism" in Colleton County, South Carolina:

> In this country the colored people own, and are successfully conducting some
> of the largest plantations. This is done under a sort of communism. A number
> of them, in some cases so many as fifty, form themselves into a society, elect
> their officers, and adopt by-laws. They have regular meetings at which the
> officers report, and a specified amount is paid into the treasury by each
> member. When sufficient is accumulated . . . a suitable plantation is selected.
> . . . the land is equally distributed by the officers elected for that purpose. . . .

[14] *BRFAL*, film M869, roll 35, "Report of Martin R. Delaney," pp. 9–15. Sterling, *The
Trouble They Seen*, pp. 87–90.

[15] *KKK Hearings*, Mississippi, vol. 11, p. 893. The St. Helena Protective Union Store, a
cooperative begun by a group of freedpeople in 1865, may have predated Delaney's Cotton
Exchange. See Pearson, p. 314.

Each is free to work as suits him, and each can dispose of his crop as he deems proper. . . . All disputes arising between the members are brought before the society.

Members of these associations were provided mutual aid services, such as health insurance. Those who voluntarily left the association were refunded their investment share with an accrual for the value of any improvements made on their alloted land. Members expelled for violation of rules forfeited their share.[16]

The major problem confronting the success of planting cooperatives was lack of an adequate amount of initial capital. The Colleton County writer explained, "These societies are principally formed from people who work for hire, fifty cents per day being the sum generally." Only a "few, if any, own any animals." This obstacle to the practical usefulness of various homestead laws was cited by an agent of the Florida bureau, who proclaimed the homestead laws useless since practically no agricultural laborer could save enough capital to feed self and family for the two years it would take to clear land and raise a salable crop. Faced with this predicament and unwilling to *contract* as laborers, many freedpeople leased or purchased anyway and found themselves, like the Patrick Glenn Colony in Liberty County, Georgia, forced to "greatly neglect their crops by absence to secure food." For example, a large and unsuccessful colony began on public lands at New Smyrna, Florida, in 1866 was farmed by methods similar to those used by homesteaders in Louisiana and elsewhere, where the capital implements consisted of "the hoe alone" and women and children tilled the fields while men sought day work to provide food.[17]

We have cast a wide geographical net in this discussion of free labor farm cooperatives, yet we have merely sighted, not touched, the tip of the iceberg. The numerical incidence of agricultural enterprise among the freedpeople was far, far greater than most historians have dared to imagine. Its historical importance lies not so much with the quantitative incidence per se but rather with its statement about the resiliency of an enslaved people, its contribution to the insecurity of social position felt by planters and poor whites, and finally the cruel glimpse of what might have been. Most of these ventures must have failed during the same five- or six-year period of disastrous crops which destroyed many southern and northern planters. For some historians it has been easy to argue that black farmers failed because as slaves they had not been prepared for independent entrepreneurship. For the great masses this inadequate endowment of human capital cannot be

[16] *BRFAL*, Louisiana, entry 1499, vol. 222, complaints Baton Rouge, "National Union Colony vs. Taney and Roberts, Merchants," Jan. 8, 1867, p. 2. *New National Era and Citizen*, Sept. 25, 1873.

[17] *New National Era and Citizen,* Sept. 25, 1873. *BRFAL*, film M752, roll 37, p. 707. ibid., film M798, roll 15, p. 249. *BRFAL, Synopsis of Reports*, entry 20, no. 136, pp. 96–98, 101, 148, 164. George R. Bentley, *Freedmen's Bureau*, p. 144. *Exodusters*, Louisiana, "Testimony of John Henri Burch," p. 218. Green and Woodson, *The Black Wage Earner*, pp. 27–29. *BRFAL*, film M1028, roll 29, Louisiana, p. 572.

disputed. But the organizational sophistication of the agricultural cooperatives implies that they were composed of the best talent available in the black population, slave and free, South and North.

Typical in its organization, if above average in leadership, was the Campbell Colony in northeast Georgia. Tunis G. Campbell, an early immigrant to Georgia from the North, managed to obtain a post with the Freedmen's Bureau in the Georgia Sea Islands. This position was lost, largely at the behest of white planters and Georgia's bureau commissioner, Davis Tillson. Chased from St. Catherine Island, where he had set up an early colony with himself as a very paternal leader, he was in fact unsuitable as a bureau agent because his extreme Jacobin positions on free labor, Negro equality, and confiscation made it impossible for him to arbitrate differences between planters and employees. Campbell organized a new colony on the mainland, renting a plantation in Mcintosh County with option to buy in 1867. Organized upon the basis of the Democratic Colleton societies, the colony impressed General O. O. Howard enough to convince him to authorize a bureau loan to support it in rations "as a test" case. Howard's action was atypical, stretching the limits of his own authority and his superiors' commitment to black land ownership. More typical was the decision made by the state bureau refusing a loan for rations to the Patrick Glenn Colony in a neighboring county because their use of it while working leased lands would "virtually amount to a loan of capital."

In July of 1867 an agent was ordered to inspect Mcintosh County with especial attention to Campbell Colony. Charles Holcombe "found the condition of the freedpeople" in the county "to be deplorable in the extreme." One of the few exceptions was "at Bellville, the colony of Reverend T. G. Campbell," which Holcombe thought "well managed." The people seemed "contented and hopeful of the future," having "a fine crop of corn and cotton growing." Holcombe's caution in stating that "should the season not prove unfavorable they will not only be able to meet existing claims against them but have more than sufficient to carry thro the next planting season" proved warranted. The crop fell victim to the cotton worm in the fall. Undercapitalized the colony was unable to continue, and Campbell turned full time to politics, where he intensified the hatred Georgia's whites held for him by getting himself elected to the state legislature.[18]

The limits of political radicalism will ultimately be defined within Kenneth Stampp's conclusion that Republican state governments did not attempt-radical experimentation in economic policy. However, that evaluation should not be carried to an extreme, for to do so forces the social historian to confine the interpretation of southern Reconstruction within the boundaries set out by the perspective of the planter class. For while hate for

---

[18] *BRFAL*, film M798, roll 32, "Report of Davis Tillson, 1865," p. 17. Leigh, *Ten Years on a Georgia Plantation*, p. 133. *BRFAL*, film M798, roll 15, pp. 86–87. Sterling, *The Trouble They Seen*, pp. 248–51, for the rules of the Campbell Colony. *BRFAL*, film M798, roll 15, pp. 157, 249; "Report of Charles R. Holcombe," June 13, 1867, p. 445. Magdol, *A Right to the Land: Essays on the Freedmen's Community,* documents many land cooperatives.

carpetbag politics, Negro suffrage, and civil equality may explain the motivation of whites with little property, it is insufficient to explain the behavior of large propertyholders. We have seen that many such men were quite willing to accept the Negro's civil rights as long as they believed he would follow their political leadership. To realize that to the propertied the problem was not that the blackman voted but why and for what he voted is to recognize that the wealthier whites, like those who were poorer, had their own economic cross to bear.

Negro agrarianism was perceived as a serious threat to the economic power of the planter class. In 1869 J. Quitman Moore, a Warren County, Mississippi, planter, voiced the ultimate fear that the "increasing disposition" of the freedpeople "to become *landholders* and *tenants* and not *hirelings*" might lead to their becoming "the cotton growers of the south leaving the planter no alternative but to sell or rent his land to them." Moore speculated that the inspiration for this spirit of accumulation might have come from the "teaching" freedpeople received in the Loyal League clubs. The speculation is suggestive. Unfortunately, we have little direct evidence about the content of the secret meetings of the Loyal or Union Leagues and the "debating societies" and "political clubs" that covered the rural counties from the tobacco fields of Virginia to the far western cotton parishes of Louisiana.[19] It is plausible that these groups of ignorant but often ably led people were the hatcheries of those seditious and mostly unsuccessful radical economic experiments which kept planters uneasy.

Undoubtedly the political activities of these clubs were not confined to black male suffrage and social equality. Even the confiscation question died hard. After the beginning of Republican campaigning in 1867 some leaders of the Negroes were again heard to "proclaim the agrarian doctrine," demanding "that the land of the leading rebels" ought to "be confiscated and apportioned" among those who had produced the wealth. Mary Jones, widow of the Reverend Charles C. Jones of Liberty County, Georgia, attended a "large meeting and registering of names" in the late spring of 1867. The speaker, "a Yankee Negro," gave "Assurances" to the blacks that in "the coming year forty acres of land would be given to each, and all our lands confiscated and given to them, to whom they justly belonged." Mary Jones considered this a "fearful state" and understandably wondered, "where will it all end?" In Sumpter, South Carolina, when the planters countered Republican politics by forming "colored Democratic clubs," a black politician suggested that the Republicans recounter with "white radical clubs," which he predicted the "poor whites would jump at" when "told they would all have land and homesteads given them."[20]

[19] Stampp, *Era of Reconstruction*, p. 170. *Southern Cultivator*, 1869, p. 302 (emphasis original). *BRFAL*, film M1048, roll 45, Virginia, 1866, p. 64. *Exodusters*, Louisiana, "Testimony of Andrew Currie," Shreveport, Louisiana, p. 78.

[20] *KKK Hearings*, Alabama, "Testimony of Benjamin F. Herr, Sumpter County, Alabama," pp. 1712–13. Myers, *Children of Pride*, p. 1382. Moore, *The juhl Letters*, Sept. 1, 1868, p. 248. *KKK Hearings*, Alabama, p. 1131. The "Yankee Negro" was almost certainly Tunis G. Campbell, whose Bellville colony had been set up in the area that spring.

To those who may be considered more conservative, or simply prudent, confiscation was infeasible, but less radical political policies were available to aid the people's attempt to undermine the plantation system or simply protect the laborer. The most straightforward attempts along these lines were the priority rankings in the crop lien laws among laborers, merchants, and landlords. General Wager Swayne's 1868 example of virtually forcing the Alabama legislature to make the laborer's claim first was followed in 1869 by a similar law in South Carolina under the governorship of former bureau assistant commissioner Robert K. Scott. Other states followed, but none went as far as South Carolina, which gave the laborer not only the primary lien but the explicit right to divide the crops in the field. Like the South Carolina truck act of 1872, which forbade the payment of wages in scrip, this legislative effort was destined to fail because it did not alter underlying conditions.[21] One obstacle to the spread of small farms was the large planters' refusal to sell their lands in small tracts. Emanuel Fortune, a black and one-time state representative from Jackson County, Florida, took the arguable position that the land was held "that way so that the colored people" could not "but it." Regardless of the motive, this was a great obstacle to the acquisition of land.

John Henri Burch, a journalist-politician from East Baton Rouge Parish, Louisiana, explained to a senate committee that the Louisiana Republican Constitutional Convention of 1868 had attempted to circumvent the unwillingness of large landholders to sell their properties in small plots by requiring that in all sheriff's sales of property for uncollected taxes, "the land should be cut up in tracts of fifty acres" in order "to give all purchasers a chance to buy." In 1868 Burch had become a shareholder in "an association of colored men" who purchased "several thousand acres of land" near Port Hudson, Mississippi, with the intention of turning a profit and colonizing "three or four hundred families who would buy on time or rent." His grasp of vital economic conditions went to the central issue of landlord-tenant relations and their struggle over the distribution of income. While a state senator in 1872, Burch introduced and helped secure the passage of an act appropriately termed "the homestead law." This law exempted $600 worth of property of any laborer from attachment for the payment of a debt. The effect of this introduction of limited liability upon agricultural laborers was to reduce the profitability and therefore extent of loans. Burch's paternalist view towards his fellow blacks was evinced when he answered the charge that credit would no longer be available with the reply that "it would be a good deal better for them" to become "more prudent and provident in their purchases." In 1877 a majority of black Republican legislators in South Carolina voted with the Democrats to repeal the act guaranteeing merchants' liens. Aware that this would mean the end of credit lines indepen-

[21] *Alabama Acts*, Oct. 10, 1868, pp. 252–53. *Statutes at Large of South Carolina, 1868–1871*, March 19, 1869, pp. 227–29. *Acts of General Assembly, State of Arkansas*, July 23, 1868, pp. 224–26. Tindall, *South Carolina Negroes*, p. 99.

dent of the landlord for many laborers, black politician Hastings Gantt from Beaufort County echoed the paternalism of Louisianian Burch, believing it was the legislature's "duty to make such laws as to bring our people up to a sense that they must take care of themselves." This legislation proved so futile in the absence of alternative credit that it was repealed within a year.[22]

In South Carolina, Negro legislators voting to weaken the merchant's lien had joined Democrats representing planter interests to form a coalition that excluded merchants, many of whom were Republicans. With Republican control quite dead in 1877, the irony of this coalition was that the last economic remnant of radical Reconstruction had been the existence of nonplanter merchants willing to offer the laborers the opportunity to become independent farmers by leasing or buying land on the strength of merchant credit. The existence of these merchants affected landlord profits in three ways. First, many merchants who acquired land by purchase or lease or as agents for absent owners decreased the supply of supervised plantation labor by leasing and granting credit to blacks, who would be allowed to farm with very little supervision. On plantations where renters and sharecroppers were able to obtain loans independent of the landlord, labor discipline was often relaxed because control over the laborer's work habits rested with the absent merchant, who controlled the supply of rations. Finally, and perhaps most importantly, the merchant's lien often took control of the distribution of the product out of the landlord's hands.

A clue to the continuing political power of the planter class during the period of redemption is given by the speed with which state legislatures regranted the landlord the first lien on crops. Every southern state granted landlords this property right before the end of the 1870s. The intent of allowing the landlords first lien was to put the financial disposition of the crop under the landlord's total authority, purportedly to guarantee payment of the rent. E. B. Borden, planter and president of a bank in Goldsborough, North Carolina, explained that his state's landlord-tenant act of 1876 prevented the laborer from giving a "mortgage" on his crop, "thereby giving the merchant a prior lien over the landlord" and possibly allowing him to "cut the landlord out of his rent altogether." Actually the counterrevolution of the landlords began earlier than the legislative repeals. Much Ku Klux Klan violence can be attributed to the attempt of resident landowners to burn the stores and goods of merchants they sought to drive out for "decreasing the supply" of plantation "labor" by leasing tenancies and granting credit independent of any landlord's control. Whippings by the Klan and sometimes hired Regulators were also used as a means of

[22] *KKK Hearings*, Florida, "Testimony of Emanuel Fortune," p. 96; "Testimony of R. Meachum," p. 101. *Exodusters*, Louisiana, "Testimony John Henri Burch," pp. 219, 218, 219-20. Similar strategies were attempted in South Carolina and Virginia. See Sterling, *The Trouble They Seen*, pp. 253-54. A. A. Taylor, p. 131. Tindall, *South Carolina Negroes*, p. 109. *Exodusters*, North Carolina, for reference to North Carolina's Homestead Exemption, p. 407.

enforcing labor discipline. The largest and most responsible planters justified their involvement on the basis of a need for social discipline. John C. Reed, reputed head of the Klan in Oglethorpe County, Georgia, described the organization: "it was a police, rather than a military force, an underground and nocturnal constabulory, detective, interclusive, interceptive, repressive, preventive—in the main—punitive only now and then, where it showed some faint resemblance to the Vehmgericht."[23]

The laborers were not always totally innocent or passive. Any economic system with haves and have-nots is forced to tolerate a certain amount of theft. Where the laboring population believes itself to be underpayed and exploited this may reach epidemic proportions. Later, historians who did not ignore the issue found it easy to buy the planters' explanation—race inferiority—or, in more modern versions, the legacy of slavery. Economic historians have found it convenient to ignore the issue, regarding it as separate from economic questions. The future dean of Gloucester, discussing the lawlessness during the industrial riots in the Wiltshire cloth industry in 1726 and 1738, was closer to the mark. The arrogant attitude of the masters toward their workmen in the atmosphere of a rigidly capitalistic work environment prompted Josiah Tucker to compare industrial relations there to that of planter and slave in the American colonies. As free people the workers were loath to brook such treatment without complaint and consequently thought "it no crime to get as much wages and to do as little for it as they possibly can, to lie and cheat and to do any bad thing provided it is only against their Master." Barn burnings, theft, and the destruction of crops have a long tradition in the annals of rural societies and have often intensified during crop failures and periods of depression. It is especially at these times that we must be on our guard to acknowledge that the political economy of the principles of income distribution often bursts the bounds of the marketplace.[24]

A year of intense political activity, 1870 was an especially violent time in the South. The ever-observant Julius J. Fleming, writing from South Carolina, eschewed the common explanation that the "very bad state of feeling" was due to "political antagonism." Fleming believed the "origin of the whole trouble" was due to the "plundering" induced by the "traffic in seed cotton" and the "rough measures" used to "arrest such traffic" by the "planters." The explanation was consistent with his earlier views. Fleming had anticipated such events at the time of the failed harvest of 1866. "Dimissals" from plantations without pay turned "out upon the country at

[23] *Exodusters*, North Carolina, pp. 209, 212. *KKK Hearings*, South Carolina, "Testimony of Leander A. Bigger," pp. 273–75, 284–85; Georgia, "Testimony of John J. Neason," pp. 42–45; Mississippi, "Testimony of Daniel H. Smith," p. 574; South Carolina, "Testimony of John Lewis," pp. 436, 440. John C. Reed quotation in Thompson, *Reconstruction in Georgia*, p. 374. The Vehmgericht was a secret society of vigilantes seeking to bring social order to late medieval German states.

[24] J. De L. Mann, "Clothiers and Weavers in Wiltshire During the Eighteenth Century," in L. S. Presnell, ed., *Studies in the Industrial Revolution* (London, 1960), p. 66.

large many who believe that they have been tortuously discharged from their places and wrongfully deprived of the fruits of their toil. These form a growing class of vagrants soured in temper and ready to make reprisals out of anything which may come to hand."

The extent of the laborers' embezzlement, the conditions surrounding it, and the landlords' more sophisticated disdain of the merchant's lien are shown by a parallel example that can be considered representative. In Chapter 9 it was seen that one of the forces leading to the leasing of land on shares or fixed rent to freedpeople was the government's willingness to subsidize planters who did so by allowing a loan of rations to such laborers. In South Carolina the Freedmen's Bureau, acting as merchant, took a lien upon the crop of those who received such loans. These loans were made upon the bonds of planters, laborers seldom receiving government credit on their own. In 1868 "a small crop" of cotton was realized which the bureau agent in Darlington District estimated to be just "enough" to pay the "lien" in "almost every case" if the "indebted" were "disposed to be honest." The problems he had are perhaps indicative of those confronted by merchants during subsequent short crop and low price years. People were very "slow" to settle but "most spoke as if they would." Those who weren't made up a substantial number:

> It is difficult for me to conceive how to stop the continual selling of seed cotton (unginned cotton) in the night by Freedmen who are determined to defraud. Planters fear to report or take action against them, fearing they will retaliate by burning their buildings or killing their stock. I have never known of so much stealing and selling of seed cotton before. . . . All night long, the roads leading to the village are dotted with Freedmen bringing in stolen cotton, and they find plenty of chance to sell the same.

The agent's opinion of the integrity of many planters was not much higher:

> Planters who rented lands to Freedmen, and then came to me and begged in the most piteous manner that I would supply the hands with provisions, promising in the most solemn manner that the lien should be paid before the rent—now have the effrontery to declare that the rent must and shall be paid first.[25]

Many petty traders, white and black, who were the suspected receivers of stolen goods were either driven out of the district by planters or convinced by nighttime visitors that discontinuing their business interests would be conducive to their continuing health. A longer-run solution was the banning of the sale of unginned cotton. The laborers, not being the owners of ginning equipment, were forced to bring their cotton to the employer's gin. Viewed solely as an expedient method of legally controlling embezzlement, the prohibition against the sale of seed cotton need not have been a hardship upon

---

[25] Moore, *The Juhl Letters*, pp. 369, 116. Botume, *First Years among the Contrabands*, pp. 260–61. *BRFAL*, film M869, roll 36, South Carolina, pp. 124, 126.

the field laborers. Coupled with the landlord's all-powerful lien, the practice had side effects which the workers considered oppressive in the extreme.

The economic conditions responsible for violent and illegal activities on all sides provide a sad epitaph for postwar Negro agrarianism. Democratic control of state legislatures would quickly put an end to any further serious threats to the plantation system by Negroes. But the role played by the agrarian spirit in the decentralization of the plantation had been essential.

15

What poor remedy is a law suit for these laborers who have neither time nor money to spare. The only security against such contracts is for the courts utterly to ignore them.

Justice Reade, Supreme Court,
State of North Carolina (dissenting)

In regard to the operation of the Civil Law in this District, I would only say, it is merely a source of power and oppression in the hands of the wealthy few. It being in this state an expensive luxury—there is no justice for poor whites or Freedmen.

Major George A. Williams

Beyond the Market

Few ruling groups relish the inconvenience of maintaining their power through the continuous application of physical violence. In the postwar South such measures were utilized to oust the radicals, gain control over the black vote, and institute race relations more attuned to the mores of whites than had prevailed during the brief interlude of radicalism. In the long run, the social exploitation of the black worker had to be buttressed by a legal system that defined the limits of that exploitation. Modern scholars seldom argue with the view that this intent existed. Disagreement has centered upon the question of whether or not exploitation actually occurred. The issue has been greatly obfuscated by debate over a subsidiary question about the existence or absence of a *competitive* market for labor in southern agriculture. Economic historians have recently argued convincingly that the southern agricultural labor market was competitive. Planters frequently attempted to depress labor earnings below the "market rate" by forming cartels or making employer agreements, but the continuing calls for employer solidarity and the need to reorganize these cartels is signal evidence of their failure to produce significant and lasting monopoly power. Actually, this argument does not represent the new interpretation or revisionist history that its latest advocates purport it to be. Earlier historical work made exactly the same points while still concluding that the worker was exploited.[1]

In neoclassical economics a competitive market is by definition devoid of exploitation. Competitive markets theoretically allocate to workers incomes equal to the marginal productivity of labor, and neoclassical economics defines worker exploitation as the payment of incomes to workers which fall short of their marginal productivity. The historians have offered no definition of exploitation, implicitly treating it as a concept explicable only within the realm of ethics. This has the strength that it allows consideration of social relations as well as income shares distributed to groups. Its weakness is that the criteria are subjective and incapable of scientific resolution. But the substitution of a precise objective definition is no improvement over an imprecise, subjective one, if the substitution itself involves a value judgment.[2] While we cannot escape the fact that exploitation is fundamentally an ethical concept, we do have a means of approaching the question that provides those who wish to make the value judgment with a clearer perspective of just what issues are involved.

We may agree that southern labor markets were competitive and still be mystified by the frequent use of the terms "competitive market" and "free market" as if they were synonymous. For as many historians have understood, the crucial question which must be answered about the southern agricultural labor market is, how free was the laborer? The question is of philosophical and material import. Competitively determined labor conditions need not equal the conditions gained by a mobile labor force free to react continuously to changing market opportunities. The contracts accepted by indentured servants may well be competitively determined, but the servant is not free and often realizes significant income losses due to missed opportunities over the duration of servitude. No group was more aware of the distinction between a merely competitive and a free labor market than the white workers of the antebellum South, whose competitively determined wages and conditions of labor were depressed by the significant numerical presence of leased and owned slave labor.

This study began with a query into the nature of free labor. It was claimed that the conception of free labor held by the former slaveowner was con-

---

[1] Higgs, *Competition and Coercion*, and DeCanio, *Agriculture in the Postbellum South* give the most extensive treatments, among those books which make the existence of competition the paramount factor in postbellum markets. Wharton, *The Negro in Mississippi, 1865-1890*, and Williamson, *After Slavery*, are excellent earlier studies that are careful to note the importance of competitive conditions as one factor in a complex picture. See also Woodward, *Origins of the New South*, chap. 8.

[2] "To choose a definition is to plead a cause, so long as the word defined is strongly emotive. . . . The purport of the definition is to alter the descriptive meaning of the term, usually by giving it greater precision within the boundaries of its customary vagueness; but the definition does *not* make any substantial change in the term's emotive meaning. And the definition is used consciously or unconsciously, in an effort to secure, by this interplay between emotive and descriptive meaning, a redirection of people's attitudes": Charles L. Stevenson, *Ethics and Language* (New Haven, 1944), p. 210.

tained within the laborer's legal right to refuse to enter the employ of any particular employer. That notion, of course, did not extend to a right to refuse the offers of all employers. Entrance into a labor contract could then be construed as a voluntary choice of a specific employer. Since no single employer exercised absolute control over the laborer's initial employment decision, labor was free and contracts competitively determined. From the beginning, the planters' trepidation in regard to free labor was based on the knowledge that a completely mobile labor force would be endowed with the power to select a day-to-day work rhythm of its own choice and, possibly worse, to abandon the employ of any planter at any time during the crop season.[3]

The "black codes" of 1865 had reflected these concerns, and had it not been for federal intervention they would have settled the Negro question in the South. The contract system enforced by the Freedmen's Bureau provided a solution, but at the heavy cost of placing limits upon the industrial relations permitted within a contract and the unacceptable cost of the political interference of bureau agents. For these reasons few planters, even among those who could command enough *money credit*, contracted for short periods with money wages. With respect to the availability of short-term wage contracts in the late 1870s, North Carolinian Green Ruffin's statement that landowners didn't "do much of that kind of hiring down there with us" may be accepted as true. As for the Southwest, the statement of a resident of Natchitoches Parish, Louisiana, that there was "no hiring by the month in" his parish can be given the same credibility.[4]

The yearly contract with a large percentage of all of the payment due at the close of the season greatly curtailed intraseason mobility but did not of itself enforce a continuous supply of labor on a daily basis. As we have seen, this was solved by the employer's or merchant's exercising control over the issue of rations to the laborer's family. Where the laborer worked for monthly wages under a year contract, this method of labor discipline increased the use of payment in scrip redeemable at plantation stores. When the employee was hired for a share of the crop, the problem was more difficult. In the late summer months a crop of corn and other foodstuffs could be used by sharecroppers to feed their families while they sought alternative day labor elsewhere or went hunting. Earlier in the season, a share laborer with possessory control over a part of the crop could hypothecate the crop to an independent merchant, obtaining a food source not controlled by the landlord. Here the lien laws and the legislation limiting the right to sell crops went beyond the landlord's attempt to protect rent from embezzlement.

---

[3] For cogent statements of these ideas by responsible southerners see the written opinions of the Supreme Courts of Alabama, Georgia, and North Carolina: *Murrell* v. *The State*, 44 *Alabama Reports*, 370–71 (1870); *Bryan* v. *The State*, 44 *Georgia Reports* 332 (1871); *Haskins* v. *Royster*, 70 *North Carolina Reports* 603–11 (1874).

[4] *Exodusters*, North Carolina, p. 382; ibid., Louisiana, p. 456.

James T. Rapier of Alabama admitted that the legislative debates did "appear" to imply that the sundown law prohibiting the sale of cotton between sundown and sunrise was "to keep the negroes from going to stores and taking off seed cotton." However, the effect of the law "went further." It stopped "a man from selling what he has raised," because landlords "prevented" laborers from leaving the plantation during working hours. As discriminatory as it was, the sundown law was less obnoxious than an 1879 act which applied only to counties in Alabama's Black Belt and forbade the sale of seed cotton "to anybody but the landlord." This granted each landlord monopoly power to dictate the selling price of all cotton grown by people on the landlord's property. By 1879 the mechanism was well established. Many planters had learned how flexible legislated monopoly power could be. The price of cotton offered did not need to be deflated below market prices. After all, the landlord also owned the cotton gin. Laborers in the great plantation belt of the Mississippi Valley had a new complaint. "In order to catch all the cotton on each place," those renting land "in the south" had to have their cotton "ginned at the gin of the owners of the land." It was "in this way," explained George Rogers, that "they take all the cotton from the colored people. I never got a bale of cotton ginned for less than $6."[5] The effect of the gin monopoly was only negligibly stronger than that of a lien law which gave the landlord control of the entire crop until rent and all advances were paid, preventing the tenant from eating any portion of the crop or getting additional credit without the landlord's consent. A renter or share laborer who ate even one ear of corn could be prosecuted for a criminal offense.

Samuel Perry, an agricultural laborer and a creditable labor organizer in North Carolina's rural districts, gave a synopsis of the white and black working people's grievances over the effect of the landlord's total lien:

> The part of it where we think it is most severe is where it gives the landlord the right to be the court, sheriff, and jury, and say when the rents shall be paid. . . . It comes hard on him [laborer] sometimes, when he wants to sell a part of his crop, to hire help, and buy meat, and get out the balance. And then they claim that it makes them sort of servants to the landlords.

Samuel Perry could speak from personal experience, with the inside knowledge of a man who worked with and knew the laborers well. Many whose knowledge was limited to observation came to similar conclusions. R. C. Badger, a Republican attorney who felt that the law itself was reasonable, admitted that the decision to "vest the whole possession of the crop in the landlord" could allow "bad men" to become "autocrat[s] with extraordinary power."[6]

[5] *Exodusters*, Alabama, "Testimony of James T. Rapier," pp. 466, 468. Ibid., Louisiana, "Testimony of J. D. Daniel," p. 50; "Testimony of George Rogers," p. 445. For charges varying between $4 and $9 a bale see ibid., pp. 42–50.

[6] *Exodusters*, Alabama, "Testimony of James T. Rapier," p. 469. Ibid., Texas, "Testimony of W. E. Horne," p. 460. Ibid., North Carolina, "Testimony of Samuel L. Perry," p. 298. Ibid., "Testimony of R. C. Badger," p. 407. *KKK Hearings*, Alabama, p. 1139.

Early in Pickens's childhood his family was trapped in debt to a landlord for a number of years. A hotel keeper in nearby Pendleton, South Carolina, offered to pay the debt and bring the family to town "if Father would be his man of all work and Mother a cook." The arrangement worked well; "wages were small but paid promptly, and there was no binding debt." In a few years the parents were persuaded by "hard times" and an "immigration agent" to "risk the future of" their "family in the malarial swamp-lands of Arkansas." The Arkansas planter advanced the train fare for their emigration. That an error had been made was clear from the moment of their arrival. No one in the family knew what was owed for transportation. The initial debt was enlarged by advances of "salt meat, meal and molasses for the first weeks of enforced idleness" due to bad weather. After fieldwork began, better food, including sugar, coffee, and flour, was supplied; the profit calculation allowed a better "maintenance in the seasons when" they "were bringing in returns." After a bountiful season for cotton production the family was told they were "deeper in debt than on the day of their arrival." Unbroken in spirit, they strove to pay off their indebtedness. To minimize the debt "no provisions were drawn from the planter." The mother cooked and took in washing while the father "felled trees in the icy 'brakes'" to pay for rations. All this was to little avail. For it is the economics of compound interest which makes an investment in debt slavery profitable to the investor. While the family was working and sacrificing, keeping the children out of school "to hoe and pick cotton," "the old debt remained, of course, and perhaps took advantage of this quiet period to grow usuriously."

At the end of the second year the family had produced a bigger and better crop, but still, by the planter's accounts, was in "enough debt to continue the slavery." The second economic lesson on the profitability of debt bondage to the bondholder was learned: "If the debt could not be paid in fat years, there was the constant danger that lean years would come and make it bigger." Add to this the political economy of race and the law: "but there was the contract—and the law; and the law would not hunt the equity, but would enforce the letter of the contract. It was understood that the Negro was unreliable, and the courts must help the poor planters."

The escape had to be executed perfectly, for capture could bring a "fine and peonage." Father Pickens, a man of above-average shrewdness and ability, managed to rent a small farm and obtain a new loan for transportation many miles away on the other side of Little Rock. The family was lucky; the escape was aided by the presence of a railroad some "twelve or fifteen miles" from their place of indebtedness. "So one night the young children and some goods were piled into a wagon and the adults went afoot," toward the railroad and back into the vortex of competitive labor markets.[14]

William Pickens's account is a classic case of debt bondage. The advance for transportation made it akin to the indentured servitude of an earlier

[14] Pickens, *Bursting Bonds*, pp. 13, 17, 24–29.

period in American history, but with one large difference. Unlike colonial servitude, the time of service was not specified; it could be compounded, with an original debt that increased via an interest rate chosen by the planter. This is the whole key to the issue of debt bondage. Once the laborer was indebted for an advance for transportation, an advance of rations and tools from a merchant, or simply a bad harvest, where was the competitive market for buying and selling old debts? Only a well-functioning market in trading such claims would prevent the planter or merchant holding the debt from charging an interest rate well above his or her own cost of borrowing. But even if a well-organized competitive market for trading labor debts and laborers existed, contrary to Higgs's assertion, legitimate profits could be made by keeping laborers in debt for long periods of time. All that is required is that the discounted stream of earnings produced by the laborer be expected to pay for the debt in some finite time.[15] Like an entrepreneur who makes an initial outlay of capital in a new plant or machine with the expectation that the discounted value of subsequent net returns will pay for the initial debt, the debt bondsman's master is not forced to make a profit through fraudulent bookkeeping.

By the end of Reconstruction a tight system of legislation had institutionalized the labor market upon the principle of one-year contracts. Even so, the landlords could still point to weaknesses in the laws. Under the anti-enticement laws laborers were still legally able to quit a contract. The law merely informed them that they could not legally find alternative employment. In the event that a runaway did find another employer there was the possibility that this second hirer might not relinquish the laborer. The only recourse for the initial employer was to sue the second for damages. Worse, the fact that the laborer could not be punished directly lowered the costs to the laborer of an unsuccessful attempt to quit a job. C. Van Woodward has noted that the "New South" was destined to find what Reconstruction had "failed to find: the measure of the emancipated slave's freedom and a definition of free labor." One of the more important components of this definition was a legal means of directly punishing a laborer who quit employment and providing for his or her return to the employer. As one planter put it, what was needed was a "law to enforce strictly the obligation of contracts" upon both sides. No effort was made to conceal the motive. Laborers in industrial areas sometimes broke contracts for the benefits of higher wages paid in "money every Saturday night." Businessman W. H.

---

[15] Let D_o be the value of initial debt and r_t the market rate of interest in year t. If R_t is the *net* revenue produced by the laborer in year t, and T the expected number of years the creditor believes he can hold the laborer, the discounted sum of net returns must exceed or equal the initial debt. That is,

$$\sum_{t=0}^{T} \frac{R_t}{(1+r_t)^t} \geq D_o$$

must hold. For any given year t prior to the last year T, the net return R_t may be positive, negative, or zero, and in fact the discounted returns up until t may be negative as long as they are non-negative at T.

Gardiner of Mobile explained that if there were "a supply of labor in excess of the demand" no such laws would be required, for then "it would be easy to get another to supply his place." As Alabama's Senator Pugh explained, "the difficulty" of writing such a law was "that the constitution of Alabama prohibits imprisonment for debt."[16]

Pugh's own state circumvented this by clever legislation which, while it did not legalize debt peonage formally, did make it a legal and flourishing practice. Alabama in 1885 made it a criminal offense to "obtain property by false pretenses":

> Any person who . . . obtains from another any money or personal property, or any person who has entered a written contract, with, at the time the intent to defraud, to do or to perform any act or service and in consideration thereof obtains from the hirer money or other personal property, and who abandons the service of such hirer without just cause, without first repaying such money or other advances, and with the intent to injure or defraud, must, on conviction be punished as if he had stolen it.

The crucial aspect of this type of legislation was that the penalty of imprisonment was applied not to the breach of contract but to the acceptance of and advance of money with "intent to defraud." The Supreme Court of South Carolina, recognizing that its law in fact enforced compulsory labor contracts, felt compelled in a written opinion to give a moral justification. In a curious mixture of voluntary contract and state interference the court was "unable to discover any feature of involuntary servitude." Any person "must take the consequences affixed by the law to the violation of a contract into which he has voluntarily entered."[17]

With the passage of this law the state had legally seized about as much control of its agricultural labor force as was possible in the absence of total ownership. Any unsuccessful attempt to escape from unwanted employment could end in imprisonment if the employer chose to prosecute. The employer now had a stronger motive to expend funds tracking down runaways, since the costs could be added onto their contract, and they could work it off when they sensibly chose returning to their employer over going to prison. The New South's definition of free labor and a feasible solution of what we termed in Chapter 1 a fundamental problem of emancipation

[16] Woodward, *Origins of the New South*, p. 205. *Senate Hearings on Labor and Capital*, vol. 4, pp. 30-31, 69.

[17] *Acts of the General Assembly of Alabama* (Montgomery, 1885), Feb. 17, 1885, p. 142. State v. *Williams*, 32 *South Carolina Reports* 123 (1889). See the discussion of moral hazard and the Roman law of slavery, an exact analogue of this legislation, in Chapter 3, and compare n. 12 in that chapter.

The practice of this legal ploy preceded its legislative sanction. Various planters during Reconstruction had attempted to bind contractees on the basis of the previously existing criminal codes covering fraud. State supreme courts during this period uniformly reversed convictions of laborers under this code. The lower courts were not so disposed, however, and it seems reasonable that many planters were able to use the threat long before the legislature altered the laws to the courts' satisfaction. See *Colly* v. *The State*, 55 *Alabama Reports* 85 (1876); *Ryan* v. *The State*, 45 *Georgia Reports* 128 (1872).

had been found. Planters desirous of credit based upon the expected returns from a future cotton crop could now give creditors reasonable assurances that they would be able to keep a sufficient supply of labor to produce the crop throughout the entire year. This was the real basis of the need for year-long labor contracts. Trapped by the seasonal nature of the credit system, poverty-stricken laborers were unable to offer any collateral for their credit advances other than their future labor. To guard against the moral hazard problem of laborers' quitting in the middle of a contract, defaulting on their labor payment, and thus undermining the entire credit system that supported laborer and planter, the state had to sanction the selling of future labor—debt bondage of at least a year's duration. For this the credit lien system was essential.

Since nearly every laborer and tenant would have to work for a long period before any money was paid them, they all required advances of credit for subsistence and household goods as well as farming implements. The landlord's or merchant's liens reached far enough to cover even the families' cooking utensils, paltry though they usually were. In such cases, when an indebted family was "cleaned out," deprived of all their worldly goods, they stood on the brink of total dependency upon the creditor to whom they were indebted. To begin anew required a loan larger than the average, since most familes in any given year had probably retained their stock of household goods. The creditor who had cleaned out a family held a monopoly advantage with respect to this family and other creditors. He could simply return to the family their household goods, setting it down in his books as a new advance of credit. In this way unpaid debts for a given year could be extended to a sequence of years. The practice of continuing the debts of one season into the next in this and other ways began immediately after emancipation.

Although the lien laws were explicitly worded to be effective against the current crop only, the practice of continuing a debt into succeeding years, and therefore effectively allowing the crop lien to cover multiple seasons, was maintained through a variety of legal procedures, the most common being the legalization of precisely the case just discussed. If a creditor returned repossessed goods to debtors or simply allowed them to retain any property, like seed, the value of the goods was considered an advance for the new season.[18]

The discussion between Senator Pugh of Alabama and his constituents emphasized one aspect that they considered desirable in a contract law but that was absent in the anti-enticement statutes. The latter laws gave the

---

[18] *BRFAL*, film T142, roll 69, Tennessee, "Lien Contract, A. O. Stuart and Company with Five freedmen," Hempstead County, Arkansas, July 4, 1867. *BRFAL*, Louisiana no. 1499, vol. 223-1/2, "Complaints Baton Rouge," pp. 2-6. Harold D. Woodman, "Post-Civil War Southern Agriculture and the Law," *Agricultural History*, vol. 53, no. 1, Jan. 1979, pp. 331-33. *BRFAL*, film M752, roll 37, Florida, 1866, pp. 687-88. Ibid., film M979, roll 44, "Lien Contract, 1868," Arkansas. Ibid., film M826, roll 50, "Labor Contract, 1868," Holmes Co., Mississippi, pp. 1015-16.

employer protection against the loss of a laborer but did not "compel the laborer when he is engaged to do a stipulated amount of work to do that work." The acts making it criminal fraud to violate a labor contract on which advances in kind or money had been obtained included under the definition of violation the nonfulfillment of "reasonable" stipulations made by the employer, whether written or oral. The tendency of competitive market forces to determine the degree of labor discipline and industrial relations was severely biased in favor of employers. Late in the century, when a group of South Carolina laborers began a strike on the Louden plantation in Marlboro County, the owner had six of the "ringleaders" arrested and jailed for fraudulent violation of their contracts. The laborers, understanding the untenability of their position, agreed to pay all court costs and returned to work.[19]

During the first few years following the war, the federal government enforced the rule that a worker contracting for a share of the crop as compensation for services held ownership rights in that crop. Given the development of family labor units, the logical progression of that ruling would have been a form of tenancy, guaranteeing the landlord rent and vesting the labor with legal tenant's rights. This outcome was resisted by the landlords. In 1869 the Arkansas Supreme Court found that sharecroppers "were not members" of a "copartnership" but received "a mere share in the nature of wages." "There is an obvious distinction between a cropper and a tenant," declared Georgia's highest court. The tenant has "possession of the premises, exclusive of the landlord," while the cropper had only the right "to plant, work and gather the crop." Other state courts concurred that the southern sharecropper was legally defined to have the status of a "day laborer" working "not a farm, but a section of a well-ordered unit." In a system of free enterprise, the landlord as owner of the land and capital certainly had the right to hire laborers to work land rather than lease to renters. But in offering a share of the crop, with the consequent risk, landlords relinquished the right to call themselves capitalists. Under capitalism the conservative validation of private ownership and profit has always been based upon the capitalist's willingness to bear risk while securing for the less enterprising laborers a guaranteed wage.[20] The southern sharecropper bore all the burdens of an entrepreneur but was dispossessed of freedom of choice in making managerial decisions. If sharecroppers had been renters with the rights of tenants, even the contract and lien laws might have had some justification. The independent renter can be conceived as an entrepreneur setting up in business. To accept an advance would have been to receive a business loan, and not an advance upon wages. No government

[19] *Senate Hearing on Labor and Capital*, vol. 4, pp. 69, 30–31. Tindall, *South Carolina Negroes*, p. 113. *Haskins v. Royster*, 70 *North Carolina Reports* 601 (1874).

[20] *Christian v. Crocker et al.*, 25 *Arkansas Reports* 330 (1869). *Appling v. Odum*, 46 *Georgia Reports* 584 (1872). Brooks, *Agrarian Revolution in Georgia*, p. 61. *Randle v. The State*, 49 *Alabama Reports* 14–15 (1873). *Haskins v. Royster*, 70 *North Carolina Reports* 611 (1874).

which allows its laboring population to mortgage its labor by enforcing debt peonage can claim to have free labor.

The development of labor legislation in the postbellum South is a superlative example of the gross inadequacy of studying real-world economic phenomena on the basis of economic models alone. In the postbellum South the inseparable interactions of politics, racism, and economics are inescapable to anyone who cares to look. The competition of convict labor, the legal crushing of a free market for mobile wage labor, the landlord-merchant lien, and control of the productive and distributional process to the exclusion of laborers, who had all the obligations of entrepreneurs but few of the benefits, all confined the southern laborer within a system that bred exploitation. The landlords and employers, competitors for the services of free laborers, were not so naive as to miss the point that the ultimate cost of that labor depended upon the material conditions, legal and social, that would define the empirical extent of that freedom.

In 1880 the social organization of the South had been pretty well demarcated. What remained for the next twenty years was the practical implementation and legal codification of a system of political, social, and economic segregation. The general success of this vision should not be allowed to disguise the fact that at times the perpetuation of the social ordering was visibly threatened. The Negroes' position at the bottom of the industrial order as a cheap mass of raw labor lowering the cost of all labor, white and black, could be continued only by keeping the blacks uneducated and, above all, in an antagonistic competition with white working people. This rested upon the political disfranchisement of the black working population and the stability of the social psychology of racial segregation.

An attempt to put a quantitative measurement on the benefits of freedom to the blacks as early as 1880 would in my opinion be so fraught with statistical guesswork it would create more harm than enlightenment. The great mass of blacks still lived on plantations, and though they were in debt too often, part of that resulted from the fact that they were consuming a larger variety of goods. Even so, at this early date the benefits of emancipation were probably dominated by these social intangibles that defy quantification. This is proper, for far be it from me to attempt to measure the value of the souls of a people.

For the most part the freedpeople considered emancipation a benefit. Perhaps the criteria they applied were too lenient. "We got a little money, but we got room and board and we didn't have to work too hard. It was enough difference to tell you was no slaves anymore," recalled one old Oklahoman. Ambus Gray had no problem remembering the primary benefit of freedom: "I didn't see much difference for a year or more. We gradually quit gettin provisions up at the house and had to take a wagon and team and go buy what we had. We didn't have near as much. Money then like it is now, it don't buy much. It made one difference. You could change places and work for different men." "Me an my mammy," recalled Mingo White of Franklin County, Alabama, "we kep movin an makin share crops twell us saved up nough money to rent us a place an make a crop fer ourselves."

Our friend the right good Reverend Thomas Gisborne had observed the behavior in his own sphere. "Another trait, in the character of a [nineteenth-century English] collier" was "his predilection to change of situation." "Annual changes," complained the reverend, were "almost as common with the pitman, as the return of the seasons." Gisborne's statement that "not unfrequently, the succeeding year finds" a collier "in the same situation, which he quitted twelve months before" leads one to suspect that English colliers, like Negro field hands, seldom roamed far in their search for a new situation. With disdain Gisborne noted the collier's disposition to consider all past favors "as all cancelled by the refusal of a single request."

Tennesseean Absolom Jenkins would have explained the behavior in simple terms: "Folks roved around for five or six years trying to do as well as they had done in slavery. It was years fore they got back to it." Planter-historian Phillip Bruce, remarking upon the similar tendencies of Virginia tobacco hands, would, like Gisborne, fail to understand the indigent laborers' nomadic quest for greener pastures made possible by the capital embodied in their newly awarded right of passage. "One ob de rights ob bein free," declared Mississippian James Lucas, "wuz dat we could move around en change bosses." The nonagricultural workers used it to their advantage in season as well as out:

> Cap'n did you hear bout
> All yo men gonna leave you,
> Nex pay day,
> Lawd, Lawd nex pay day? [21]

Jefferson B. Algood, a planter from Noxubee County, Mississippi, said about as much as anyone can when asked about the social relations existing between would-be masters and their former charges: "a great many negroes will take off the hat and call you master now, but that thing is pretty much played out." Marie Hervey, born in the 1870s, was part of a new generation:

> You had to call everybody "Mis" and "Mars" in those days. All the old people did it right after slavery. They did it in my time. But we children wouldn't. They sent me and my sister up to the house once to get some meal. We said we wern't goin' to call them no "Mars" and "Mis." Two or three times we would get up to the house, and then we would turn 'round and go back. We couldn't make up our minds how to get what we was sent after without sayin' "Mars" and "Mis." Finally old man Nick noticed us and said, "What do you children want?" And we said, "Grandma says she wants some meal." When we got

[21] Rawick, vol. 7, *Oklahoma Narratives*, p. 18. Ibid., vol. 9, part 3, *Arkansas Narratives*, p. 78. Ibid., vol. 6, *Alabama and Indiana Narratives*, pp. 420–21. *Society for the Poor Reports*, vol. 1, p. 253. Rawick, vol. 9, part 4, p. 48. Bruce, pp. 177–78, 199–200, quoted in Michael S. Wayne, p. 63. Odum and Johnson, *Negro Workaday Songs*. Pickens, *Bursting Bonds*, pp. 7, 11, 29–30.

back, grandma wanted to know why we took so long to go and come. We told her all about it.[22]

Bound between two cultures, a fading memory and an emerging reality, the black worker was to remain an enigma for American democracy.

[22] *KKK Hearings*, Mississippi, "Testimony of Jefferson B. Algood," p. 508. Rawick, *Arkansas Narratives*, vol. 9, part 3, p. 232. *Senate Hearings on Labor and Capital*, vol. 4, pp. 49–50, 70–71, 153, 156, 386.

Appendix A
Description of Methods and Data: BRFAL Collection

I will not attempt a comprehensive description of the material available in the immense archive of documents of the Bureau of Refugees, Freedmen, and Abandoned Lands, Record Group 105 in the National Archives. The National Archives has published a number of documents which do this. The two major sources I have utilized are labor contracts and the field reports of Freedmen's Bureau agents. Here I will present a description of the sources I have relied upon most heavily, with a discussion of possible biases and correctives based upon corroboration from alternative and independent sources.

The labor contracts represent the single most valuable primary source for ascertaining plantation labor organization, methods of labor payment, and work and social relations between employer and employee and among workers during the years 1865 through 1868. There are four main candidates for bias in these records. The first three are the absolute numerical size of the sample, the sufficiency of data in terms of geographical diversity, and discrepancies between plantations and smaller farms. Finally, there is the possibility of differences in contractual relations in verbal and written contracts *not* approved by the bureau and those contained in the sample. A fifth problem, how to interpret the idealized stipulations of written contracts as a source for actual practices, is treated separately.

The size of the sample utilized in this study is indicated by table A.1. This sample represents all complete contracts in original form, fair copies of contracts, and brief summaries of contracts by bureau agents that I read. The cyclical reduction in contract numbers from 1866 through 1868 is probably the greatest weakness in the data base. The weakness bears directly upon the fourth possible source of bias mentioned above—the representativeness of bureau-approved contracts in relation to non approved contracts. For this reason I will discuss this last problem later. The geographical distribution of contracts by states is also displayed in table A.1.

The data show a fairly large bias toward the western states of Mississippi, Tennessee, and Arkansas. There are, however, good reasons why this does not represent a serious problem. Most important is the fact that representation of geographical distribution by state is itself an arbitrary line of demarcation. To illustrate the point, consider the state of Mississippi. As shown in table A.2, the distribution of contracts within the state covers every geographical area. It is not at all clear that contracts from eastern counties bordering Alabama should be classified with contracts represent-

Table A.1   Sample Distribution of Contracts by State

	1866	1867	1868
Total Contracts	3,274	618	395
Percentage of Total Contracts			
Alabama	3%	2%	2%
Arkansas	12%	33%	42%
Georgia	3%	5%	16%
Kentucky	2%	0%	9%
Louisiana*	3%	8%	3%
Mississippi	28%	24%	8%
N. Carolina	3%	16%	10%
S. Carolina	7%	6%	10%
Tennessee	30%	5%	'0%

ing western or even central counties in such a large state. In fact, contracts from counties like Noxubee, Lauderdale, and Lowndes, bordering Sumter, Pickens, and Lamar counties in Alabama, probably tell us much more about the state of the labor market in Alabama's plantation belt, whose western frontier was composed of the latter three counties. More importantly, quantitative geographical bias would only be an important problem if the contracts were appreciably different in their primary stipulations. Such differences do not appear as long as the major crop produced is the same.

The number of counties within each state represented in the sample of contracts utilized for this study is presented in table A.2. This sample, while clearly not geographically uniform across the South, is rich. A more important possible source of bias, in my opinion, would be a lack of data

Table A.2   Number of Counties Represented in Contract Sample

	1866	1867	1868
Alabama	10	3	2
Arkansas	31	30	27
Georgia	9	4	14
Kentucky	5	5	5
Louisiana*	3	4	5
Mississippi	31	28	11
North Carolina	32	24	11
South Carolina*	15	19	19
Tennessee	14	3	0

*This does *not* include hundreds of very brief descriptions of contract terms from Louisiana and South Carolina which were studied.

representing small farms. As is well known, in 1850–60 roughly a quarter of all slaves lived and worked on farms with fifteen or fewer slaves. In the postbellum period a significant fraction of the black population continued to work for landowners whose holdings were small compared to those of large planters. Since all employees of a single employer do not necessarily appear on the same contract, it is not possible to obtain an exact measure of the distribution of employees by farm from the contracts. The distribution that can be obtained from the contract sample provides a lower bound upon the size of labor forces and in many cases is no doubt equal to or very close to the exact number. In any case, this distribution provides good information about the amount of data available from which inferences about labor arrangements can be made about large plantations and small farms, and is presented in table A.3.

Table A.3   Distribution of Number of Employees by Contract, 1866

	% of total 2–10	% of total 11–50	% of total 51–100	% of total >100
Arkansas	47.0	38.2	05.8	08.8
Mississippi	69.6	27.8	02.3	00.3
North Carolina	77.7	22.3	00.0	00.0
South Carolina	32.5	60.4	04.6	02.3
Georgia } Alabama }	75.3	23.0	01.5	00.0

A point that no doubt needs little emphasis for historians is that statistical weaknesses in the contract sample itself need not be important if independent sources are available to corroborate details and fill in missing areas. The monthly narrative and statistical reports of local bureau agents and the letters of planters, freedpeople, and others are excellent and indispensable sources for this purpose. As the text itself demonstrates, thousands of pages have been read and utilized for just these purposes. This material, along with other sources, provides the primary basis upon which judgments about the actual enforcement of contractual arrangements must be made. I hope the text makes it clear that these sources show that a naive reading of the contracts, which takes their prebargained conditions as a historical record of what actually occurred after the execution of the contracts, would be an error of some magnitude. As one simple example, the analyses in Chapters 8, 9, and 12 show that the wages and other payments included in contracts cannot be used as a measure of the incomes or welfare imputed to laborers. Alternatively, these data are extremely useful as evidence of conditions in the labor market, such as regional differences in the demand for labor.

Finally, we must discuss the most serious problem with the contracts. Were the bureau-regulated contracts representative of all contractual labor arrangements, written or verbal? As mentioned above, the seriousness of this potential problem is illustrated by the decline in the sample size over time, a product of a general reduction in bureau-approved contracts after 1866. Independent sources, newspapers, travel accounts, and plantation records strongly indicate that the contents of bureau-approved contracts were no different from labor agreements made independently of the bureau. More importantly, the reports of bureau agents themselves are useful on this point. Agents were required to make inspection tours of the plantations and farms in their districts, and many agents complied with their instructions. Their reports of labor arrangements, terms of contracts, and developing problems are consistent with those found in the sample used here. Agents were far from insensitive about the decline in the demand for their services as labor agents and contract arbitrators. One simple reason for that reduction was that neither employers nor employees desired to pay the fee charged by the bureau to cover its administrative costs.

Another major reason for the decline was that neither of the parties to contracts was disposed to bind themselves in writing to payment terms they might find difficult to observe. While this did not affect the terms of contracts, it did influence the possibilities of breaking agreements on payment obligations and work performance at the end of the year. These problems were addressed in the text. The laborers, because their concept of freedom required absence of external constraints on mobility, were generally averse to entering binding contracts. As many agents of the bureau noted, the bureau's inability to enforce payment of contracted wages in many cases "embarrassed" the bureau and destroyed the primary incentive of freedpeople to overcome their fears and enter bureau contracts. A Florida agent cited a legal reason why freedpeople preferred no written contract when he noted that a laborer who was under written contract had to go to criminal court with a jury to sue for wages. A laborer without a written contract could avoid white juries by going to civil court, where suits were tried before a justice of the peace. Since the bureau attempted a judicial enforcement of contractual obligations upon both parties, it found itself in the position of satisfying neither party in the majority of its decisions. This result is hardly surprising, given the preconceived beliefs of contractors about the altered nature of labor relations and their belief that their opinion alone was just.[1]

There is, on the whole, little evidence to support a biased selection problem in the *terms* of contracts retained in the records of the bureau. The decline was not because contracts were different but because contractors desired to avoid enforcement costs. Enforcement was left largely to social and market forces, as we have seen.

Estimation of the Proportion of Share Contracts and Confidence Intervals, Chapter 3.

Estimates of the proportion of laborers and employers contracting on shares and confidence intervals for these estimates were calculated using the technique of stratified random sampling.[1] The parishes of Louisiana, com-

---

[1] *BRFAL*, film M809, roll 18, p. 0177, Alabama. Ibid., film M752, roll 37, p. 826, Florida; roll 40, p. 118, Tennessee; film M826, roll 30, p. 0154, Mississippi.

posing each of the six reported regions shown in table 3.2, were chosen on the basis of geographical closeness. For each region and for Shelby County, Tennessee, the calculations were performed as follows:

Let

$$N = \text{\# farms and plantations in the region.}$$
$$N_a = \text{\# farms and plantations in parish } a.$$
$$n_a = \text{\# farms and plantations sampled from parish } a.$$
$$P_a = \text{\% of sampled units in parish } a \text{ paying shares.}$$
$$Q_a = 1 - P_a$$

Then if P equals the estimated proportion of units paying shares in the region,

$$P = \sum_a \frac{N_a}{N} \cdot P_a$$

To construct confidence intervals, the variance and standard error of P must be estimated. The appropriate formula[2] for the variance V is

$$V = \frac{1}{N^2} \sum_a N_a (N_a - n_a) \frac{P_a \cdot Q_a}{n_a}$$

The standard error of P is calculated by taking the square root of V, and then the confidence interval for the given size sample is derived through standard methods.[3]

The Plantation inspection reports for Louisiana and the sample of contracts for Shelby County, Tennessee, list the name of each employer, the number and sex of employees, and the type of payment offered. Therefore, each variable needed to calculate the estimates is available except for N and the N_a's. However,

$$N = \sum_a N_a$$

so the only problem involves the need for an independent value for each N_a. These were found by consulting the *U.S. Census of Agriculture for 1860* and setting each N_a equal to the number of farms and plantations in the parish that year.

The percentage of laborers contracting on shares in a parish was derived in the same way, except that P_a now equals the ratio of laborers paid shares to total laborers in the sample. The variance was calculated with the same formula, N, N_a, and n_a, the reason for this being that the type of payment received by any two workers is not necessarily independent. The conditional probability that two laborers received the same form of payment given that they worked for the same employer was nearly equal to one. Therefore the appropriate N_a and n_a are the number of employers, not laborers.

[2] See William G. Cochran, *Sampling Techniques* (New York, 1953), p. 91.
[3] Cochran, pp. 53–54.

Estimation of Female Labor Force Participation Rates; Chapter 12
The Plantation inspection reports of the Louisiana bureau provide a monthly record of the number of working hands on each individual plantation for the years 1866–68. The itemization of working hands was recorded by sex and compartmentalized for adults and children. Therefore, the number of adult males and of adult females were reported separately. Unfortunately, the number of abled-bodied nonworking hands was not reported separately. These individuals were either not reported or included under the category "aged and helpless." This means that a direct estimate of adult female labor force participation was not possible. The indirect estimation utilized is as follows.

Let M and F equal the number of adult (sixteen years and over) males and females working on plantations. Let $\bar{M}$ and $\bar{F}$ equal the total population of adult males and females, so that $n = \bar{F}/\bar{M}$ is the population ratio of adult females to males. Then, if $\hat{F} = F/\bar{F}$ and $\hat{M} = M/\bar{M}$ are the female and male labor force participation rates, we have:

$$\hat{F} = \frac{F \cdot \hat{M}}{M \cdot n}$$

n can be computed from the U.S. population census. F and M are provided by the Plantation inspection reports. Therefore $\hat{F}$ is determined if $\hat{M}$ is given.

The reported estimates of $\hat{F}$ are based upon the assumption that $\hat{M} = 1$. This assumption seems very reasonable for agricultural labor and is completely supported by the reports of bureau agents. It seems appropriate, however, to inquire into the magnitude of error in the estimated $\hat{F}$ induced by errors in $\hat{M}$. Since $\hat{M}$ appears in the equation as a proportional, it is easy to see that any percentage change in $\hat{M}$ alters $\hat{F}$ by exactly the same percentage. Therefore, a 5 percent error in $\hat{M}$ would imply reductions in $\hat{F}$ of 5 percent.

The adult female-to-male population ratio has also been set equal to one. This is purposely calculated to underestimate $\hat{F}$. Although black females of all ages have outnumbered black males since 1860 in Louisiana, the reverse has been the case for rural areas, where males have constituted the majority. This, combined with the fact that the female majority was due to greater numbers of female children, makes $n = 1$ a conservative choice. To verify this, population statistics of the number of males and females over the age of five by parish show that males were in the majority in nearly all Louisiana parishes in 1879. Exceptions to this are conspicuous by the existence of cities or large towns.[4]

[4] *Bureau of the Census: Negro Population, 1790–1915* (Washington, 1918), pp. 147–70. *Compendium of Census, 1880*, vol. 1, pp. 579–80.

Estimation of Rates of Absenteeism, Chapter 12

The records of time lost by employees were reported and filed upon a monthly basis. These records report for each employee the total number of days worked and the days absent when work was *required by contract*. The record also gives the employee's name (and thus sex) and type and rate of pay. Since the record was kept by the bureau for purposes of preventing freedpeople from being defrauded of their pay, and the records were maintained by the employer, there was no incentive to overreport days worked or underreport days not worked. The data is quite precise, since many employers reported days worked and lost to a close degree by recording many fractional days of varying length. The data provide a lower bound on total days worked by month.

The estimates were made from the data presented in table A.4 as follows. Separating workers by sex, let

$$D_{ij} = \text{\# of days worked by the ith laborer in month } j.$$
$$N_j = \text{\# of sampled workers in month } j.$$
$$\bar{D}_j = \frac{1}{N_j}\sum_{i=1}^{N_j} D_{ij} = \text{average days worked in month } j.$$
$$\bar{D} = \frac{1}{12}\sum_{j=1}^{12} \bar{D}_j = \text{average monthly work days.}$$

The estimate of total days worked during the year is

$$D = \sum_{j=1}^{12} \bar{D}_j = 12\bar{D}.$$

Denoting L. C. Grey's estimates of days worked per year for adult males and females by D_M and D_F, the estimate[5] of the percentage reduction in days worked is

$$R = 1 - \frac{D}{D_k} \quad k = M, F$$

The estimate of hours worked per week was straightforward. The standard contractual working day was ten hours and the days contracted per week were five and one-half. Each sex, therefore, contracted to work fifty-five hours per week. Total hours worked during an average month were therefore

$$H = 10 \cdot \bar{D}$$

The number of weeks in an average month is

$$\frac{365}{12} \cdot \frac{1}{7} = 4.34$$

Thus the average hours worked per week is calculated by

$$h = \frac{H}{4.34} = 50.62$$

[5] Ransom and Sutch, p. 233, based on Grey, calculate D_M between 268 and 000 and D_F between 261 and 284. My D_M and D_F use the mean of the two end points.

Table A.4   Time Lost by Arkansas Freedpeople, 1866

Month	# Observations Male	# Observations Female	$\bar{D}_j$ M	$\bar{D}_j$ F
1	37	38	21.72	20.07
2	43	44	20.52	17.97
3	47	42	23.07	18.85
4	70	57	21.39	19.76
5	75	55	21.88	19.60
6	50	53	23.25	21.79
7	29	24	23.45	21.65
8	29	23	20.44	20.82
9	41	39	22.08	21.68
10	41	30	22.82	21.76
11	37	30	21.65	19.25
12	21	19	18.26	15.37
Year	520	454	21.96	19.88

for males. Calculating h by yearly hours, $10 \cdot 12 \cdot \bar{D}$ divided by 52 weeks would yield an estimate $h = 50.69$. The calculation of absentee rates is similar. The percentage of time lost due to absenteeism is

$$A = 1 - \frac{h}{55}$$

These calculations were done for each sex and each type of payment as reported in Chapter 12.

The average hours worked per week by textile workers was reported by Berglund, et al.[6] This source also reports average hours required of full-time textile workers. Absentee rates for textile workers were calculated using the formula for A described above.

Confidence intervals for the precision of the estimates can be calculated. The sample was taken by month rather than by workers or employers, so the appropriate methods are those of stratified sample theory, where the strata are the twelve months. The variance of $\bar{D}$ is a weighted sum of the variances of the independent monthly means $\bar{D}_j$. If V^i is the variance of $\bar{D}_i$ and N, N_i, and n_i the respective total working population, working population in month i, and size of the sample in month i, then the variance of $\bar{D}$ is

$$V = \sum_{i=1}^{12} \left[\frac{N_i}{N} \right]^2 \cdot \frac{V^i}{n_i}$$

N and N_i are unknown, but note that N_i may be assumed to be approximately equal to $\frac{N}{12}$ in each month i. If labor force participation of adults was con-

[6] Berglund et al., *Labor in the Industrial South*, pp. 72–87.

stant throughout the year this relationship holds exactly. We estimate

$$V = \frac{1}{144} \sum_i \frac{V^i}{n_i}$$

Reported confidence intervals were then constructed with the standard error shown in table A.4.

Table A.5   Aggregates for the Year 1866

# Observations = 974	Male	Female
Standard error	.161	.264
99% confidence interval for $\bar{D}$	(21.54,22.37)	(19.19,20.56)
99% confidence interval for D	(258.48, 268.44)	(230.28,246.72)
99% confidence interval for h	(49.63,51.54)	(44.21,47.37)
99% confidence interval for A	(.06,.09)	(.139,.197)

Appendix B
Productivity and Organizational Efficiency on Postbellum Cotton Plantations

Table B.1 Description of Symbols and Terms

Q = Dollar value of crop outputs produced, plus value of wood cut on plantation and improvements and maintenance on farm. Measured in farmgate prices as reported by census enumerator

Q_c = Dollar value of cotton produced

ℓ = Labor input measured in man years

k_A = Dollar value of animal workstock inputs

k_M = Dollar value of machinery, tools, and work implements inputs

k_F = Dollar value of fertilizer input

A = Land input measured in acres

W = Dollar value of yearly wages plus rations

ϱ = Rental price of a unit of capital

r = Rental price of an acre of land

M = Gross quantity of capital input

k = Adjusted capital input

P_k^i = Value of a unit of capital on unit i

L = Dollar value of labor input

K = $\varrho \cdot k$ = Rental value of capital input

T = $r \cdot A$ = Dollar value of land input

G = Base index of total factor productivity

G_i = Plantation-specific total factor productivity

f = Relative index of total factor productivity

$\hat{a}$ = Number of acres tilled by the average adult male farmworker

c = Dollar value of cotton output per acre measured in 475-lb. bales valued at ten cents a pound

$Q[L,K,A,]$ = Revenue (production) function on a cotton plantation

Q_i = Marginal revenue product of input i

$\bar{A}$ = $A + k_F/r$ = Quality adjusted land input

H = Stock of work animals, computed in horse price equivalents

$\hat{P}_H$ = Average price of a horse

$\hat{P}_k$ = Average value of a unit of capital

$\hat{}$ = Signifies an average value of the variable indicated

---

A comparison of the relative productive efficiency of wage- and share-operated cotton plantations has been carried out by methods similar to those introduced to historical analysis by Douglass North and also by Robert Fogel and Stanley Engerman in their study of productivity under slave agriculture in the antebellum South. The present approach utilizes disaggregated calculations of total factor productivities on *individual* plantations to approach the problem at the fundamental microeconomic unit—the firm. This procedure allows us to avoid the major criticisms leveled at the methods of Professors Fogel and Engerman.[1]

The major problem to be surmounted in the calculation of productive efficiency is an index problem common to all applied economics. If all firms under consideration produced one homogeneous output with one identical input, meaningful comparisons could be made readily by comparing the magnitude of output per unit of input at different firms. This is the intuitively simple idea behind the various measures of partial factor productivities of output per unit of labor Q/L, capital Q/K, and land Q/T often encountered in economic work.

The problem, of course, is that no one or two of these partial indexes of productivity can be used to make meaningful comparisons across production units. A larger Q/K at one firm may well be accompanied by smaller Q/L and Q/T indexes. The universally applied solution to index or aggregation problems of this nature is to weight each element to be compared by some factor and then sum or multiply to obtain a composite index of output per unit input where output and input are now defined in dollar terms. Their ratio will thus represent a certain quantity of *composite* output per unit labor. This is an abstraction, but a scientifically useful one.

The productivity index most often used by economists is the geometric index of total factor productivity;

$$G = \left(\frac{Q}{L}\right)^{\omega_L} \left(\frac{Q}{K}\right)^{\omega_K} \left(\frac{Q}{T}\right)^{\omega_T}$$

ω_i represents the weight of factor i in the index.[2] If the ω_i's sum to one this index is equivalent to

$$G = \frac{Q}{L^{\omega_L} K^{\omega_K} T^{\omega_T}} = \frac{Q}{I}$$

[1] Robert William Fogel and Stanley L. Engerman, *Time on the Cross: Evidence and Methods*, vol. 2 (Boston, 1974), appendix B, pp. 126–44. An extensive critique is given by Paul A. David and Peter Temin in P. A. David, H. G. Gutman, Richard Sutch, Peter Temin, and Gavin Wright, *Reckoning with Slavery* (New York, 1976). Fogel and Engerman reply in "Explaining the Relative Efficiency of Slave Agriculture in the Antebellum South," *American Economic Review*, vol. 67, no. 3 (June 1977). The critics retort en masse in *American Economic Review*, vol. 68, no. 1 (March 1979), pp. 213–26. Fogel and Engerman reply again in the same issue. No attempt to assess the outcome of this extensive debate will be made here. Douglass North, "Sources of Productivity Change in Ocean Shipping," *Journal of Political Economy*, vol. 76, 1968.

[2] For a good discussion, see Evsey D. Domar, "On Total Factor Productivity and All That," *Journal of Political Economy*, vol. 70 (Dec. 1962), pp. 597–608; M. Ishaq Nadiri, "Some Approaches to the Theory and Measurement of Total Factor Productivity: A Survey," *Journal of Economic Literature*, vol. 8, No. 4 (Dec. 1970), pp. 1137–77.

or composite output per unit of composite input, as described above. Productivity across plantations may now be compared, but with some fairly obvious cautions. Is the index, or equivalently the constructed composite, a meaningful representation of what we mean by output per unit input? There exists no pat answer to this query. But one can say that there exist weights that are at least as good as any alternative ones, and compelling in their own right. For example, if each index of partial productivity were *always* equal to a constant $1/\alpha_i$ for $i = L, K, T$, the share of output attributed to any factor, say K, would also equal a constant $K/Q = \alpha_K$. If in addition $\Sigma_i \, \alpha_i = 1$ it would be difficult to conceive of a better set of index weights than $\omega_i = \alpha_i$.[3]

Southern agricultural laborers and employers, for whatever reasons, behaved in just such a manner that the distribution of output satisfied these conditions. Under the sharecrop and share rent systems, the proportions of revenue produced due as earnings to each factor input were constant and general throughout the cotton states for over a century. The few contractual terms generally agreed to (see table B.2) specify the input and output shares to be contributed and received by employer and employee. We may write down a system of three linear equations in three unknowns from which we may solve for the share of output imputed to each input by the labor market.

$$\frac{1}{2} = \frac{Q}{L}$$

$$\frac{3}{4} = \frac{Q}{L+K}$$

$$1 = \frac{Q}{L+K+T}$$

Table B.2   Input and Output Shares for Employer and Employee

	Sharecropping	Share Renting
Landlord inputs	Land, Cabin Work stock Implements	Land, Cabin
Landlord share	$Q/2$	$Q/4$
Labor inputs	Labor Food rations	Labor Food rations Work stock Implements
Laborer share	$Q/2$	$3Q/4$

[3] The popularity of the Cobb-Douglass specification of production technologies is in large part due to the result that under profit maximization in a competitive environment these conditions are satisfied if technology is Cobb-Douglass.

Solving the system, we find that the incomes imputed to labor, capital, and land are $\ell = 1/2$, $k = 1/4$, $A = 1/4$.

Individuals operating in cotton agriculture who took these income shares as parametrical constants were behaving as if they produced under constant returns to scale with the Cobb-Douglass technology

$$B.1: \; Q = G \cdot K^{1/4} A^{1/4} \ell^{1/2}$$

The geometric index of total factor productivity becomes

$$G = \frac{Q}{(W)^{1/2} \, (\varrho \cdot k)^{1/4} \, (r \cdot A)^{1/4}}$$

From the constancy of the factor shares,

$$G = \frac{Q}{(Q/2)^{1/2} \, (Q/4)^{1/4} \, (Q/4)^{1/4}} = 2.82$$

This represents our base index. The factor productivity G_i of any particular plantation i is measured relative to G.

$$f_i = \frac{G_i}{G}$$

The major task is to find good estimates of the dollar measure of labor, capital, and land on each sampled plantation.

The Measurement of Labor

The measurement of L, equal to $w \cdot \ell$, on wage plantations presents no special problems. For each such plantation the 1880 census manuscripts provide a tally of the value of total wages plus in-kind ration payments dispensed by the plantation operator for the year 1879. On plantations utilizing sharecroppers or renters a procedure for estimating the value of labor input must be devised.

The estimate of labor input is based upon the standard allotment of acreage to a first-class male hand. Let $\hat{a}$ equal the amount of land the average first-class male hand was expected to cultivate. Then a plantation cultivating A acres is estimated to have used the equivalent of $A/\hat{a}$ units of first-class male labor. There are three cases to consider. Imagine a plantation cultivating a portion of its acreage by leasing to renters paying a certain sum of money or cotton per acre leased, another portion with sharecroppers, and the rest with wage hands. Both types of tenants would generally have to hire extra labor in addition to their families at some time during the cultivating season, and especially during harvest. The dollar value of wages and rations paid to this additional labor on each tenancy is reported by the census enumerator as total wages plus rations paid, just as on a wage plantation. Let this amount be denoted W^i. All that is left is to estimate the value of the labor supplied by the tenants themselves.

If the average wage in the state of residence is $\hat{w}$, our estimate of the value of labor input on tenancy i is

$$L^i = W^i + \hat{w} \cdot A^i / \hat{a}.$$

There is a third case. A few census takers reported the share paid to share-croppers as wages paid, just as if they were wage hands.[4] In these cases total wages paid always appear to have nearly exhausted the gross value of output. To use reported wages paid in these cases would have been a serious overestimation of the labor input. The needed correction is to value only the additional labor hired by the sharecropper and that part of his share of output that should properly be imputed to labor income exclusive of any earned profit. Therefore in these cases

$$L^i = W^i - Q^i/2 + \hat{w} \cdot A^i / \hat{a}$$

On the fraction of the plantation worked by wage hands only

$$L^i = W^i$$

The total valuation of labor on the plantation is the sum across each tenancy and wage-worked section,

$$L = \sum_i L^i$$

The crucial empirical step in this estimation procedure is the determination of $\hat{a}$. I have based the choice of $\hat{a}$ for each region on the typical acreage allotments reported by agricultural experts to the Department of Agriculture and the Bureau of the Census. Where possible these numbers have been compared to the reports of additional sources and statistical data. The general allotment of acreage to the hand producing short-staple cotton on reasonably flat land was a total of fifteen acres in cotton and corn. This rule was universal throughout the western areas of the sampled states: Louisiana, Arkansas, and Mississippi.[5] A statistical sample from labor contracts verifies the pronouncement that $\hat{a}$ equaled 15 for these states.

Alabama, Georgia, and North Carolina showed more variation in the acreage allotment. A general explanation of the variation was given by a census enumerator for the state of Georgia who explained that in the north-western portion of the state, where the soil quality was metamorphic, the usual allotment was fourteen or fifteen acres to the hand. The same correspondent explained that in the southern half of the state the rule was closer to twenty acres to the hand. Consistent with these two regional acreages, planter David Dickson of Hancocke County, Georgia, an assiduous and astute student of and authority on cotton culture, confirmed these two figures by reporting in 1874 that the *average* in Georgia was eighteen acres to the hand, twelve in cotton, and six in corn. Use of similar evidence and the type of soil and terrain in a sampled county lead me to make the

---

[4] The only plantations used in my sample where this method was adopted were in Concordia Parish, Louisiana.

[5] *Tenth Census of the United States, 1880: Cotton Production, I*, pp. 84–85, 104, 106, 155.

estimates of $\hat{a}$ exhibited in table B.3.[6] In all cases I have used a high estimate, since this results in a low estimate of labor utilized and therefore biases upwards the calculation of factory productivity on share plantations.

The wage rate used to value the estimate of labor input is taken from the Department of Agriculture's record of the average value of wages plus rations by state in 1879. In cases like Bolivar County, Mississippi, situated on the alluvial flood plain of the Mississippi River and just across the river from high-wage counties in Arkansas, it was deemed more appropriate to use the higher average wage given for Arkansas than the Mississippi average.

The wage valuation of labor input on share-worked plantations was reduced further by allowing for an average rate of absenteeism of 20 percent during the year. This figure was derived as follows. As reported by Roger Ransom and Richard Sutch, in 1875 the Georgia Department of Agriculture calculated from a survey of Georgia planters that share hands worked an average of 4.6 days per week. Taking as standard a five and one-half day work week for wage hands, this computes on the basis of a fifty-week work year to a yearly absentee rate of 16 percent. Again, to bias my estimates in favor of share plantations the adjustment for the valuation of labor input is made by reducing the appropriate yearly wage by 20 percent before multiplying by the labor estimate.[7]

Estimation of Land Rent

With the exception of one minor point, the estimation of a rental price on land for each plantation is relatively straightforward. I treat fertilizer as a land-augmenting resource. That is, a pound of fertilizer increases output per acre and can be thought of as an increment to the quantity cum quality of land. Within this framework, fertilizer costs should be treated as a part of the total rent of land. This is justified in the theoretical derivation of land rent.

Under the assumption of a competitive market with profit-maximizing landlords the derivation of the *direct rent* on each actual acre of cultivated land is straightforward. Starting from the specification of the Cobb-Douglass production function, taken as given by cotton producers, profit maximization requires that the landlord j equate the marginal product of land to its rental price. Consider the revenue (production) function $Q = Q[L,K,A]$. Profit maximization implies $Q_A = r^j$, but also the income share of land is equal to one quarter total output, so

$$Q_A \cdot A^j = r^j \cdot A^j = \frac{Q_c^j}{4}$$

[6] Ibid., vol. 2, pp. 174–75. *Report of the Commissioner of Agriculture, 1874* (Washington, D.C.), p. 219. Ibid., C. W. Howard, "Condition of Agriculture in the Cotton States," p. 222. *U.S. Census, 1880, Cotton Production*, vol. 2, pp. 77, 156, 174. Trowbridge, pp. 496–97. Lewis Cecil Gray, *History of Agriculture in the Southern United States to 1860* (Gloucester, Mass., 1958), vol. 2, p. 708.

[7] *Report of the Commissioner of Agriculture, 1887* (Washington, D.C., 1888), p. 582. Ransom and Sutch, p. 235.

Table B.3 Estimated Average Acreage and Wages per Hand, by Region

Location	Soil Type	â	w Unadjusted	w Adjusted for Absenteeism	c	r
Concordia, L.	alluvial	15	$196.80	$157.44	.79	9.38
Lowndes, Miss.	black prairie	15	159.72	127.77	.34	4.03
Noxubee, Miss.	black prairie	15	159.72	127.77	.31	3.68
Yazoo, Miss.	alluvial (Yazoo bottoms)	15	159.72	127.77	.58	6.88
Tunica, Miss.	alluvial	15	205.44	164.35	.60	7.12
Hinds, Miss.	black prairie	15	159.72	127.77	.46	5.46
Desha, Ark.	alluvial-prairie	15	205.44	164.35	.86	10.21
Saint Francis, Ark.	alluvial-prairie	15	205.44	164.35	.50	5.93
Jefferson, Ark.	alluvial-pine woods	15	205.44	164.35	.76	9.02
Lowndes, Ala.	black prairie	15	158.40	126.72	.30	3.56
Marengo, Ala.	black prairie	15	158.40	126.72	.29	3.44
Greene, G.	metamorphic	20	128.76	103	.31	3.68
Houston, G.	sand hills–pine hills	20	128.76	103	.26	3.08
Harris, G.	metamorphic–pine hills	20	128.76	103	.29	3.44
Burke, G.	pine hills–redlands	20	128.76	103	.33	3.91
Edgecomb, N.C.	pine woods	25	134.28	107.42	.51	6.05
Wake, N.C.	metamorphic	25	134.28	107.42	.50	5.93

The *direct rent* on land[8] is therefore

$$r^j = Q_c^j/A_c^j \cdot \frac{1}{4} = c^j/4, \ c^j = Q_c^j/A_c^j$$

The indirect component of the rental on land is just $P_F \cdot F$, cost of fertilizer used. The rent on A acres of land is then

$$T = r \cdot A + P_F \cdot F$$

This is derived in the following manner.

Under the assumption that fertilizer is land-augmenting, we have acreage cultivated as an increasing function of fertilizer use, $A = A(F)$. Profit maximization again requires the equation of marginal products and prices, so

$$P_F = Q_A \cdot \frac{dA}{dF}$$

But $Q_A = r$, so $\frac{P_F}{r} = \frac{dA}{dF}$. Total acreage after an incremental dosage of fertilizer is just $A + dA$. The last term can be approximated by

$$dA \cong \frac{dA}{dF} \cdot F = P_F \cdot F/r$$

Thus the rental cost of land is

$$r[A + dA] \ = \ r \cdot A + P_F \cdot F.$$

An alternative way of seeing this is as follows. Under constant returns the well-known condition of Euler's theorem is that total product must be exhausted by the imputed return to each input.

$$Q = Q_L \cdot L + Q_K \cdot K + Q_A \cdot A + Q_A \cdot \frac{dA}{dF} \cdot F$$

or

$$Q = Q_L \cdot L + Q_K \cdot K + Q_A \left[A + \frac{dA}{dF} \cdot F \right]$$

The rental valuation of land is the last term,

$$Q_A \left[A + \frac{dA}{dF} \cdot F \right] = r[A + P_F \cdot F/r] = T$$

To compute T for a given plantation we require data on the value of output per acre, c^i, A^i, and $P_F \cdot F$. The last two are supplied by the manuscript census returns for 1879. c^i is more problematical. It is possible from the census data reported to compute c^i for each plantation. But the use of plantation specific output per acre would bias the estimation of the independent return to land, because a high or low c^i on any particular plantation could be due to high or low inputs of capital and labor as well as to the quality of the land input. What we desire is the average output per acre on land of a given quality. This is approximated by using average output per acre in the plan-

---

[8] In the cotton South land rents were based upon land's productivity in terms of cotton.

tation's home county. This has the advantage of controlling for land quality and averaging over a large number of production units with varying input combinations.

Cotton producers valued land rents on the basis of the output of cotton per acre alone. This estimate was made on the basis of an average price of ten cents a pound. Taking a standard bale at 475 pounds of lint cotton, this means that the direct rent on land for any county is given by

$$r^j = (\text{bales/acre}) \cdot (475) \, \frac{10\cancel{c}}{4}$$

The average number of bales per acre for each sample county was obtained from the U.S. census of cotton production.

The Rental Price on Capital Services

From both a conceptual and practical view the estimation of the rental price on capital is the most difficult problem. My approach is explained here in all its gory detail.

I begin with a theoretical derivation of the rental price of a unit of capital as a function of variables that can be empirically estimated. The derived rental price is then estimated with statewide data from the 1880 census and *Reports of the Department of Agriculture*. The major conceptual difficulty in the empirical estimate is the necessity of defining what is meant by a unit of capital. My approach to this is to use an analogue of a fixed coefficient model of production to construct a composite capital input. While the conceptual theoretical difficulties of this method will be of special concern to followers of Piero Sraffa, I suggest that the closeness of my estimate to two independent itemizations of the cost of capital on cotton farms by 1879 planters justifies it as a practical method.

Begin by considering that the rental price of a mobile input like capital, unlike unmovable land, should in a competitive economy be reasonably constant across geographic space.[9] After making this assumption, we may either assume profit maximization with competitive markets, so that the rental price of capital equals capital's marginal product, or drop that theoretical assumption and simply write down the equation representing capital's share of total income. Either way we get the same result for every production unit or region[10]

$$\varrho \cdot k^j = \frac{Q_c^j}{4}$$

Supposing j to index the major cotton-producing states, we divide through by total acreage cultivated in state j, weigh each state by $1/N$, and sum over

---

[9] The same argument might have been made for the price of labor services. However, in that case available wage data allow no method of correcting for differences in the quality of labor services across space. The best index for these differences is simply the location-specific wage.

[10] Dividing through by k^j we find that the rental price ϱ equals capital's marginal product.

states. This gives

$$\frac{\varrho}{N}\sum_{j=1}^{N} k^j/\bar{A}^j = \frac{1}{4N}\sum_{1}^{N} Q^j/\bar{A}^j$$

equivalently,

$$\varrho(\hat{k}/\bar{A}) = \hat{c}/4$$

or

$$\varrho = \frac{\hat{c}\cdot(\bar{A}/k)}{4}$$

with N = number of states and $(\bar{A}/k)$ = average acreage per unit of capital. If k can be defined in measurable terms, published census data will permit the estimation of ϱ.

The basic components of capital on cotton farms were tools, machinery, and work animals. Cotton producers defined the basic production unit in terms of work animals—for example, the "one-horse farm." Begin by defining a unit of capital to be the capital requirements of a one-horse farm. Then the value of a unit of capital is just the dollar value of one horse plus the value of the machinery utilized by a horse on a one-horse farm. If k'_M is the dollar value of machine capital and P^i_H the value of the horse, the value of such a unit of capital is $P^i_H + k'_M$. On any farm i the mean value of a unit of capital is simply

$$\frac{\sum_{j=1}^{H^i} P^j_H + \sum_{j=1}^{H^i} k^j_M}{H^i} = \hat{P}_H + \frac{k^i_M}{H^i} = P^i_k$$

If we were to average over all farms we would obtain the mean value of a unit of capital, $\hat{P}_k = \hat{P}_H + (\frac{\hat{k}_M}{H})$. The last two values P^i_k and $\hat{P}_k$ are our basic units.

Suppose we wish to know how much capital is on a one-horse farm? Observe that the percentage of capital (in dollars) embodied in the horse is P_H/P^i_k. We may then say that a horse equals P_H/P^i_k units of capital, or

$$(P_H/P^i_k)\cdot M^i = H^i$$

The gross units of capital on the farm are

$$M^i = (P^i_k/P_H)\cdot H^i$$

Because of the varying age, size, and quality of work stock and machinery on different plantations, a further adjustment to gross capital stocks must be made. For example, we would not wish to define a mule of highest quality equipped with the finest new machinery on one farm as an identical unit of capital to an old worn-out work ox equipped with inferior and aged machinery. The only practical way to handle this problem is to take market prices as indicative of the quality of goods. Thus our estimation of the quantity of capital on any production unit or in any region is not its gross capital but the amount of capital in base units.

$$K^i = M^i \cdot k^i \text{ where } k^i = P_k^i / \hat{P}_k.$$

To estimate the quantity of capital used on a farm or plantation or in a region I used the sequence of computations now described. Utilizing average prices of horses, mules, and work oxen for 1879 as reported in the *Report of the Commissioner of Agriculture, 1880*, mules and oxen were converted into horse equivalents by multiplying the ratio of their respective prices and the price of a horse times the number of oxen or mules. The sum of total horses, mule equivalents, and oxen equivalents was defined to equal the total stock of animal capital. This was done for each state, using state prices. The gross units of capital on a farm or in a state were then computed by multiplying animal capital H by the inverse of the ratio of the price of a horse and the value of a unit of capital in the corresponding state, as described above. Gross capital M was then adjusted to net capital k by multiplying M by the value of a unit of capital on the respective farm or state, P_k^i, and indexing or deflating by the average value of a unit of capital, $\hat{P}_k$. This procedure was carried out for every farm (plantation) and for the seven major cotton-producing states (see table B.4).

To estimate ϱ using this procedure, we require for each state its quantity of work stock on farms, the value of machinery on farms, total cultivated farm acreage, farm work stock values, and cotton output per acre. Finally, in order to adjust acreage inputs in the manner described in the previous section, total fertilizer costs are needed.

Since the estimate of ϱ is based upon averages of cotton output per acre and land per unit of capital, the reliability of the estimate would be expected to increase with the number of states used. To calculate the average acreage per unit of capital, $(\bar{A}/k)$, I used the seven major cotton-producing states (table B.4). To calculate $\hat{c}$ I added the remaining cotton-producing states, excluding the border states. Calculation of the data for Louisiana was adjusted by excluding all parishes producing sugar as the major crop.

The resulting estimate is $\varrho = \$73.65$. How do we know if this estimate is at all reasonable? At the bottom line the estimate of ϱ, just like the estimates of each of the other variables, must be judged on the basis of the estimated total factor productivities they generate. The relative productivity index $f_i = \frac{G_i}{G}$ has been constructed so that if the production function in cotton agriculture were correctly specified by Equation B.1, then in a statistical sense the mean value of the sampled productivity indices should not stray too far from unity. In fact, from a theoretical perspective, given a correct specification of the production function, the only source of variation from unity would be due to errors in the measurement of outputs and inputs. But this theoretical criterion is too strong. For one thing, the requirement that f_i equal unity is to ask that a production unit be producing at maximum theoretical efficiency on the frontier of its production possibilities. In practice this may be too much to expect. Furthermore, since one of the hypotheses to be tested asserts that share-worked plantations were not efficient, that hypothesis actually requires that for that set of plantations $\hat{f}$ be a statistically significant distance below unity.

Table *B.4* Estimates of Variables, by State

State	H	P_k	$[P_H/P_k]^{-1}$	M	$I/\hat{P}_k$	K	$\bar{A}$	$\bar{A}/k$	c	Q^i
Alabama	258,698	73.88	1.25	323,373	.96	310,438	6,471,545	20.85	14.25	74.27
Arkansas	269,085	66.59	1.35	363,265	.86	312,408	3,440,685	11.01	27.55	75.83
Georgia	288,382	90.41	1.26	363,361	1.17	425,132	8,871,520	20.87	14.72	76.80
Louisiana	183,864	71.99	1.26	231,669	.93	215,452	1,714,059	7.96	28.02	55.75
Mississippi	275,448	90.50	1.24	341,556	1.17	399,621	4,947,204	12.38	21.85	67.59
North Carolina	231,087	93.61	1.39	321,211	1.21	388,665	6,394,328	16.45	20.90	85.86
South Carolina	140,011	105.92	1.28	179,214	1.37	245,523	4,394,498	17.90	18.05	80.73
Tennessee	482,006	72.76	1.35	650,708	.94	611,666	7,733,187	12.64	21.85	69.05
Virginia									20.90	
Texas									17.57	
Florida									10.45	

Table B.5   Average Estimate of Variables in Table B.4

$(\bar{A}/k)$	15.00
$\hat{c}$	19.65
$\hat{P}_k$	77.28
ϱ	73.65

An independent test of the reasonableness of the estimated rental cost of capital services is its quantitative nearness to the cost of capital as reported by practicing cotton planters in 1879. I have located two fully itemized estimates for the year 1879 of my equivalent to ϱ in the published census of cotton production of 1880. A planter in Crittenden County, Arkansas, offered an estimate on *capital* of $5 per acre. Using the average $(\bar{A}/k) = 15$, this computes to $\varrho = \$75$. A second cotton planter, located in North Carolina, estimated for the years 1878 and 1879 that the interest cost on capital for a twenty-two- acre farm was $72.[11] The estimates of ϱ and P_k provide an estimate of the interest rate on capital available to sharecroppers.

$$\frac{\varrho}{P_k} = .95$$

This nearly 100 percent rate of interest may seem excessive to those unfamiliar with conditions in the cotton South. Actually it is within the bounds of reported interest rates on consumption items purchased by sharecroppers. Roger Ransom and Richard Sutch report interest rates for credit purchases of corn and bacon in Georgia and Louisiana over a number of years. Average rates ranged from 59 percent to 92.8 percent, many years exceeding 100 percent in Georgia. For fairly obvious reasons the interest rate on capital which must be returned to its owner after use should be expected to be higher than that on borrowed food. A final corroboration of the validity of the estimation procedure is the calculated data exhibited in table B.4. The resulting estimates of $\bar{A}/k$ and ϱ^i for states seem very reasonable.

A direct test of the hypothesis that wage plantations were more efficient than share plantations requires a comparison of the total factor productivities on both types of plantations. One way to accomplish this would be to simply take all share plantations and, after aggregating their outputs and inputs, compute an aggregate index of total productivity. After doing the same for wage plantations a comparison could be made. A major problem with this procedure is that it does not allow a determination of the independent effect of wage versus share organizations on the productivity of a plantation. For instance, it may be the case that cotton was a more profitable crop in 1879 than corn or other products. If wage plantations planted a higher fraction of their total acreage in cotton than share plantations of identical size, this difference in the cotton-corn mix might be responsible for some or even most of any computed differences in productivity. In that case

---

[11] *U.S. Census, 1880, Census of Cotton Production*, vol. 1, p. 107, vol. 2, p. 78.

it could be argued that wage plantations were allocatively more efficient, but the evidence would not allow a statement concerning relative productive efficiency. Analogously, it may be that significant returns to scale existed in postbellum cotton agriculture. Again, if for some reason those planters who chose the share organization were also, on average, owners of larger estates, a calculation of aggregate productivities might lead to an erroneous conclusion that the share *organization* was more efficient when the result was really due to returns to scale.

In order to circumvent these problems and to make a direct test of the *independent* effect of organizational form on productive efficiency, I utilize the method of ordinary least squares (ols) regression. The fact that total factor productivities were calculated for each individual plantation in the sample makes this feasible. As explained earlier, many cotton plantations were organized by allocating part of the total acreage cultivated to wage labor and part to sharecroppers. Define the continuous variable A_w to be the percentage of total cultivated acres on a plantation worked by wage labor. Similarly, define A_c to be the percentage of cultivated acreage planted in cotton. Recall that A and Q denote total cultivated acres and produced revenues respectively. Two direct tests of the relative efficiency of share and wage plantations are given by regression equations B.2 and B.3:

B.2: $$f_i = \alpha + B_1 \cdot A_w^i + B_2 \cdot A_c^i + B_3 \cdot A^i$$

B.3: $$f_i = \alpha + B_1 \cdot A_w^i + B_2 \cdot A_c^i + B_3 \cdot Q^i.$$

The B_i coefficients represent estimates of the independent quantitative influence of their corresponding variables on the total factor productivity (TFP) of cotton plantations. Thus, in Equation B.2 B_1 represents the average change in productivity induced by an increase (decrease) in the fraction of total acres worked by wage hands, while holding total acres worked and crop mix unchanged. The statistical results from estimating these equations are exhibited in table B.6. B_1 is positive, well away from zero and statistically significant at the 99 percent level. We may conclude that, holding the other variables constant, a 1 percent increase in wage-organized acreage will increase TFP by one-third of 1 percent. Equivalently, a 100 percent increase in wage acreage (conversion from complete sharecropping to a complete wage plantation) will, on average, increase TFP by one-third of the TFP of a sharecropped plantation. The crop mix variable is positive in one equation and negative in the other, but not statistically significant in either. We may conclude that for the year 1879 the crop mix choice had no significant effect upon productive efficiency. Acreage cultivated and revenues produced represent two means of testing for returns to scale. Both have positive but extremely small coefficients. Since the coefficient of A is statistically insignificant, we may conclude that early postbellum cotton plantations exhibited no scale economies due to increasing productive capacity or plantation size. The coefficient for Q is practically zero but

Table B.6 Total Factor Productivity Regressions

	Constant	A_w	A_c	A	Constant	A_w	A_c	Q	f
Value of regression coefficient	.669	.356	.153	.164E-03	.837	.327	-.172	.132E-04	
T ratio	5.15	4.69	.728	1.65	7.00	4.85	-.882	5.37	
Mean	.999	.588	.596	362.88	.999	.588	.596	7673.14	1.02

Equation B.2: $f = \alpha = B_1 A_w = B_2 \cdot A_c = B_3 \cdot A$

$R^2 = .288$

Equation B.3: $f = \alpha + B_1 \cdot A_w + B_2 \cdot A_c + B_3 \cdot Q$

$R^2 = .392$

NOTE: Significance level = 1%. Sample size = 98 plantations.

statistically significant. Since there were no increasing returns due to plantation size it seems reasonable to interpret this last result as meaning that there were small-scale returns to marketing crops.

Alternatively, the negligibility of Q's coefficient may be interpreted as a confirmed test of the freedom from bias of the productivity index. Suppose we knew that there existed no increasing or decreasing returns, as is confirmed by Equation B.2, then a test for bias in f_i would be the requirement that Q's coefficient equal zero. The geometric index of input measure should not favor small or large farm revenues.

As stated earlier, the construction of the geometric index of total factor productivity implies that any statistically significant variations in measured productivity should be due to errors of measurement. The differences between share and wage plantations can be understood more fully by reconsidering the estimation techniques used. The only input for which estimation differed across the types of plantations was labor. In the estimation of the labor input on sharecropped farms, each unit of male equivalent labor was converted into dollar values by imputing to it an absenteeism-adjusted *yearly* wage. This procedure implicitly assumed that the total quantity (in terms of work intensity) of labor supplied by a sharecropper was equal to that supplied by a wage hand. In view of the productivity differential we must assume that this implicit assumption is unwarranted. The labor supply of sharecroppers, even after adjusting for a higher rate of absenteeism, has been overestimated. But this is precisely the result which economic theory implies. Sharecroppers, due to work disincentives at the margin, supplied a lower level of work effort than wage hands. While the choice was voluntary, it cannot be characterized as Pareto efficient. The wedge between the laborer's marginal product and marginal disutility of work effort produced the standard dead-weight loss in welfare. Both laborer and landlord could have been made better off with a different payment system.

One final comment should be made. The presumed reduction in supervisory costs allowed by the share contract has been accounted for by the technique used. Any supervisory services purchased by wage plantations should be included as part of the total wages paid by such plantations. Any *possible* services of the owner have not been included, but this is true for both types of plantations. Since the labor elasticity of total factor productivity

$$\frac{dG}{dL} \cdot \frac{L}{G} = \frac{1}{2}$$

any differential in the quantity of these services would have to amount to 70 percent of the wage plantation's labor bill to eliminate the productivity differential.

Appendix C

Table C.1 The Immediate Incidence of Share Payments

"The Freedmen are working for a portion of the crop, varying on different plantations from 1/10 to 1/4"	Western Alabama	1865	*BRFAL*, film M752, no. 22
". . . nearly all receive a portion of the crops"	Alabama	Feb. 12, 1866	*New York Times*
"The Freedmen share in the general misfortune, many having worked for a share of the crop" (Gen. Wager Swayne)	Alabama	1866	*BRFAL, Synopsis of Reports*
"A majority of the planters give a share in the crop"	Arkansas	June 1866	*BRFAL*, film M979, no. 23
"They are generally cultivating the lands of their employers on shares of from one-tenth to one-fourth of the crop produced."	Arkansas	Oct. 6, 1865	*New York Tribune*
"About one half the laborers are working for a share of the crop, from one half to one third" (Gen. J. G. Foster)	Florida	Sept. 30, 1866	*BRFAL, Synopsis of Reports*
"The most of them fair, giving one-fourth (1/4) as the share of the crop for the laborer" (General C. H. Howard)	Tallahassee, Fla.	1865	*Joint Committee on Reconstruction*
"most of them will work for a share of the crop"	Georgia	1866	Davis Tillson Report *BRFAL*, film M798
"I could see there was no money for them. . . . Most of them . . . a part of profits" (Clara Barton)	Georgia	1865	*Joint Committee on Reconstruction*
"I found generally in this section that the contracts for the last year [1865] had been	Georgia	1865	*Joint Committee on Reconstruction*

at the rate of from one-sixth
to one-tenth of the crop, and
the latter for the greater
number" (C. H. Howard)

"As a general thing, the compensation . . . has been an interest in the crop"	Georgia	Dec. 11, 1866	*New York Times*
"the contracts on large farms, generally are made for from one quarter (¼) to one half (½) of the crop"	Green and Oglethorpe Cos., Ga.	1866	*BRFAL, Synopsis of Reports*
"The system of remunerating the laborers by a fourth of the cotton crop . . . worked so advantageously as to be generally adopted"	Sumpter Co., Ga.	June 8, 1866	*New York Times*
"It will be observed that a majority of the contracts for labor are a 'share of the crop'"	Catahoula Parish, La.	1866	*BRFAL*, film M1027, no. 28
"The Plantations were mostly worked on shares"	Mississippi	Mar. 11, 1866	*New York Times*
"The great majority of the laborers hired in this portion of the state have preferred to work on shares"	Mississippi	Jan. 31, 1866	*Mississippi Intelligence*, quoting *Columbus Sentinel*
"Nearly all the contracts for labor are in writing. Four fifths for a share of the crop instead of fixed money wages"	Raliegh, N.C.	1866	*BRFAL*, film M843, no. 22
"Even for this year they [planters] have been obliged to resort to some arrangement for a division of crops"	N.C.	1866	*Joint Committee on Reconstruction*
"In nearly all cases of contracts made in this District the negro agrees to labor upon the plantation receiving in payment a proportion of the crops"	Darlington, S.C.	1865	*BRFAL*, film M869, no. 34
"Freedmen expect a reward for their labor generally by receiving a specified portion or share from the crops which they cultivate"	Aiken, S.C.	July 1866	*BRFAL*, film M869, no. 34

"The operations of the year, on almost every plantation, have been conducted upon the plan of sharing the crop"	South Carolina, Sea Islands	Oct.2, 1866	*New York Times*
"Some pay from six to ten dollars per month. Some contract for a year from $96 to $150, but the larger portion are made on shares in the coming crop"	Orangeburg, S.C.	Jan. 29, 1866	*BRFAL*, film M869, no. 34
"Nearly all Freedpersons contracted to work on shares with the planters, for the period of one year"	Monks Corner, S.C., Berkeley Co.	Nov. 1, 1866	*BRFAL*, film M869, no. 34
"The county reports show that the Freedmen through-out the state are generally working lands on shares. . . . Comparatively few are working for wages"	Tennessee	Aug. 11, 1866	*New York Tribune*
"The great majority are working for a portion of the crop"	Cheatham Co., Tenn.	1866	*BRFAL*, film T142, no. 38
"about two thirds (2/3) of these are at work for an interest in the crop"	Warren Co., Tenn.	1866	*BRFAL*, film T142, no. 38
"The [Freedmen's Bureau] supt. gave it as his opinion that about 500 were at work for wages and about 1000 received an interest in the crop"	Wilson Co., Tenn.	1866	*BRFAL*, film T142, no. 38
"Four-fifths (4/5) of the parties are at work for an interest in the crop"	Haywood Co., Tenn.	1866	*BRFAL*, film T142, no. 38
"about three quarters of the Freedmen are at work for an interest in the crop"	Tipton Co., Tenn.	1866	*BRFAL*, film T142, no. 38
"In about two thirds (2/3) of the contracts in the county the Freedmen are working for an interest in the crop"	Fayette Co., Tenn.	1866	*BRFAL*, film T142, no. 38

"Three fourths (¾) of the Freedmen are at work for an interest in the crop"	Hardeman Co., Tenn.	1866	*BRFAL*, film T142, no. 38
"Four-fifths (4/5) are at work for an interest in the crop"	Lauderdale Co., Tenn.	1866	*BRFAL*, film T142, no. 38
"A large majority of the Freedmen in the county are working for an interest in the crop"	Carroll Co., Tenn.	1866	*BRFAL*, film T142, no. 38
"About two thirds (2/3) of the Freedmen of the County are at work for a share of the crop"	Gibson Co., Tenn.	1866	*BRFAL*, film T142, no. 38
"About one half for a share of the crop and the balance receive from $70, to $200 per year"	Henderson Co., Tenn.	1866	*BRFAL*, film T142, no. 38
"A large majority of the Freedmen in the county are at work for an interest in the crop"	Madison Co., Tenn.	1866	*BRFAL*, film T142, no. 38
"Most of the Freedmen are working for a portion of the crop"	Texas	1866	*BRFAL*, film M752, no. 40
"At least three fourths of the Freedmen work for a share of the crop" (General Kiddo)	Texas	1866	*BRFAL, Synopsis of Reports*
"The Freedmen in these counties are mostly working for part crop"	Texas	Aug. 20, 1866	*New York Tribune*
"The majority of contracts made under my supervision and otherwise have been made for portions of the crop"	Fredericksburg, Va.	1866	*BRFAL*, film M1048, no. 44
"The Negroes being few and scattered and most of them having rented land for a share of the crop. . . . out of two hundred and	Virginia	1866	*BRFAL*, film M1048, no. 46

thirteen farmhands and
mechanics in Scott Co.
I did not find over ten who
were working for stated
wages"

"The Freedmen in this county, most of them are employed for a share of the crop"	Dinwiddie Co., Va.	Nov. 30, 1866	*BRFAL*, film M1048, no. 46, p. 96.
"Most of the colored people have been working on shares	Amherst, Livingston, Nelson Co., Va.	Nov. 30, 1866	*BRFAL*, film M1048, no. 46
"The majority of the Freed-people entered into con-tracts to labor for a certain portion of the crop"	New Kent, C.N., Va.	Nov. 25, 1866	*BRFAL*, film M1048, no. 46
"the most of them this year are working the farms on shares"	Essex Co., Va.	Feb. 24, 1866	*BRFAL*, film M1048, no. 44
"Very few of them are working for stated wages; the majority are renting land or working for a share of the crop"	Scott, Wise Cos., Va.	Nov. 25, 1866	*BRFAL*, film M1048, no. 46
"Nearly all the heads of families in the Sub-District say they have raised enough in cropping this year to support them until the crop next year"	Arlington, Va., Washington, Russell, Buchanan Cos.	Nov. 25, 1866	*BRFAL*, film M1048, no. 46

Index

African Methodist Episcopal Church, 257–58, 267, 273–75, 280, 284
agrarian reform, 6, 10, 14, 245–46, 293, 296
Aiken, David Wyatt, 191–93, 195, 207, 209, 212–13
American Cotton Planters Association, 35–36, 41–42, 70
Andrew, John A., 289
apprenticeship system, 3
Ashley, William J., 26
Ashton, Thomas S., 49, 89–90, 174
Atkinson, Edward, 40

Babbage, Charles, 196, 199
balance of trade, 9
Bancroft, George, 255
bankruptcy: planters, 145, 150, 156; merchant confiscation of crops, 150, 152, 312
Banks, Nathaniel P., 13–14; free labor, 68–69, 71
Barrow, David C., Jr., 159–60, 167, 188, 214
Baylor, Charles, G., 39–40, 41
Beecher, Henry Ward, 23
Benham, George, 81, 82–83, 91n., 144, 260
Bentham, Jeremy, 20–23, 61, 82
Berkeley, George, 97
black autonomy, 114–15, 159–60, 162–64, 173, 180, 182–83, 187, 189, 201–3, 207–9; and labor contracts, 220–21, 240–41
black codes, 16–17, 18, 58–60, 303
black press: views on contract system, 73, 288–89; views on black self-help and employment, 64–65, 66, 285–86, 289–91, 294, 296–97; views on free labor, 73,192

Blassingame, John: slave family, 69–70
Boston Board of Trade, 8, 19, 24
Botume, Elizabeth Hyde, 86
Breman, Jan, 105, 124
Bright, John, 10
Britain, 3–5; economy, 26–29
British West Indies, 4, 12; sugar plantation, 17, 19, 20
Brock, W. R., 6n., 7n., 9
Brooks, Robert P., 26n.
Burch, John Henri, 296
Bureau Refugees, Freedmen and Abandoned Lands, 16, 19, 20, 21, 23, 32, 45–48, 294; labor policy, 62–74, 107, 133–36, 149–57, 133, 203–9, 226–7, 285–86
Burke, Sir Edmund, 20
Butty-Gang, 183

Campbell, Sir George, 6n.
Campbell, Tunis G., 289, 294, 295
capital: European, 35–36, 41; northern, 23, 24, 35, 36, 48, 238, 269–70, 272; stock, 25, 26, 29; working, 26; shortage, 142–43, 156, 224–26, 244, 246–47, 293
Carlyle, Thomas, 11–12, 13
Chase, Salmon P., 13, 163
civil war, 24, 25, 30
Clapham, Sir J. H., 23, 184
coal mining, 270–71, 278
Cobb, Howell, 89, 98
Cocke, John Hartwell, 107–9, 209–10
colonies, 3, 4
colonialism, 59–60; Chamberlain, Neville, 60
Commons, John R., 233, 282

347

Wilberforce, William, 61, 102, 249, 260, 262–64
Williams, Eric, 3n., 18n.
Wilson, Henry, 7, 17
Woodward, C. Vann, 78, 310
worker-collective, 158, 162, 174, 182–84, 199, 201, 276, 284, 290–92, 296, 300; egalitarianism, 178, 183, 202–03; demo-
cratic, 183, 189, 204–205, 213, 288–89, 292–94
work ethic: freed people, 57–58, 96–98, 194, 237, 287; slaves, 69–70, 76–77, 194, 236
Woodman, Harold D., 35n., 42
Wright, Gavin, 245–46